BODY AREA COMMUNICATIONS

BODY AREA COMMUNICATIONS

CHANNEL MODELING, COMMUNICATION SYSTEMS, AND EMC

Jianqing Wang
Nagoya Institute of Technology, Japan
Qiong Wang
Dresden University of Technology, Germany

IEEE
WILEY

This edition first published 2013

Registered office
John Wiley & Sons Singapore Pte. Ltd., 1 Fusionopolis Walk, #07-01 Solaris South Tower, Singapore 138628

For details of our global editorial offices, for customer services and for information about how to apply for permission to reuse the copyright material in this book please see our website at www.wiley.com.

Library of Congress Cataloging-in-Publication Data

Wang, Jianqing.
Body area communications : channel modeling, communication systems, and EMC / Jianqing Wang, Qiong Wang.
pages cm
Includes bibliographical references and index.
ISBN 978-1-118-18848-4 (cloth)
1. Body area networks (Electronics) 2. Medical telematics. I. Wang, Qiong. II. Title.
TK5103.35.W36 2013
621.382–dc23

2012027876

Set in 11/13 pt Times by Thomson Digital, Noida, India.

Contents

Preface

The past decades have witnessed wide demand and applications for wireless communications in the human body area, that is, in the immediate environment around a human body. These demands and applications especially focus on wireless transmission and networking of personal information for user identification, healthcare and medical applications. Body area communication techniques make this more feasible. Compared with classical wireless communications, the human body acts as a communication medium either passively or favorably in body area communications. This is no doubt a kind of novelty in the context of classical wireless communications. As expected, the human body exhibits various different features from the typical wireless medium, that is, air, when it is used as a communication medium. Nowadays, more and more people are trying to uncover its intrinsic mystery so as to profit from the expected communication purposes.

As noticed by many researchers in this area, body area communications is an emerging technique which exhibits very good prospects not only in entertainment and user identification but also in healthcare and medical applications. The required knowledge ranges from wireless communications to bio-electromagnetics. In fact, we have been working in wireless communications for many years. The first author spent the first six years on completion of his doctoral work in developing wireless transceivers/systems for mobile communications and personal computer communications in both industry and research institutions, and then began biomedical electromagnetic compatibility research after moving to a university. In 2005, when the first author became a full professor, it was felt that his background in both wireless communications and bio-electromagnetics would be very suitable to pursue this new research area. This triggered our full-scale research on body area communications.

Until now, however, there has been no textbook which systematically and completely covers this area and provides an introductory course for newcomers to this field. Two years ago, we started a course in this area for graduate students at the Nagoya Institute of Technology and for undergraduate diploma/master students at Dresden University of Technology, respectively. During the preparation of the lecture materials for these courses, we thought about how to gather our research to produce a systematic and introductory work. Fortunately, Mr James Murphy, a senior commissioning editor at John Wiley & Sons provided us with such a valuable opportunity. Following his kind invitation, we met him at an international conference in 2010 and discussed the contents of this book.

This book attempts to provide an introductory course for graduate students and newcomer engineers/researchers who need to know about or intend to be involved with body area communications. The book starts with an introduction to basic electromagnetic properties and modeling methods of the human body in various frequency bands that are available for body area communications. Next, the representative analysis techniques for body channel modeling are introduced. Based on the basic knowledge, the book focuses on three major areas: channel modeling, modulation/demodulation communications performance and electromagnetic compatibility considerations in body area communications. Most of the contents is based on the research in our laboratories.

The following topics will be described in detail from the viewpoint of an introductory course:

- Available frequency bands and expected applications for body area communications.
- Electromagnetic properties of the human body in various frequency bands, and how to model them.
- Major propagation mechanisms in various frequency bands.
- Channel models for on-body communication and in-body communication.
- Modulation and demodulation schemes used in body area communications and their analytical error probabilities.
- Link budget, bit error rate performance, RAKE and diversity reception for various on-body and in-body modulation/demodulation schemes in the available frequency bands.
- Specific absorption rate analysis for human safety evaluation in the available frequency bands.
- Modeling of electromagnetic interference with implanted cardiac pacemakers, based on both electromagnetic field approach and electric circuit approach.
- Useful computer codes for channel model generation, bit error rate analysis and specific absorption rate evaluation.

The completion of this book has been achieved by selfless contributions and help from many people. First, we would like to express our deep gratitude to the commissioning editor, Mr James Murphy, for providing us with this valuable writing opportunity, and to the project editor, Ms Shelley Chow, for her great assistance and patience in handling our manuscript. Without their support and effort it would have been impossible to publish this book. Secondly, we would like to take this opportunity to acknowledge the five anonymous referees who provided many constructive comments and suggestions, which helped us to finally form the framework of this book. Thirdly, we would like to thank our colleagues and students in our laboratory at the Nagoya Institute of Technology, especially Dr Daisuke Anzai and Dr Jingjing Shi for their contributions to the analytical and experimental data in this book. Finally, we also want to thank our families, Sufang, Moe and Mizuki in the first author's family, and Hui and Nina in the second author's family. We are sorry that we had to spend most of our spare time in writing this book rather than being able to spend our time with them. We are grateful to them for their support and understanding.

We sincerely hope that this book is both of interest and useful as an introduction for newcomers to this emerging and exciting research area.

Jianqing Wang
Nagoya Institute of Technology
Qiong Wang
Dresden University of Technology

1

Introduction to Body Area Communications

The ever-advancing miniaturization and low-power consumption of electronic devices, combined with recent developments in wireless communication, is leading to a rapidly increasing demand for wireless communications in the human body area. In a scenario of human body area communications, various communication devices may locate on, in or near the human body to form a wireless link or a small-scaled network to share data, reduce functional redundancies, and allow for new services. As an emerging communication technique, it is especially expected to be useful in medical, healthcare and consumer electronics applications. It may provide new possibilities in high-quality medical and healthcare services by linking various on/in-body vital sensors to establish a body area network (BAN) of personal health information. It may also allow high convenience and security in consumer electronics and user identification systems.

1.1 Definition

Body area communications is a short range wireless communication technique in the vicinity of, or inside, a human body. Differing from other short range communication systems such as Bluetooth and Zigbee, it focuses on communications just in the human body area, that is, the immediate environment around the human body which only includes the nearest objects that may be one part of the body.

As a most promising scenario for body area communications, BAN is attracting much attention especially for medical and healthcare applications. The concept of BAN was first proposed by Zimmermann (1996), and the definition of BAN is given by the IEEE 802.15.6 task group (IEEE802.15.TG6). BAN operates in the body area with radio frequencies to provide a wireless network of wearable and implanted sensors and/or devices in the human body. It can take a continuous measure and transmit

Body Area Communications: Channel Modeling, Communication Systems, and EMC, First Edition.
Jianqing Wang and Qiong Wang.

a vital sign or body physiological data to facilitate remote monitoring for the purposes of healthcare services, assistance for people with disabilities, and entertainment or user identification. Since it operates on or in the human body and focuses on personal information, requirements such as support for the quality of service to keep a highly reliable communication link, extremely low consumption power for long term use, and high data rate for real time transmission, should be considered. Moreover, body area communications uses the human body as a transmission medium. The transmitter and receiver are also in close vicinity in the body area. This means human body effects on the transmitting and receiving antennas have to be considered. As a transmission medium, the human body not only creates a completely different channel characteristic but also introduces a safety issue. The safety of the human body in body area communications has a higher priority than for other wireless communications.

In addition, BAN can be divided into wearable BAN and implant BAN according to its location on or in the human body where it operates. Wearable BAN includes all of the communication devices worn on the body, while implant BAN has some in-body devices which communicate with on- or off-body devices. The different operating environments lead to some differences between a wearable BAN and an implant BAN or on- and in-body communications. First, on-body communication may mainly suffer from a shadowing due to the body shape and structure or a multipath fading due to the body movement, while in-body communication mainly undergoes severe signal decay during the transmission through the lossy human tissue (Hall and Hao, 2006). These result in different requirements for the operating frequency bands. Secondly, in-body communication devices are generally more power limited and sometimes require smaller or specific shape due to their locations inside the body. Thirdly, both need to consider the bio-electromagnetic compatibility issue or transmitting power restriction to ensure human safety.

1.2 Promising Applications

There are different categorizations for application and usage models of body area communications. Table 1.1 gives the categorization (Astrin, Li, and Kohno, 2009):

Table 1.1 Categorization of applications of body area communications

Medical and healthcare applications	Assistance to people with disabilities	Consumer electronics and user identification
Medical check-up	Blind person	Wireless headphone
Medical diagnosis and treatment	Speech disability	Audio/video streaming share
Physical rehabilitation	Artificial hands and legs	User identification
Physiological monitoring	Accident prevention for elderly people	Automatic payment

(1) medical and healthcare applications; (2) assistance to people with disabilities; and (3) consumer electronics and user identification.

1.2.1 Medical and Healthcare Applications

1.2.1.1 Healthcare Monitoring in Hospital and at Home

Today's aging population is leading to a wide-scale demand for more advanced and efficient medical and healthcare treatment using wireless communication techniques. For example, the demand for wireless health-state monitoring for both in-hospital and at-home patients is growing dramatically. This is because wireless patient monitoring can effectively reduce the inconvenience of wire links, and save time and resources when people are monitored remotely at home.

Body area communications provides a wide range of possibilities in supporting such medical and healthcare services (Li, Yazdandoost, and Zhen, 2010). It may cover three areas: medical check-up; physical rehabilitation; and physiological monitoring. As a typical usage model, the body area communication device is a transceiver together with a health information sensor or a set of health information sensors. For medical check-up, such devices can collect electroencephalogram (EEG) data for monitoring brain electrical activity, electrocardiogram (ECG) data for monitoring heart activity, breathing data for monitoring respiration, as well as blood pressure, heart rate and body temperature. For physical rehabilitation, tilt sensors for monitoring accidental falls, foot sensors for monitoring steps, movement sensors for monitoring activities, breathing sensors for monitoring respiration, as well as blood pressure sensors, heart rate sensors and body temperature sensors are all possible candidates. For physiological monitoring, acceleration sensors for monitoring instant behavior, foot sensors for monitoring steps, breathing sensors for monitoring respiration, as well as blood pressure sensors, heart rate sensors and body temperature sensors may be required.

A typical application of these sensor data is the real time monitoring of patient state in a hospital. Another typical application is the real time monitoring of the health state of elderly people at home. By attaching such devices to patients or elderly people, vital healthcare data are automatically collected, and then forwarded to medical staff in a hospital or medical center for medical and healthcare administration (Bonato, 2010). Figure 1.1 shows the concept of the two applications. The sensor data are collected at an on-body server as shown by the circle in the center of the body, and then sent to a hospital or medical center. The wireless link to the on-body server needs a body area communication technique, while the data transmission to a hospital or medical center can employ cellular systems or local area networks (LANs). This usage mode reduces the work load of the medical staff and results in increased efficiency of patient or at-home elderly people management. Moreover, as a simple extension, the patient monitoring system in a hospital can be also used in a sports center to monitor physiological information. The data from the physiological

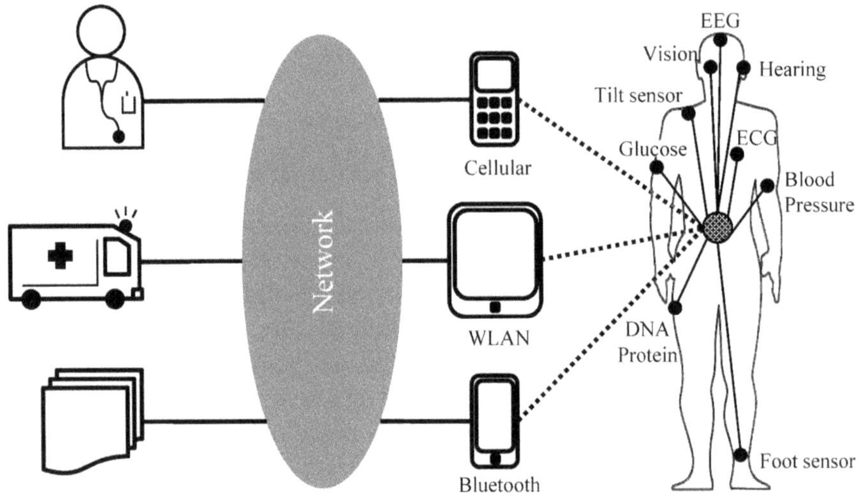

Figure 1.1 Concept of body area communications for healthcare application

sensors are collected by using the on-body communication technique and are then sent to trainers for analysis and administration.

1.2.1.2 Healthcare Monitoring in a Car

The application of body area communications to healthcare monitoring has broad possibilities. A promising use is to monitor a driver's health as a means of in-car communication. In this scenario, as shown in Figure 1.2, some vital sensors are placed on the driver's body to collect healthcare data such as ECG, blood pressure

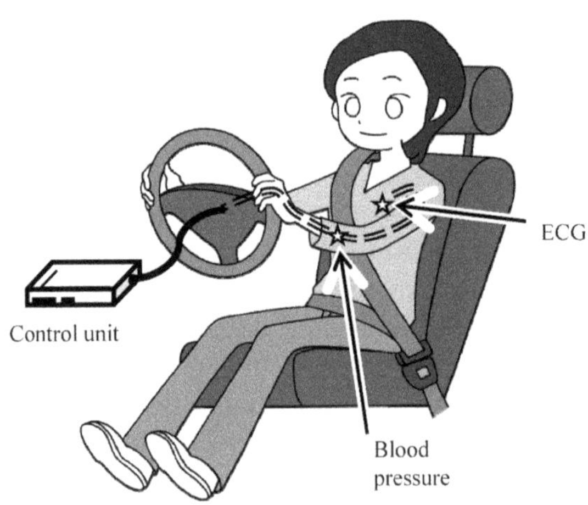

Figure 1.2 Concept of an in-car healthcare monitoring system

and pulse rate. The vital sensors may be embedded in the driver's seat, seat belt, or the steering wheel so that the driver unconsciously wears the sensors. These parts of the car are chosen as they are always in contact with the driver's body when driving. Such a system makes it easy to collect the driver's healthcare data and send them to a control unit with the body area communication technique. The control unit can then analyze the driver's health and generate warning signs or take automatic control of the car, if necessary, for driving safety.

1.2.1.3 Medical Diagnoses and Treatment

Body area communications can also be used in medical diagnoses and treatment. In this scenario, an in-body device should consist of a sensor, a transceiver and an operation unit. The sensor data are sent to an on- or off-body control unit by the wireless transceiver. The control unit makes a medical measurement and sends the corresponding command for medical treatment to the operation unit. The operation unit then carries out medical treatment based on the received command. One example of this scenario is an automatically controlled cardiac pacemaker (Bradley, 2007). A cardiac pacemaker is an electronic device which helps people with irregular heart beat problems. As shown in Figure 1.3, first, the pacemaker collects sympathetic nerve signals using sensors and sends them to a control unit. Then, the control unit calculates the correct heart beat rate and instructs the pacemaker. Finally, the pacemaker helps to adjust the heart beat to the correct beating rhythm.

Another example of body area communications for medical diagnosis is the capsule endoscope. The ingestible capsule consists of a camera and a transceiver. It takes pictures during its course through the digestive tract after being swallowed, and transmits the pictures or video data in real time from the in-body transceiver to

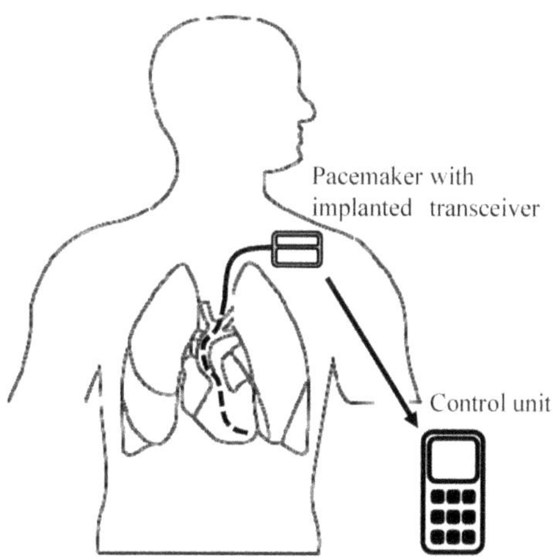

Figure 1.3 In-body to on-/off-body communication scenario for a cardiac pacemaker

Table 1.2 Required data rates for medical and healthcare information

Health information	Data rate
On-body	
ECG	36 kbps
EEG	98 kbps
Pulse rate	2.4 kbps
Respiratory rate	1.0 kbps
Blood pressure	1.92 kbps
Heart rate	1.92 kbps
Body temperature	2.4 bps
In-body	
Capsule endoscope	10 Mbps

off-body medical instruments. Figure 1.4 shows the concept of this scenario, which can effectively promote the noninvasive diagnosis.

In addition, automatic insulin injection for diabetes patients is also a possible application of body area communications. Using the data from a glucose sensor under the skin, an injection control unit linked by the body area communication technique can decide the correct amount of insulin to be injected. Then, an insulin pump carries out the injection according to the instruction from the control unit.

The required data rate for various medical and healthcare data can be calculated with the number of used channels N_C, the sampling rate f_s and the number of quantization bits N_b by $f_b = N_C f_s N_b$. On-body sensor data are usually sampled with a sampling rate f_s between 0.2 and 256 Hz and quantized with a 12- or 16-bit analog to digital converter. On the other hand, the raw data for the capsule endoscope may need a data rate of 76 Mbps for real time transmission. Even if an image compression technique is employed, a data rate as high as 10 Mbps may be still necessary for a high-quality picture or video transmission. Table 1.2 summarizes some estimated data rates required for body area transmission of medical and healthcare information data (Misic and Misic, 2010). A data rate ranging from several bps to 10 Mbps should be supported by body area communications.

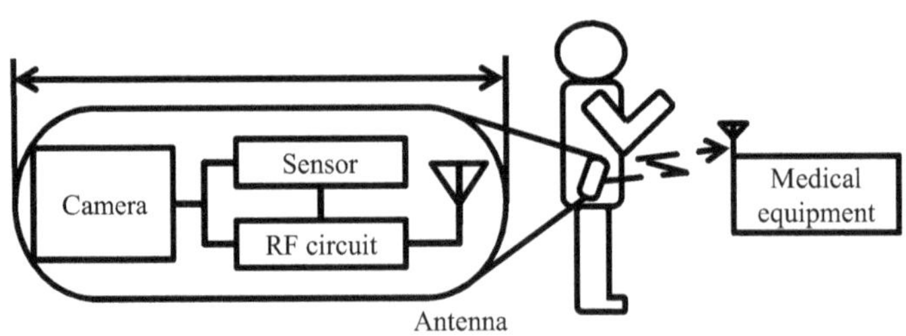

Figure 1.4 In-body to on-/off-body wireless capsule endoscope

1.2.2 Assistance to People with Disabilities

In the second category of applications, there are also plenty of possible scenarios (Li, Takizawa, and Kohno, 2008). One of the typical scenarios is the application for assisting people with visual disabilities. In this application, a wireless body area link is formed among the sensors attached to the belongings of a person and a transceiver worn by the person. A reasonable range between these sensors and the transceiver is set in advance. When the person forgets his or her belongings and leaves them over the pre-set range, a warning signal will be generated automatically from the transceiver. Moreover, in advanced applications, a camera with on-body communication function can be attached to a person with visual disability. Pictures taken by the camera are sent to an on-body control unit, where they are converted to audio signal to provide guidance to the person. A similar principle can also be used for assisting people with speech disability. Here, sensors to catch finger and hand movements are used, and the obtained information is converted into speech.

Accident prevention or rescue for elderly people is also a promising application of body area communications. For example, an elderly person can wear a foot sensor for monitoring steps and some tilt sensors for monitoring accidental falls. The sensor data are continuously sent to an on-body receiver by means of the body area communication technique. If these sensors detect anything unusual, the receiver can give a warning signal to the elderly person or emit a warning sound.

1.2.3 Consumer Electronics and User Identification

The third category of applications is for consumer electronic connectivity. A typical example is to connect a headphone to a music player without wires. By using body area communications, we can not only increase convenience by removing wires but also provide a method of source sharing in audio or video streaming. In these applications, for example, two or more people can share the same music player by using wireless headphones. People can also exchange their business card information by handshake.

On the other hand, body area communications also provides a paradigm shift toward intuitive applications on user identification and user–machine interface (Baldus *et al.*, 2009). By wearing an on-body transceiver, we can communicate with a machine by touching it to establish a body area link. For example, we can embed a user identification function in the on-body transceiver and an entrance door. Then, when we touch the door knob, an on-body communication link is established and the door will be unlocked. A similar system is to embed the user identification function in the mouse of a personal computer. Instead of entering a password, we only need to touch the mouse and the user identification will be automatically made. In a more advanced application, we may embed an automatic payment function in the on-body transceiver. Such a system can be used for passing through an automatic ticket gate. As long as the person touches the ticket gate, the gate will open and a charge is paid.

1.3 Available Frequency Bands

Based on these promising application scenarios, it is found that body area communications may utilize a broad data rate range. Very low consumption power is also a remarkable feature. Figure 1.5 makes a comparison of the requirements between body area communications and other short range communications as well as wireless LANs. To meet such a broad range of data rates, different frequency bands may be required to best fit theses requirements.

1.3.1 UWB Band

The ultra wideband (UWB) technique generally employs very narrow or short duration pulses as modulation signal, which results in a very large frequency bandwidth coverage. The UWB signal is defined as the fractional bandwidth (FBW) being greater than 0.20–0.25 or the whole occupied bandwidth being greater than 500 MHz. Within the defined fractional bandwidth the UWB signal is restricted by the power spectrum density (PSD). The PSD is the ratio of the transmitting power P_T to the frequency bandwidth B, that is,

$$PSD = \frac{P_T}{B} \tag{1.1}$$

In 2002, the US Federal Communications Commission (FCC) approved the first rules regarding the UWB transmission system (FCC, 2002). The formula proposed for calculating the fractional bandwidth is

$$\text{FBW} = 2\frac{f_H - f_L}{f_H + f_L} \tag{1.2}$$

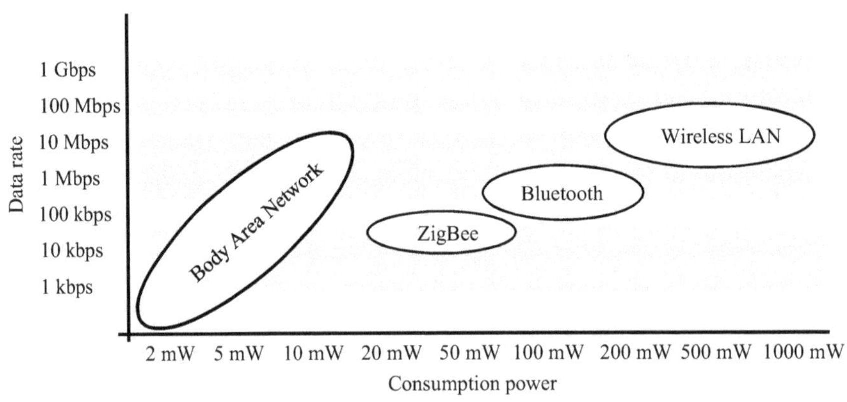

Figure 1.5 Comparison of the requirements between body area communications and other short range communications

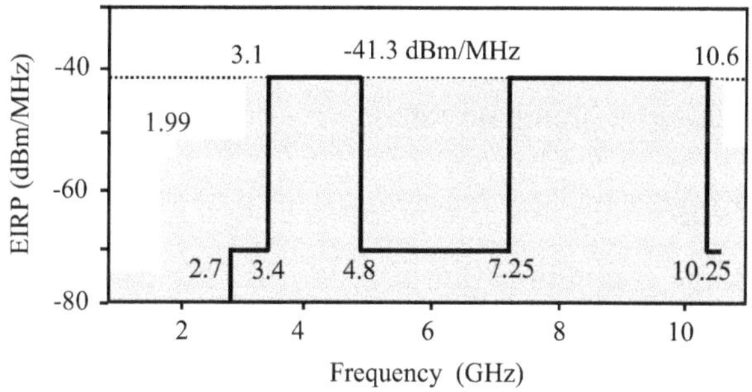

Figure 1.6 FCC-regulated (gray scale) and Japanese (bold line) UWB PSD mask

where f_H is the upper frequency of the -10 dB PSD point and f_L is the lower frequency of the -10 dB PSD point. The center frequency of the UWB transmission is defined as the average of the upper and lower -10 dB points, that is, $(f_H + f_L)/2$. Figure 1.6 shows the FCC-regulated PSD mask for UWB transmission. The equivalent isotropic radiated power (EIRP) of a UWB transmitter is required to meet this PSD mask. The EIRP is the amount of power that a theoretically isotropic antenna would emit to produce the peak power density observed in the direction of maximum antenna gain. With P_T as the transmitting power and G_T as the antenna gain, we have

$$\mathrm{EIRP} = P_T \cdot G_T \tag{1.3}$$

From 3.1 to 10.6 GHz, the measured EIRP is not allowed to exceed −41.3 dBm/MHz or 74.13 nW/MHz. This yields an maximum allowable power of 0.556 mW in the full UWB band.

On the other hand, in Europe and East Asia, the UWB bands are further divided into low band and high band (EC, 2009). This is mainly for removing the 5 GHz band where wireless LANs are being used. The low band may range from 3.1 to 4.8 GHz, and the high band may range from 6.0 to 10.2 GHz, although there are some minor differences among regions and countries. Moreover, in the UWB low band, low duty cycle or detected-and-avoid (DAA) algorithm is used to avoid possible interference with other systems.

The UWB technique has shown an extensive potential for applications on either high data rates over short ranges or low data rates over largely attenuated channels. The extremely low PSD of the UWB signal can be thought to have less influence on medical equipment than other communication systems such as Bluetooth, wireless LAN and cellular phones. Its wideband nature also permits a fine time resolution which is particularly beneficial for health monitoring, human body probing and real time diagnosis (Staderini, 2002). In addition, the hardware miniaturization and low consumption power can be expected for UWB transceivers because of its simpler modulation and demodulation schemes. All of these features make UWB a promising candidate for wireless body area communications, especially for providing a high data rate.

1.3.2 MICS Band

The medical implant communication service (MICS) band is specified between 402 MHz and 405 MHz for communication with medical implants (ITU-R SA.1346, 1998). It allows bi-directional wireless communication with a pacemaker or other electronic implants. With a MICS band technique, one can establish a wireless link between an in-body medical device and on- or off-body monitoring and control equipment. In order to reduce the risk of interfering with other users of the same band, the permissible maximum transmitting power in the MICS band is very low, with an EIRP = 25 μW or −16 dBm. The maximum bandwidth is also limited to 300 kHz, thus a high data rate is difficult. The main advantage of the MICS band signal is its low attenuation compared with the UWB signal when the signal propagates through the human body. This feature makes it a promising choice, especially for in-body communications.

In Japan, MICS devices are classified as specific low power equipment permitted to have an emission up to 0.01 W. An adequate modification for the permissible maximum transmitting power may significantly increase the usefulness of this frequency band. In addition, the wireless medical telemetry system (WMTS) is assigned to 420–430 and 440–450 MHz. WMTS devices are also classified as specific low power equipment, which exhibit potential for use in body area communications.

1.3.3 ISM Band

The industrial, scientific and medical (ISM) bands are mainly for the use of radio frequency (RF) energy for industrial, scientific and medical purposes other than communications. The ISM band does not need end-user licenses up to 1 W, but may be subjected to authorization of local administrations. Despite the intent of the original allocations, in recent years many short range and low power communication systems have been used in these bands.

Suitable ISM bands for body area communications may include the 430 MHz band and the 2.4 GHz band. The former is currently available in Europe and the maximum effective radiated power may be up to 10 mW. The latter is supporting the fast growth in various short range communication services such as Bluetooth, Zigbee, and wireless LAN. A drawback of the 2.4 GHz band is the lack of any protection against interference from other communication services in the same band. The coexisting issue with current communication services also limits the introduction of body area communications.

1.3.4 HBC Band

Human body communication (HBC) uses the human body as a communication route to transmit data. It usually operates in the range of dozens of kHz to dozens of MHz, since at these frequencies the propagation loss along the human body is smaller than that through air. Based on the consideration in IEEE 802.15.6, we here refer to the frequency band from 10 to 50 MHz as the HBC band. HBC provides a new possibility for low data rate on-body communication. Its low propagation loss may yield superior communication performance compared with UWB and ISM bands, and the low radiation outward the human body also leads to high security.

There is not a common regulation in this frequency band, which actually covers several bands including wireless card and amateur radios. In Japan, a transceiver in this frequency band may be classified as extremely low power radio equipment in which the radiated electric field from the transceiver is asked to not exceed 500 μV/m at a distance of 3 m. When this requirement is met, no license is needed for the transceiver. In view of the low radiation feature of HBC, an HBC-based on-body communication link can be established under extremely low power radio regulation.

1.4 Standardization (IEEE Std 802.15.6-2012)

The IEEE 802 Standards Committee is an international organization which develops international standards on wireless communications. As one of the working groups under IEEE 802, IEEE 802.15 focuses on wireless personal area networks. A task group referred to as IEEE 802.15.6 was set up in December 2007 for defining new physical (PHY) and media access control (MAC) layers for wireless BANs. Its scope is to cover not only medical healthcare applications but also consumer electronics applications. The large scope of applications and wide range of technical requirements, however, mean the standard allows multiple PHY layers. In order to provide a common platform for the multiple PHY layers, a common MAC layer, both a beacon mode and a nonbeacon mode, is then defined. The typical number of devices is assumed to be 6 nodes but should be scalable up to 256 nodes.

The IEEE Standard for local and metropolitan area networks – Part 15.6: Wireless Body Area Networks (IEEE Std 802.15.6-2012) was approved in February 2012. In the introduction it is stated that:

> IEEE Std 802.15.6-2012 is a standard for short-range, wireless communications in the vicinity of, or inside, a human body (but not limited to humans). It uses ISM and other bands as well as frequency bands in compliance with applicable medical and communication regulatory authorities. It allows devices to operate on very low transmit power for safety to minimize the specific absorption rate (SAR) into the body and increase the battery life. It supports quality of service (QoS), for example, to provide for emergency messaging. Since some communications can carry sensitive information, it also provides for strong security.

In IEEE Std 802.15.6-2012, the main PHY proposals include narrowband PHY in MICS band, WMTS band and ISM band, UWB PHY as well as HBC band PHY. First, the basic modulation schemes for narrowband PHY are $\pi/2$-shifted differential binary phase shift keying (DBPSK), $\pi/4$-shifted differential quadrature phase shift keying (DQPSK) and Gaussian minimum shift keying (GMSK) at a data rate of 50 kbps to 1 Mbps. The use of Bose–Chaudhuri–Hocquenghem (BCH) error correction code, which belongs to a class of cyclic codes with efficient multiple-error-correcting features, is further recommended to improve communication performance. Secondly, the basic modulation/demodulation schemes for UWB are impulse radio UWB (IR-UWB) and wideband frequency modulation UWB (FM-UWB) with noncoherent detection or differentially coherent detection at a data rate of 0.2–12 Mbps. An automatic repeat request (ARQ) algorithm is suggested for maintaining a high quality of service. Finally, the HBC band PHY employs 21 MHz to implement a baseband transmission of digital signals over the human body at a data rate of 164 kbps to 1.3 Mbps.

Since this book focuses on the PHY layer of body area communications, we here only summarize the basic PHY specifications in IEEE Std 802.15.6-2012.

1.4.1 Narrow Band PHY Specification

In order to transmit a PHY service data unit (PSDU), one first needs to transform it into a PHY protocol data unit (PPDU). As shown in Figure 1.7, the PPDU is composed of a PHY layer convergence protocol (PLCP) preamble, a PLCP header and a PSDU. The PLCP preamble is used to aid the receiver in packet detection, timing synchronization and carrier recovery, and the PLCP header is used to convey the necessary PHY parameter information to aid the decoding of the PSDU. The PLCP head also includes BCH parity bits for improving its robustness.

Based on this definition, a packet is transmitted in the order of the PLCP preamble, the PLCP header and the PSDU. The available operating frequency bands and modulation parameters are summarized in Table 1.3.

According to Table 1.3, the binary bit stream $b(n)$ in the PPDU will be mapped onto either a rotated and differentially encoded constellation referred to as DPSK or a corresponding frequency deviation referred to as GMSK. In GMSK, the frequency deviation is equal to the product of the symbol rate and a modulation index of 0.5 divided by two. The Gaussian pulse shape is used to

PLCP preamble	PLCP header	PSDU

Figure 1.7 Narrow band PPDU structure

Table 1.3 Modulation parameters in narrow band PHY layer. Reproduced with permission from IEEE Std 802.15.6-2012 (2012): IEEE Standard for Local and metropolitan area networks – Part 15.6: Wireless Body Area Networks

Packet component	Frequency band (MHz)	Modulation	Symbol rate (kbps)	Code rate (k/n)	Data rate (kbps)
PLCP header	402–405	$\pi/2$-DBPSK	187.5	19/31	57.5
PSDU		$\pi/2$-DBPSK	187.5	51/63	75.9
		$\pi/2$-DBPSK	187.5	51/63	151.8
		$\pi/4$-DQPSK	187.5	51/63	303.6
		$\pi/8$-D8PSK	187.5	51/63	455.4
PLCP header	420–450	GMSK	187.5	19/31	57.5
PSDU		GMSK	187.5	51/63	75.9
		GMSK	187.5	51/63	151.8
		GMSK	187.5	1/1	187.5
PLCP header	863–870	$\pi/2$-DBPSK	250	19/31	76.6
PSDU	902–928	$\pi/2$-DBPSK	250	51/63	101.2
	950–958	$\pi/2$-DBPSK	250	51/64	202.4
		$\pi/4$-DQPSK	250	51/65	404.8
		$\pi/8$-D8PSK	250	51/66	607.1
PLCP header	2360–2400	$\pi/2$-DBPSK	600	19/31	91.9
PSDU	2400–2483.5	$\pi/2$-DBPSK	600	51/63	121.4
		$\pi/2$-DBPSK	600	51/64	242.9
		$\pi/2$-DBPSK	600	51/65	485.7
		$\pi/4$-DQPSK	600	51/66	971.4

filter the symbol and shape the spectrum. In DPSK, the encoded information is carried in the phase transitions between symbols. Denoting the mapped sequence by a complex value expression

$$S(k) = S(k-1)e^{j\phi_k} \qquad k = 0, 1, \ldots, N/\log_2(M) - 1 \tag{1.4}$$

the phase change ϕ_k will be determined in terms of the bit stream $b(n)$ as given in Table 1.4.

1.4.2 UWB PHY Specification

The UWB PHY specification includes two types of techniques: one is IR-UWB and the other is FM-UWB.

Table 1.4 Relationship between the bit stream and phase change

$\pi/2$-DBPSK ($M=2$)	$b(n)$			ϕ_k
	0			$\pi/2$
	1			$3\pi/2$
$\pi/4$-DQPSK ($M=4$)	$b(2n)$		$b(2n+1)$	ϕ_k
	0		0	$\pi/4$
	0		1	$3\pi/4$
	1		0	$7\pi/4$
	1		1	$5\pi/4$
$\pi/8$-D8PSK ($M=8$)	$b(3n)$	$b(3n+1)$	$b(3n+2)$	ϕ_k
	0	0	0	$\pi/8$
	0	0	1	$3\pi/8$
	0	1	0	$7\pi/8$
	0	1	1	$5\pi/8$
	1	0	0	$15\pi/8$
	1	0	1	$13\pi/8$
	1	1	0	$9\pi/8$
	1	1	1	$11\pi/8$

A UWB PPDU is composed of a synchronization header (SHR), a PHY header (PHR) and a PSDU, as shown in Figure 1.8. The SHR is further divided into two parts. The first part is the preamble used for timing synchronization, packet detection and carrier recovery, and the second part is the start-of-frame delimiter for frame synchronization. The PHR contains information about the data rate, pulse shape, burst mode, and so on.

To transmit the bits of the PPDU as a RF signal, there are three possible modulation schemes: on-off keying (OOK), DBPSK/DQPSK or a combination of continuous-phase BFSK and wideband frequency modulation, FM-UWB. The operating frequency bands are listed in Table 1.5. A UWB device must operate in either channel 1 or channel 6. The other channels are optional.

The shape of pulses to be transmitted is not mandatorily defined, but it must fulfill the transmit power spectral mask given as follows:

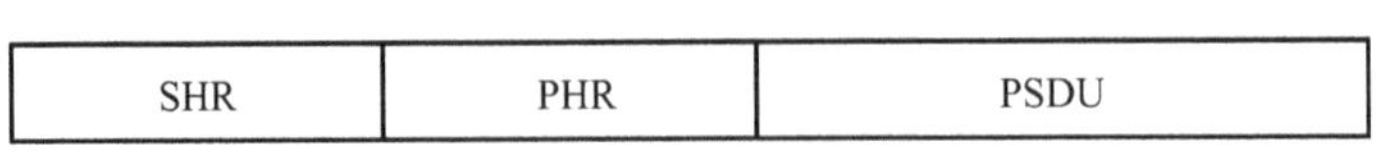

Figure 1.8 UWB PPDU structure

Table 1.5 UWB operating frequency bands

Band group	Channel number	Central frequency (MHz)	Bandwidth (MHz)
Low band	0	3494.4	499.2
	1	3993.6	499.2
	2	4492.8	499.2
High band	3	6489.6	499.2
	4	6988.8	499.2
	5	7488.0	499.2
	6	7987.2	499.2
	7	8486.4	499.2
	8	8985.6	499.2
	9	9484.8	499.2
	10	9984.0	499.2

$$M(f) = \begin{cases} 0 & |f - f_c| < \dfrac{0.5}{T} \\ -60[|f - f_c|T - 0.5] & \dfrac{0.5}{T} \leq |f - f_c| < \dfrac{0.8}{T} \\ -10[|f - f_c|T - 0.8] - 18 & \dfrac{0.8}{T} \leq |f - f_c| < \dfrac{1}{T} \\ -20 & |f - f_c| > \dfrac{1}{T} \end{cases} \quad \text{(dB)} \qquad (1.5)$$

where $T = 1/499.2$ MHz.

The date rates depend on the PSDU symbol duration T_s and modulation schemes. For uncoded transmission, the data rates are $1/T_s$ for OOK and DBPSK, and $2/T_s$ for DQPSK, respectively. When the channel coding is made, a channel coding rate, defined as the number of information bits divided by the number of coded bits, will be introduced. Then the coded data rates are given by the channel coding rate/T_s for OOK and DBPSK, and twice the channel coding rate/T_s for DQPSK, respectively. Table 1.6 summarizes the data rates for IR-UWB with OOK and Table 1.7 summarizes that for IR-UWB with DPSK. R0, R1 and R2 are defined in the PHR frame for specifying a data rate.

1.4.3 HBC PHY Specification

A HBC packet is composed of a PLCP preamble, a start frame delimiter (SFD), a PLCP header, and a PSDU, as shown in Figure 1.9. During packet reception, the receiver achieves the packet synchronization by detecting the PLCP preamble, and then finds the starting point of each frame by detecting the SFD. The PLCP header

Table 1.6 Data rates for IR-UWB with OOK modulation. Reproduced with permission from IEEE Std 802.15.6-2012 (2012): IEEE Standard for Local and metropolitan area networks – Part 15.6: Wireless Body Area Networks

R0, R1, R2	Symbol duration T_s (ns)	Uncoded data rate (Mbps)	Channel coding rate	Coded data rate (Mbps)	Number of pulses in one pulse waveform
000	2051.300	0.487	0.81	0.395	32
100	1025.600	0.975	0.81	0.790	16
010	512.820	1.950	0.81	1.579	8
110	256.410	3.900	0.81	3.159	4
001	128.210	7.800	0.81	6.318	2
101	64.103	15.600	0.81	12.636	1

Table 1.7 Data rates for IR-UWB with DPSK modulation. Reproduced with permission from IEEE Std 802.15.6-2012 (2012): IEEE Standard for Local and metropolitan area networks – Part 15.6: Wireless Body Area Networks

R0, R1, R2	Symbol duration T_s (ns)	Modulation	Uncoded data rate (Mbps)	Channel coding rate	Coded data rate (Mbps)	Number of pulses in one pulse waveform
000	2051.300	DBPSK	0.487	0.5	0.243	32
100	1025.600	DBPSK	0.975	0.5	0.457	16
010	512.820	DBPSK	1.950	0.5	0.975	8
110	256.410	DBPSK	3.900	0.5	1.950	4
001	128.210	DBPSK	7.800	0.5	3.900	2
101	128.210	DQPSK	15.600	0.5	7.800	2
011	1794.900	DBPSK	0.557	0.5	0.278	4
111	1794.900	DQPSK	1.114	0.5	0.557	4

records the transmission parameters such as data rate, burst mode, PSDU length and so on. Such a packet is transmitted based on a frequency selective digital transmission scheme, in which the data are spread in the frequency domain using frequency selective spread codes.

Figure 1.9 HBC packet structure

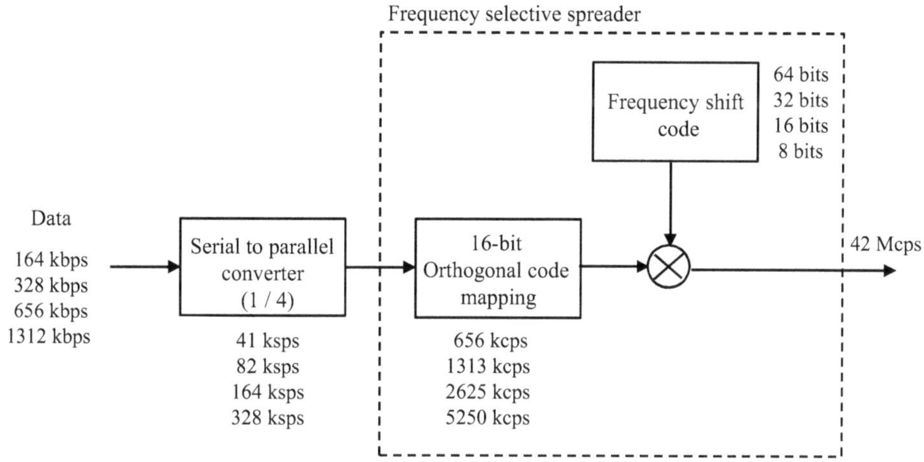

Figure 1.10 Block diagram of serial to parallel converter and frequency selective spreader. Reproduced with permission from IEEE Std 802.15.6-2012 (2012): IEEE Standard for Local and metropolitan area networks – Part 15.6: Wireless Body Area Networks

Referring to Figure 1.10, the data to be transmitted are created by mapping 4 bits (a symbol) from serial-to-parallel converter to a 16-bit chip. The 16-bit chip is then spread by applying a frequency shift code. This modulation scheme is referred to as a frequency selective spreader. It is composed of orthogonal coding and frequency shift coding. The frequency shift code may be a repeated [0,1] code. The number of repeated times is defined as the spreading factor. For a spreading factor of 8, for example, the frequency shift code will be [0,1, 0, 1, 0, 1, 0, 1], that is, 8 bits. In Figure 1.10, the final chip rate at the output of the frequency selective spreader is thus the same regardless of the input data rate. Since the HBC frequency band is centered at 21 MHz, the operating clock frequency is fixed at 42 MHz. Table 1.8 shows the main modulation parameters at 21 MHz in the HBC PHY layer.

Table 1.8 Modulation parameters for the PLCP header and PSDU in the HBC PHY layer

Packet component	Modulation	Symbol rate (ksps)	Data rate (kbps)	Operating clock frequency (MHz)
PLCP header	Frequency selective	41	164	42
PSDU	spreader	41	164	42
		82	328	42
		164	656	42
		328	13 125	42

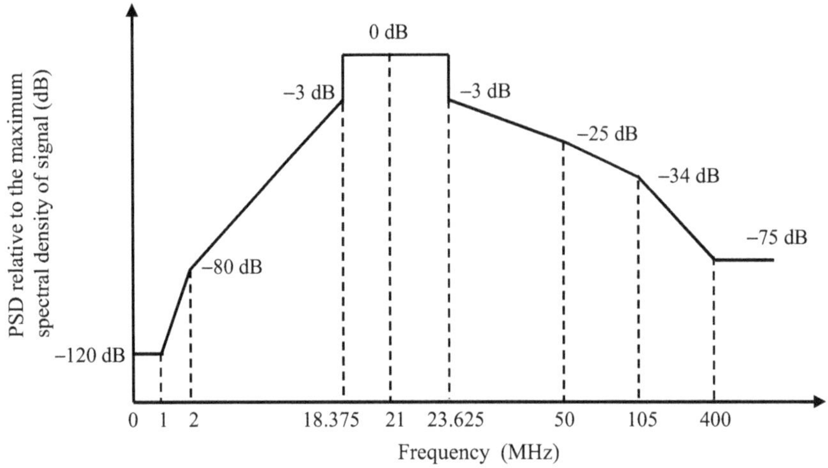

Figure 1.11 Transmit power spectral mask

In addition, the transmit power spectrum is required to be under the mask in order to avoid possible interference with other bands, especially in the 400 MHz band. The transmit power spectral mask is defined in Figure 1.11 within the channel bandwidth of 5.25 MHz at the central frequency of 21 MHz. The vertical axis is the PSD (in units of dB) relative to the maximum spectral density of the signal. The maximum transmit power in a HBC transmitter is limited by local regulations which usually define a radiated electric field level at a specified distance from the transmitter in free space.

IEEE Std 802.15.6-2012 is a new international standard for body area communications, especially for wireless BANs. In this book, although we treat many parts of it, we will go beyond it using more fundamental and broader viewpoints of body area communications.

References

Astrin, A.W., Li, H.-B., and Kohno, R. (2009) Standardization for body area networks. *IEICE Transactions on Communications*, **E92-B** (2), 366–372.

Baldus, H., Corroy, S., Fazzi, A. *et al.* (2009) Human-centric connectivity enabled by body-coupled communications. *IEEE Communications Magazine*, **47** (6), 172–178.

Bonato, P. (2010) Wearable sensors and systems for enabling technology to clinical applications. *IEEE Engineering in Medicine and Biology Magazine*, **29** (3), 25–36.

Bradley, P.D. (2007) Implantable ultralow-power radio chip facilitates in-body communications, RF Design, June issue, pp. 20–24.

EC (2009) Commission Decision of 21 April 2009 - amending Decision 2007/131/EC on allowing the use of the radio spectrum for equipment using ultra-wideband technology in a harmonised manner in the community, 2009/343/EC.

FCC (2002) Revision of Part 15 of the Commission's rules regarding ultra-wideband transmission system: first report and order, Technical Report FCC 02-48 (Adopted February 14, 2002; Released April 22, 2002).

Hall, P.S. and Hao, Y. (eds) (2006) *Antennas and Propagation for Body-Centric Wireless Communications*, Artech House, Norwood, MA.

IEEE Std 802.15.6-2012 (2012) IEEE Standard for local and metropolitan area networks – Part 15.6: Wireless Body Area Networks.

ITU-R SA.1346 (1998) Sharing between the meteorological aids services and medical implant communication systems (MICS) operating in the mobile service in the frequency band 401–406 MHz.

Li, H.-B., Takizawa, K., and Kohno, R. (2008) Trends and standardization of body area network (BAN) for medical healthcare. Proceedings of European Wireless Technology Conference, pp. 1–4.

Li, H.-B., Yazdandoost, K.Y., and Zhen, B. (2010) *Wireless Body Area Network*, River Publishers, Aalborg.

Misic, J. and Misic, V.B. (2010) Bridge performance in a multitier wireless network for healthcare monitoring. *IEEE Wireless Communications*, **17** (1), 90–95.

Staderini, E. (2002) UWB radars in medicine. *IEEE Aerospace and Electronic Systems Magazine*, **17** (1), 13–18.

Zimmerman, T.G. (1996) Personal area networks: near-field intrabody communications. *IBM System Journal*, **35** (3&4), 609–617.

2

Electromagnetic Characteristics of the Human Body

2.1 Human Body Composition

The elements of the human body can be classified into different levels such as the atom, molecule, cell, tissue, and internal organs. When the human body is exposed to an external electromagnetic field, its electromagnetic characteristics are generally treated at the celluar or tissue level.

At the cellular level, electrical properties of the human body are characterized by the cell membrane and the conductive intracellular fluid and extracellular fluid. The cell membrane consists of a phospholipid bilayer with embedded proteins, and separates the interior of the cell from the outside environment. Although the membrane thickness is 10 nm at most, both its resistance and capacitance are large. Since life originated from the sea, the extracellular fluid has a composition similar to that of seawater. The composition of the intracellular fluid is about 20% protein and this is where metabolic activity occurs. The electrical properties change according to the intracellular fluid composition.

On the other hand, the resistance of the extracellular fluid is considered to be smaller than that of the intracellular liquid. The cells unite in the extracellular fluid to compose tissue. The tissue is therefore roughly distinguishable according to the method of uniting so that the electrical properties of the human body can also be characterized by the different moisture content and composition at the tissue level. One kind of tissue is known as low water content tissue which includes skin and bone. The cells unite closely, and there is little intracellular fluid and extracellular liquid. Another kind of tissue is known as high water content tissue, such as blood and muscle, in which there is more extracellular fluid. As for fat, the moisture content varies. The fat in the abdomen is similar to muscle, whereas in other parts of the

Body Area Communications: Channel Modeling, Communication Systems, and EMC, First Edition.
Jianqing Wang and Qiong Wang.

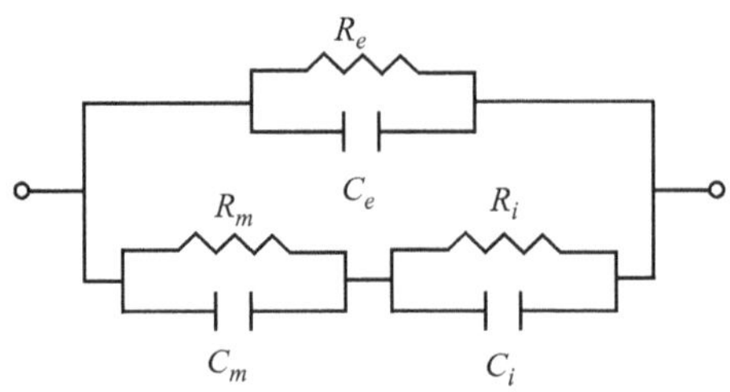

Figure 2.1 Equivalent electric circuit of biological tissue

body it may be similar to bone. Since each tissue is fundamentally the same structure, an equivalent circuit of the electrical properties at the tissue level can be shown as in Figure 2.1 (IEE of Japan, 1995), where the subscripts *m*, *e* and *i* denote the membrane, the extracellular fluid and the intracellular fluid, respectively.

At low frequencies, the equivalent circuit in Figure 2.1 only needs to consider R_e and C_e, because the impedance of the cell membrane is very high and the current will flow in the extracellular fluid. As the frequency increases, the impedance of the cell membrane becomes small so that the membrane can be disregarded by a short-circuit and the current then flows in the cell. In this case, the equivalent circuit should consist of the parallel circuit of R_i, C_i and R_e, C_e. However, C_i and C_e can be disregarded, as long as it is not in a very high frequency band, for example in the microwave band. In the intermediate frequency band, it is necessary to consider both the current which flows in the extracellular fluid and the current which flows in the cell. This means that the equivalent circuit is Figure 2.1 itself in the intermediate frequency band. In short, the equivalent circuit shows a strong frequency dependency.

2.2 Frequency-Dependent Dielectric Properties

Polarization is the most important effect arising from the interaction of an electromagnetic field with the human body. It occurs when internal charge moves in response to an external electromagnetic field. This yields both displacement and conduction currents. For this reason, the human body is classified as a lossy dielectric material. In general, dielectric properties of the human body are expressed as complex permittivity $\dot{\varepsilon}$ with relative permittivity ε_r and conductivity σ

$$\dot{\varepsilon} = \varepsilon_0 \dot{\varepsilon}_r = \varepsilon_0 (\varepsilon_r' - j\varepsilon_r'') = \varepsilon_0 \left(\varepsilon_r - j\frac{\sigma}{\omega \varepsilon_0} \right) \tag{2.1}$$

where ε_0 is the permittivity of free space, and ω is the angular frequency. In Equation 2.1, ε_r'' is a loss factor associated with the conductivity σ. The conductivity σ is given by

$$\sigma = \sigma_d + \sigma_0 \tag{2.2}$$

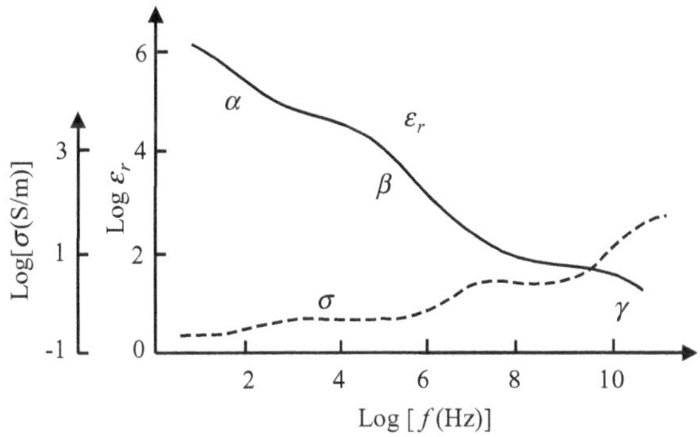

Figure 2.2 Dispersions in the dielectric spectrum of biological tissue

where σ_d is the displacement conductivity and σ_0 is the ionic conductivity. This frequency dependence is determined mainly by three interaction mechanisms, each governed by its own characteristic. When we systematically illustrate the complex permittivity as a function of frequency for human tissue, we can observe three main dispersion regions from the dielectric spectrum, as shown in Figure 2.2 (Schwan, 1957). These dispersions are identified experimentally in the hertz, MHz and GHz frequency regions, and are known as the α, β and γ dispersions, respectively.

The low frequency α dispersion is associated with an ionic diffusion process at the cell membrane. The β dispersion is mainly due to the polarization of cell membranes which act as barriers to ionic flow between the interior and exterior of the cell. Other contributions to the β dispersion come from the polarization of protein and other organic macromolecules. With respect to the γ dispersion in the GHz region, the polarization of water molecules, which are the main constituent of the human body, plays a major role.

To characterize the human body at tissue level, dielectric properties are usually used. The dielectric properties are expressed as ε_r' and ε_r'' values, or ε_r and σ values, as shown in Equation 2.1, as a function of frequency.

2.3 Tissue Property Modeling

Human body modeling should be anatomically accurate at the tissue and organ levels in body area communications. Current high resolution computer models of the human body are based on medical imaging data. The level of detail is such that over 30 tissue types are used, and the resolution is of the order of several millimeters. The application of such models requires the dielectric properties to be allocated to various tissues and organs at the considered frequencies so that the electromagnetic fields can be analyzed with Maxwell's equations. An analytic expression for the dielectric properties, that is, the complex permittivity as a function of frequency, is therefore highly useful in body area communications.

The database on the dielectric properties of biological tissue is mainly based on Gabriel's measurement data (Gabriel, 1996). The dielectric measurements were performed in the frequency range from 1 MHz to 20 GHz for over 20 tissue types. The tissue samples came from animals (mostly ovine from freshly killed sheep), human autopsy materials, and human skin and tongue. All animal tissues used were as fresh as possible, mostly obtained within 2 h of death, and human tissues were obtained 24–48 h after death. The principle is based on scattering parameter or S-parameter measurement with a two-port swept frequency network analyzer or an impedance analyzer. An open-ended coaxial probe is used to interface the input port of the measurement equipment with the tissue samples. At first, the reflection coefficient S_{11} at the input port is obtained. Then, under the assumption that the tissue sample has a sufficiently large cross-section, the complex permittivity is derived from the reflection coefficient as a function of frequency (Misra, 1987). Gabriel carried out a comprehensive survey of past published dielectric data, and found that her data fell well within them and bridge the gaps between them.

The basis used to model the frequency dependence of dielectric properties is the dispersion phenomena in the dielectric spectrum of tissue. The dielectric spectrum is characterized by several dispersion regions. Each one can be characterized by a single relaxation time constant τ with the following frequency dependence

$$\dot{\varepsilon}_r(\omega) = \varepsilon_\infty + \frac{\Delta\varepsilon}{1 + j\omega\tau}. \tag{2.3}$$

This is known as the Debye expression. The magnitude of the dispersion is described by $\Delta\varepsilon$, and ε_∞ is the permittivity when the frequency approaches infinity.

The dielectric properties of the human body can be therefore expressed as a summation of terms corresponding to various dispersion mechanisms. For a frequency range of Hz up to 10 GHz, four Debye-type dispersion region expressions provide good modeling for most tissues. The corresponding expression, named a 4-Cole-Cole expression, is

$$\dot{\varepsilon}_r(\omega) = \varepsilon_\infty + \sum_{n=1}^{4} \frac{\Delta\varepsilon_n}{1 + (j\omega\tau_n)^{1-\alpha_n}} + \frac{\sigma_0}{j\omega\varepsilon_0} \tag{2.4}$$

where each term is described in terms of a modified Debye expression because α_n is introduced to describe the deviation from Debye behavior. It should be noted that the 4-Cole-Cole expression corresponds to the α, β and γ dispersions fundamentally but adds a new dispersion term in higher frequencies. It is actually a semi-empirical multiple Cole-Cole expression. An ionic conductivity term σ_0 is also added to the expression for modeling a conduction current at zero frequency. In each dispersion region, τ_n is the relaxation time constant, and $\Delta\varepsilon_n$ is the magnitude of dispersion. With an appropriate choice of parameters for

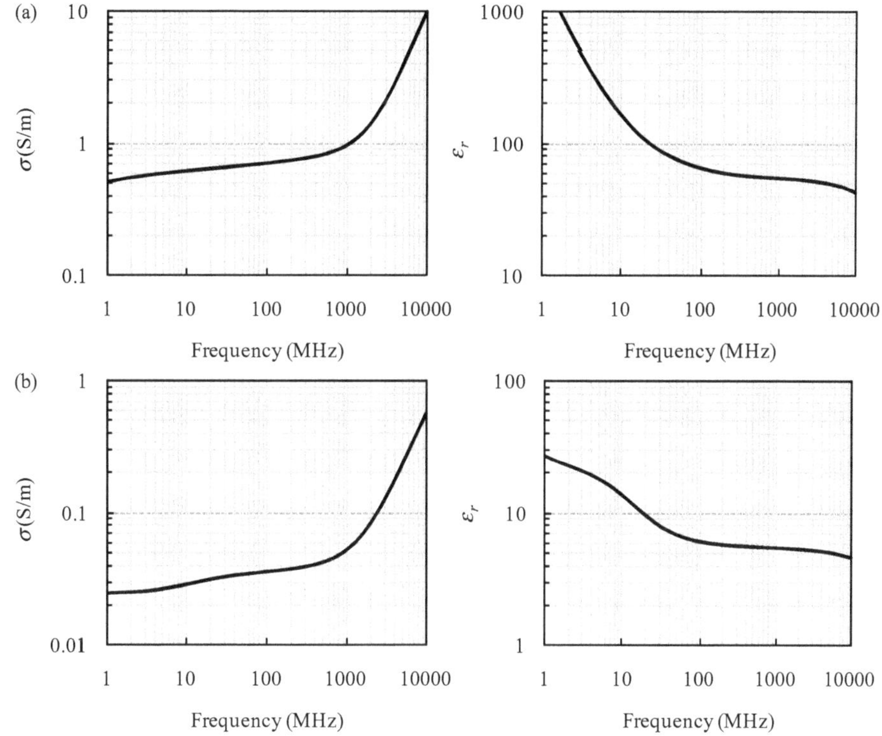

Figure 2.3 Frequency dependence of conductivity and relative permittivity of typical (a) high water content muscle and (b) low water content fat

each tissue, Equation 2.4 can give the dielectric properties as a function of frequency over a desired frequency range. The parameters in the 4-Cole-Cole expression are determined to closely fit the measurement data for each tissue. However, due to the difficulty in measurement at lower frequencies, the determined parameters in the 4-Cole-Cole expression only have confidence for frequencies above 1 MHz.

Table 2.1 gives the corresponding parameters in Equation 2.4 for obtaining the complex permittivity at any frequencies of interest below 10 GHz (Gabriel, 1996). Higher permittivity corresponds to more water content. Such tissues are called high water content tissues. Conversely, lower permittivity corresponds to less water content. Such tissues are called low water content tissues. Figure 2.3 shows the frequency dependence calculated from Equation 2.4 for typical high water content tissue of muscle and low water content tissue of fat. The relative permittivity decreases and the conductivity increases with the frequency. Significantly higher permittivity and conductivity are shown in the high water content tissues compared with the low water content tissues. Tables 2.2–2.5 give the conductivity σ, relative

Table 2.1 Parameters in equation 2.4 for obtaining the complex permittivity at frequencies of interest below 10 GHz

Tissue	ε_∞	$\Delta\varepsilon_1$	τ_1 (ps)	α_1	$\Delta\varepsilon_2$	τ_2 (ns)	α_2	σ_0 (S/m)	$\Delta\varepsilon_3$	τ_3 (μs)	α_3	$\Delta\varepsilon_4$	τ_4 (ms)	α_4
Skin dry	4.0	32.0	7.234	0.00	1100.0	32.481	0.20	0.00	0.00E + 00	159.155	0.20	0.00E + 00	15.915	0.20
Skin wet	4.0	39.0	7.958	0.10	280.0	79.577	0.00	0.00	3.00E + 04	1.592	0.16	3.00E + 04	1.592	0.20
Fat	2.5	9.0	7.958	0.20	35.0	15.915	0.10	0.04	3.30E + 04	159.155	0.05	1.00E + 07	15.915	0.01
Muscle	4.0	50.0	7.234	0.10	7000.0	353.678	0.10	0.20	1.20E + 06	318.31	0.10	2.50E + 07	2.274	0.00
Bone cancellous	2.5	18.0	13.263	0.22	300.0	79.577	0.25	0.07	2.00E + 04	159.155	0.20	2.00E + 07	15.915	0.00
Bone cortical	2.5	10.0	13.263	0.20	180.0	79.577	0.20	0.02	5.00E + 03	159.155	0.20	1.00E + 05	15.915	0.00
Bone marrow	2.5	9.0	14.469	0.20	80.0	15.915	0.10	0.10	1.00E + 04	1591.55	0.10	2.00E + 06	15.915	0.10
Brain gray matter	4.0	45.0	7.958	0.10	400.0	15.915	0.15	0.02	2.00E + 05	106.103	0.22	4.50E + 07	5.305	0.00
Brain white matter	4.0	32.0	7.958	0.10	100.0	7.958	0.10	0.02	4.00E + 04	53.052	0.30	3.50E + 07	7.958	0.02
Vitreous humor	4.0	65.0	7.234	0.00	30.0	159.155	0.10	1.50	0.00E + 00	159.155	0.00	0.00E + 00	15.915	0.00
Blood	4.0	56.0	8.377	0.10	5200.0	132.629	0.10	0.70	0.00E + 00	159.155	0.20	0.00E + 00	15.915	0.00
Heart	4.0	50.0	7.958	0.10	1200.0	159.155	0.05	0.05	4.50E + 05	72.343	0.22	2.50E + 07	4.547	0.00
Kidney	4.0	47.0	7.958	0.10	3500.0	198.944	0.22	0.05	2.50E + 05	79.577	0.22	3.00E + 07	4.547	0.00
Liver	4.0	39.0	8.842	0.10	6000.0	530.516	0.20	0.02	5.00E + 04	22.736	0.20	3.00E + 07	15.915	0.05
Lung deflated	4.0	45.0	7.958	0.10	1000.0	159.155	0.10	0.20	5.00E + 05	159.155	0.20	1.00E + 07	15.915	0.00
Lung inflated	2.5	18.0	7.958	0.10	500.0	63.662	0.10	0.03	2.50E + 05	159.155	0.20	4.00E + 07	7.958	0.00
Stomach	4.0	60.0	7.958	0.10	2000.0	79.577	0.10	0.50	1.00E + 05	159.155	0.20	4.00E + 07	15.915	0.00
Colon	4.0	50.0	7.958	0.10	3000.0	159.155	0.20	0.01	1.00E + 05	159.155	0.20	4.00E + 07	1.592	0.00
Small intestine	4.0	50.0	7.958	0.10	10000.0	159.155	0.10	0.50	5.00E + 05	159.155	0.20	4.00E + 07	15.915	0.00

Table 2.2 Dielectric properties of some main tissues and organs at 10 MHz

Tissue	Conductivity (S/m)	Relative permittivity	Loss tangent
Skin dry	0.20	361.66	0.98
Skin wet	0.37	221.81	2.97
Fat	0.03	13.77	3.81
Muscle	0.62	170.73	6.49
Bone cancellous	0.12	70.78	3.12
Bone cortical	0.04	36.77	2.09
Bone marrow	0.01	19.27	1.02
Brain gray matter	0.29	319.67	1.64
Brain white matter	0.16	175.72	1.62
Vitreous humor	1.50	70.01	38.56
Blood	1.10	280.03	7.04
Heart	0.50	293.47	3.07
Kidney	0.51	371.15	2.46
Liver	0.32	223.12	2.55
Lung deflated	0.44	180.32	4.37
Lung inflated	0.23	123.66	3.27
Esophagus	0.78	246.43	5.72
Stomach	0.78	246.43	5.72
Colon	0.49	271.45	3.24
Duodenum	0.78	246.43	5.72
Small intestine	1.34	488.46	4.95

permittivity ε_r and loss tangent (defined as $\sigma/\omega\varepsilon_0\varepsilon_r$) at 10 MHz, 400 MHz, 2.45 GHz and 5 GHz, respectively, for some main tissues and organs of the human body. The chosen frequencies cover main candidate bands for body area communications, and the dielectric properties play a dominant role in characterizing the body area channels. A comprehensive database for the dielectric properties of various tissues can be found from the FCC website http://transition.fcc.gov/oet/rfsafety/dielectric.html or the Italian National Research Council website http://niremf.ifac.cnr.it/tissprop/htmlclie/htmlclie.htm.

Although Equation 2.4 gives a good expression for the dielectric properties of tissue, it is sufficiently complicated so as not to be easily incorporated into actual electromagnetic analysis. As shown in Figure 2.2, however, the dielectric properties are dominated by different dispersions in different frequency bands. For example in Figure 2.3, the variations of conductivity and relative permittivity from 100 MHz to 1 GHz are relatively small, which suggests that only one Cole-Cole term, that is, a simpler expression for Equation 2.4 may be sufficient in a specific frequency band. In these cases, the parameter α_n is nearly zero and the complex relative permittivity

Table 2.3 Dielectric properties of some main tissues and organs at 400 MHz

Tissue	Conductivity (S/m)	Relative permittivity	Loss tangent
Skin dry	0.69	46.79	0.66
Skin wet	0.67	49.90	0.60
Fat	0.04	5.58	0.33
Muscle	0.80	57.13	0.63
Bone cancellous	0.23	22.44	0.47
Bone cortical	0.09	13.15	0.31
Bone marrow	0.03	5.67	0.23
Brain gray matter	0.74	57.44	0.58
Brain white matter	0.44	42.07	0.47
Vitreous humor	1.53	69.00	1.00
Blood	1.35	64.18	0.94
Heart	0.96	66.10	0.66
Kidney	1.09	66.42	0.74
Liver	0.65	51.24	0.57
Lung deflated	0.68	54.58	0.56
Lung inflated	0.37	23.81	0.71
Esophagus	1.00	67.49	0.67
Stomach	1.00	67.49	0.67
Colon	0.86	62.59	0.62
Duodenum	1.00	67.49	0.67
Small intestine	1.90	66.14	1.29

can be approximated by the Debye expression. The first-order Debye expression is given by

$$\dot{\varepsilon}_r(\omega) = \varepsilon_\infty + \frac{\Delta\varepsilon_1}{1 + j\omega\tau_1} \tag{2.5}$$

or

$$\dot{\varepsilon}_r(\omega) = \varepsilon_\infty + \frac{\Delta\varepsilon_1}{1 + j\omega\tau_1} + \frac{\sigma_0}{j\omega\varepsilon_0} \tag{2.6}$$

with an ionic conductivity term σ_0 in the human body. At high frequencies, however, the third term in Equation 2.6 is omissible. Figure 2.4 shows an example of the first-order Debye expression to approximate the dielectric properties of muscle for frequencies from 3.4 to 4.8 GHz, that is, in the UWB low band. The symbols are the actual values and the lines are the Debye approximated values with $\varepsilon_\infty = 4.0$, $\Delta\varepsilon_1 = 48.5$, $\tau_1 = 7.6$ ps and $\sigma_0 = 1.0$ S/m. A reasonable accuracy with the maximum difference smaller than 5% demonstrates the usefulness of the first-order Debye approximation.

Table 2.4 Dielectric properties of some main tissues and organs at 2.45 GHz

Tissue	Conductivity (S/m)	Relative permittivity	Loss tangent
Skin dry	1.46	38.01	0.28
Skin wet	1.59	42.85	0.27
Fat	0.10	5.28	0.15
Muscle	1.74	52.73	0.24
Bone cancellous	0.81	18.55	0.32
Bone cortical	0.39	11.38	0.25
Bone marrow	0.10	5.30	0.13
Brain Gray matter	1.81	48.91	0.27
Brain white matter	1.22	36.17	0.25
Vitreous humor	2.48	68.21	0.27
Blood	2.54	58.26	0.32
Heart	2.26	54.81	0.30
Kidney	2.43	52.74	0.34
Liver	1.69	43.04	0.29
Lung deflated	1.68	48.38	0.26
Lung inflated	0.80	20.48	0.29
Esophagus	2.21	62.16	0.26
Stomach	2.21	62.16	0.26
Colon	2.04	53.88	0.28
Duodenum	2.21	62.16	0.26
Small intestine	3.17	54.43	0.43

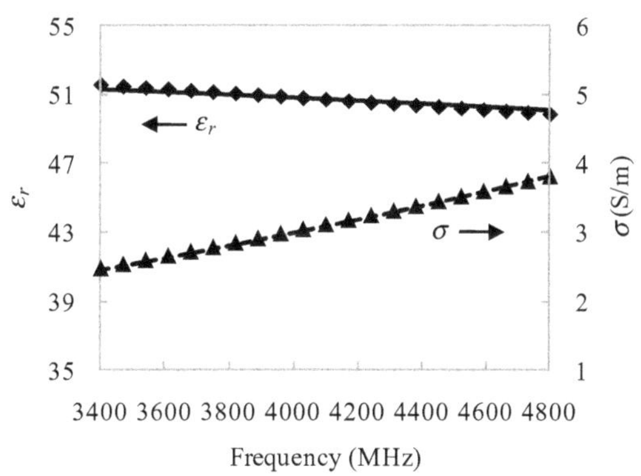

Figure 2.4 An example of the first-order Debye expression to approximate the dielectric properties of muscle in the UWB low band

Table 2.5 Dielectric properties of some main tissues and organs at 5 GHz

Tissue	Conductivity (S/m)	Relative permittivity	Loss tangent
Skin dry	3.06	35.77	0.31
Skin wet	3.57	39.61	0.32
Fat	0.24	5.03	0.17
Muscle	4.04	49.54	0.29
Bone cancellous	1.81	16.05	0.41
Bone cortical	0.96	10.04	0.34
Bone marrow	0.23	5.04	0.17
Brain gray matter	4.10	45.15	0.33
Brain white matter	2.86	33.44	0.31
Vitreous humor	5.41	65.81	0.30
Blood	5.40	53.95	0.36
Heart	4.86	50.27	0.35
Kidney	4.94	48.06	0.37
Liver	3.83	39.26	0.35
Lung deflated	3.94	44.86	0.32
Lung inflated	1.72	18.97	0.33
Esophagus	5.16	57.89	0.32
Stomach	5.16	57.89	0.32
Colon	4.58	49.72	0.33
Duodenum	5.16	57.89	0.32
Small intestine	5.75	49.98	0.41

2.4 Aging Dependence of Tissue Properties

For the complex relative permittivity expression of a biological tissue, we can rewrite Equation 2.1 as

$$\dot{\varepsilon}_r = \varepsilon_r' - j\varepsilon_r'' = \varepsilon_r - j\frac{\sigma}{\omega\varepsilon_0} = \varepsilon_r\left(1 - j\frac{1}{\omega\tau_A}\right) \tag{2.7}$$

where $\tau_A = \varepsilon_0\varepsilon_r/\sigma$. There is a useful expression, known as a Cole-Cole plot, for $\dot{\varepsilon}_r$. A Cole-Cole plot is a plot of ε_r'' on the y (or imaginary) axis versus ε_r' on the x (or real) axis. Figure 2.5 gives such a plot for measured complex relative permittivity data for aging rat tissues (Peyman, Rezazadeh, and Gabriel, 2001). It is found that all five tissues, that is, the skin, muscle, skull, brain, and salivary gland, have an almost constant ratio of ε_r'' to ε_r' or an almost constant τ_A. This finding suggests that τ_A is nearly independent of age.

On the other hand, a biological tissue can be considered as a composition of water and organic material. The amount of water inside the biological tissue should change with age, while the organic material depends only on the tissue type. Therefore, ε_r in

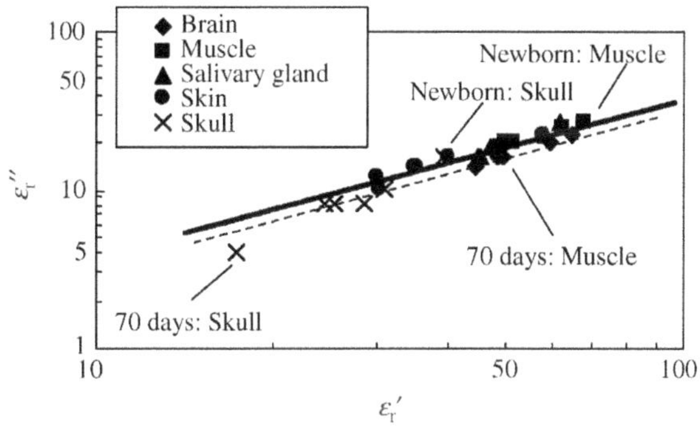

Figure 2.5 Cole-Cole plot for the dielectric properties at 900 MHz. The solid and dashed lines are the approximated dielectric properties for the high water content and low water content tissues, respectively (Wang, Fujiwara, and Watanabe, 2006). Reproduced with permission from Wang J., Fujiwara O. and Watanabe S., "Approximation of aging effect on dielectric tissue properties for SAR assessment of mobile telephones," *IEEE Transactions on Electromagnetic Compatibility*, **48**, 2, 408–413, 2006. © 2006 IEEE

Equation 2.7 should dominate the change in dielectric properties with age. In addition, it is known that Lichtenecker's exponential law holds for ε_r in composite dielectric materials (Lichtenecker, 1926). We can thus express ε_r for any tissue as

$$\varepsilon_r = \varepsilon_{rw}^{\alpha} \cdot \varepsilon_{rt}^{1-\alpha} \tag{2.8}$$

where ε_{rw} is the relative permittivity of water (ranging from 74.3 at 10 MHz to 71.6 at 5 GHz at 37 °C), ε_{rt} is the relative permittivity of organic material, and α is the hydrated rate which is related to the mass density ρ and the total body water (TBW) by $\alpha = \rho \cdot \text{TBW}$. It should be noted that ε_{rt} depends only on the tissue type and is not a function of age. Substituting Equation 2.8 into 2.7 for adult tissue, we can represent ε_{rt} using the relative permittivity ε_{rA}, relaxation time constant τ_A, and hydrated rate α_A for adult tissue. After a primary operation to eliminate ε_{rt} in Equation 2.7, we have

$$\dot{\varepsilon}_r = \varepsilon_{rw}^{\frac{\alpha-\alpha_A}{1-\alpha_A}} \cdot \varepsilon_{rA}^{\frac{1-\alpha}{1-\alpha_A}} \left(1 - j\frac{1}{\omega\tau_A}\right). \tag{2.9}$$

Equation 2.9 gives an empirical representation of the complex relative permittivity for a tissue with a hydrated rate α. It is therefore possible to derive the dielectric properties at different ages from Equation 2.9 as long as the hydrated rate α is known as a function of age and the dielectric properties are known for an adult.

Actually, the hydrated rate α is related to the TBW by $\alpha = \rho \cdot \text{TBW}$. Figure 2.6 shows the TBW as a function of age based on the data in Altman and Dittmer (1974). As can be seen, the TBW varies greatly under 3 years old, but becomes

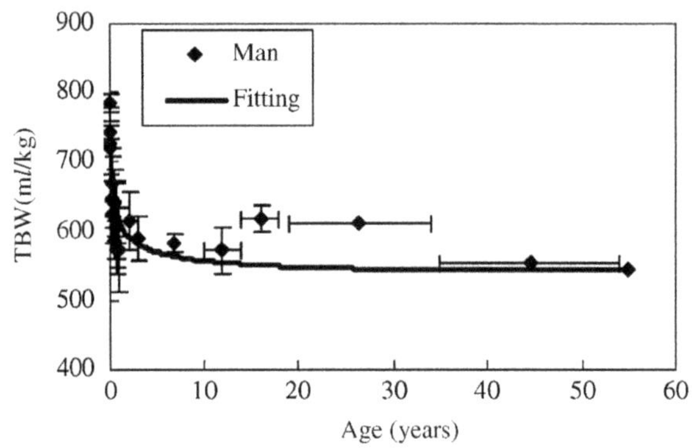

Figure 2.6 Total body water for humans as a function of age (Wang, Fujiwara, and Watanabe, 2006). Reproduced with permission from Wang J., Fujiwara O. and Watanabe S., "Approximation of aging effect on dielectric tissue properties for SAR assessment of mobile telephones," *IEEE Transactions on Electromagnetic Compatibility*, **48**, 2, 408–413, 2006. © 2006 IEEE

insignificant over 3 years old. In a total trend, the TBW may be fitted with the following equation (Wang, Fujiwara, and Watanabe, 2006)

$$TBW = 784 - 241e^{-\left[\frac{\ln(Age/55)}{6.9589}\right]^2} \quad [\text{ml/kg}] \tag{2.10}$$

and then the hydrated rate α can be obtained for Equation 2.9. It should be noted that the TBW is not guaranteed to be the same in various human tissues because the developmental change of tissues may be quite different. However, as a first step for approximating the aging dependence of dielectric properties of various tissues, Equation 2.10 is a reasonable approximation to determine the hydrated rate α. Moreover, as for the tissue density ρ, there is not enough data to take into account its change with the amount of water. We thus also assume a direct link between α and TBW for a fixed tissue density.

We can check the validity of Equation 2.9 using measured data. Figure 2.7 shows the derived ε_r and σ, respectively, for the skin, muscle, skull, and brain at 900 MHz. Also shown by symbols are measured data for rats in Peyman, Rezazadeh, and Gabriel (2001). The differences between the calculated and measured data are $\pm 20\%$ for all the tissues. The empirical formula of Equation 2.9 is therefore a reasonable representation of the complex relative permittivity for various aged tissues.

Based on the above result, we can obtain the relative permittivity and conductivity at various ages and various frequencies. The calculation steps are as follows:

1. Calculate the TBW using the approximation formula of Equation 2.10, and then obtain the hydrated rate α using $\alpha = \rho \cdot \text{TBW}$.

2. Obtain the relative permittivity ε_{rA} and conductivity σ_A from the database for adult tissues, for example, the Gabriel data from the FCC website, and also calculate $\tau_A = \varepsilon_0\varepsilon_{rA}/\sigma_A$.
3. Calculate the complex relative permittivity $\dot{\varepsilon}_r$ and then the relative permittivity and conductivity at a specified age and frequency using Equation 2.9.

Figures 2.8–2.10 show the calculated relative permittivity and conductivity as a function of age at 400 MHz, 2.45 GHz and 5 GHz, respectively. A decreasing

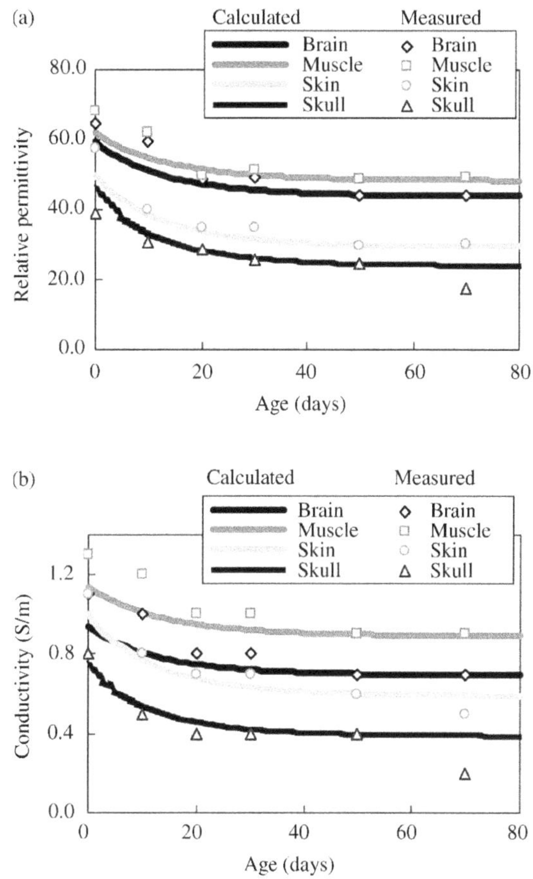

Figure 2.7 Comparison of calculated and measured dielectric properties of tissues as a function of age. (a) Relative permittivity and (b) conductivity (Wang, Fujiwara, and Watanabe, 2006). Reproduced with permission from Wang J., Fujiwara O. and Watanabe S., "Approximation of aging effect on dielectric tissue properties for SAR assessment of mobile telephones," *IEEE Transactions on Electromagnetic Compatibility*, **48**, 2, 408–413, 2006. © 2006 IEEE

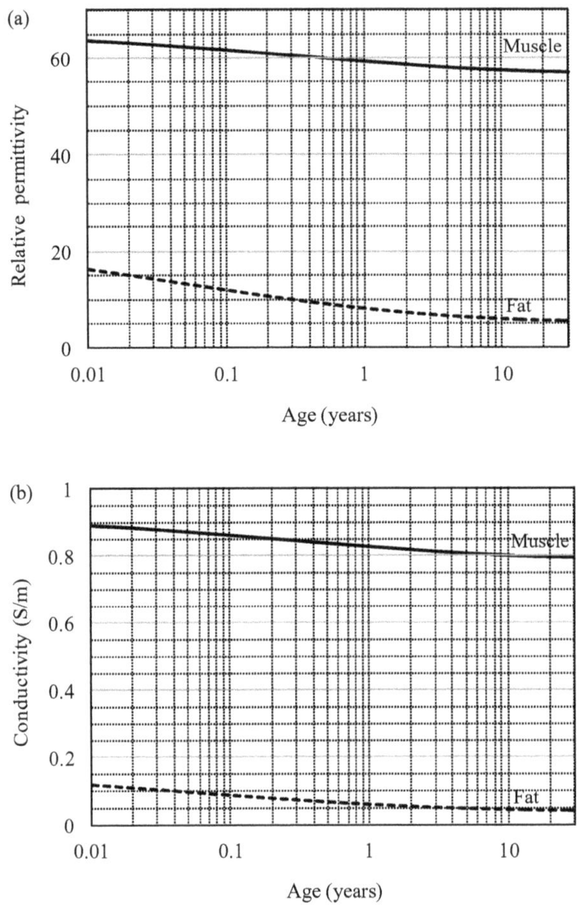

Figure 2.8 Calculated (a) relative permittivity and (b) conductivity as a function of age at 400 MHz

tendency with age can be observed in both the relative permittivity and the conductivity, which is attributed to the decrease in water content in the human body with increase in age. It can also be found that the obvious aging dependence of dielectric properties mainly occurs below 10 years old.

The approximation formula of Equation 2.9 is very useful when we have to consider the aging effect of human tissue. However, it suffers from a limitation that its validity is only confirmed when the relative permittivity of the tissue is below that of water. In reality, especially at lower frequencies such as 10 MHz, the relative permittivity of human tissue may be much larger than that of water. Whether Equation 2.9 is still valid in that case needs further validation.

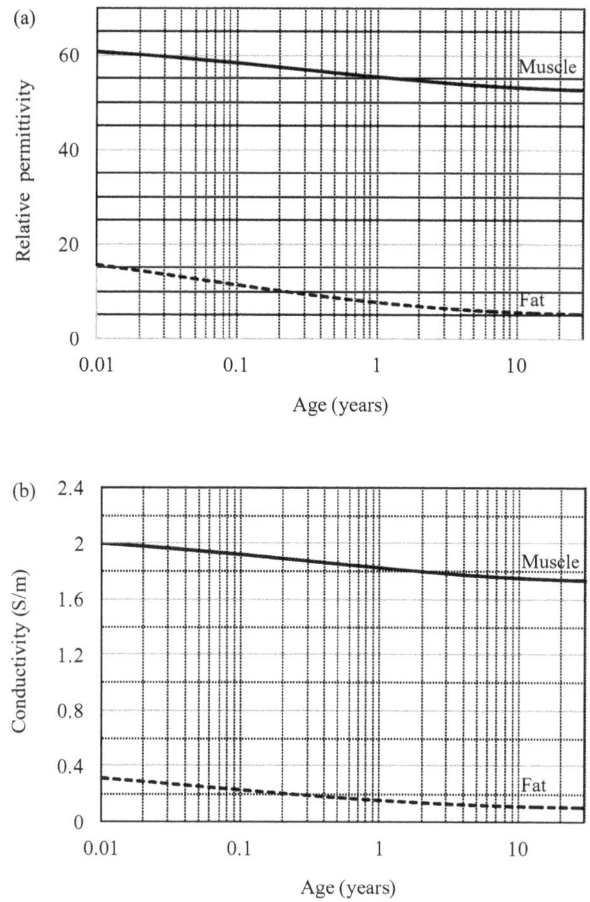

Figure 2.9 Calculated (a) relative permittivity and (b) conductivity as a function of age at 2.45 GHz

2.5 Penetration Depth versus Frequency

Since the human body is a lossy medium, the wavelength is shortened mainly due to the real part of the frequency-dependent dielectric properties. This makes the attenuation inside the human body also change with frequency. With the notation in Equation 2.1, the wavenumber

$$k = k' - jk'' = \omega\sqrt{\mu_0 \dot{\varepsilon}} = \omega\sqrt{\mu_0 \varepsilon_0 \left(\varepsilon_r - j\frac{\sigma}{\omega\varepsilon_0}\right)} \tag{2.11}$$

is complex where μ_0 denotes the permeability of free space. Let us consider a semi-infinite large plane medium of homogeneous human body tissue with a plane wave

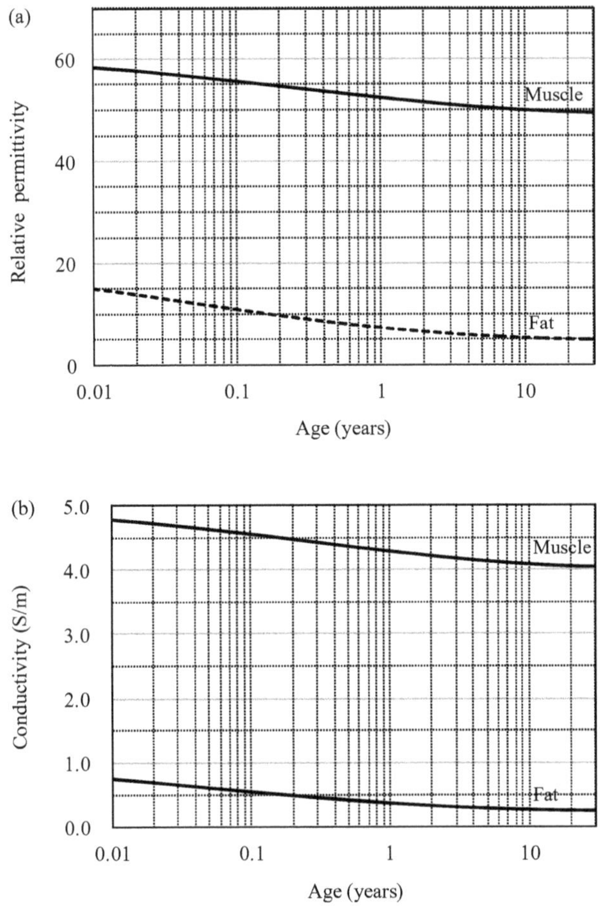

Figure 2.10 Calculated (a) relative permittivity and (b) conductivity as a function of age at 5 GHz

with normal incidence to it, as shown in Figure 2.11. The electric field intensity in human body tissue can be described as a function of the propagation distance d along the x axis by

$$E_z = E_{z0}e^{j(\omega t - kd)} = E_{z0}e^{-k''d}e^{j(\omega t - k'd)} \tag{2.12}$$

where E_{z0} is the electric field at the air–body boundary, and $e^{-k''d}$ depicts the attenuation term in the direction of propagation with

$$k'' = -\omega\sqrt{\mu_0\varepsilon_0}\sqrt{\varepsilon_r}\mathrm{Im}\left[\sqrt{1 - j\frac{\sigma}{\omega\varepsilon_0\varepsilon_r}}\right] \tag{2.13}$$

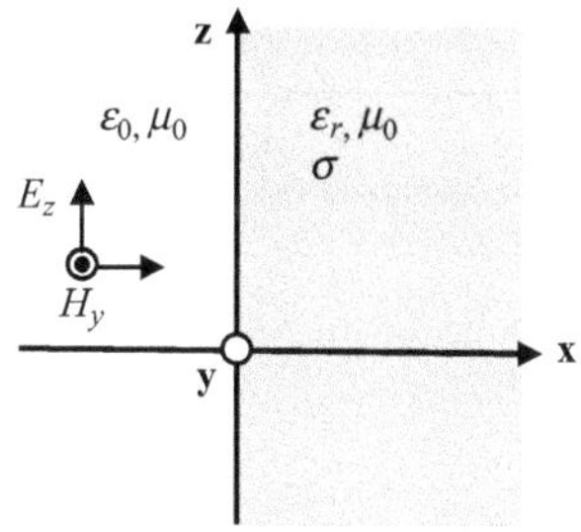

Figure 2.11 A plane wave with normal incidence to a semi-infinite large homogeneous tissue

where $\omega(\mu_0\varepsilon_0)^{1/2} = 2\pi/\lambda_0$ because the speed of light $c = 1/(\mu_0\varepsilon_0)^{1/2}$. The wavelength in human body tissue is thus

$$\lambda = \frac{\lambda_0}{\sqrt{\varepsilon_r}\mathrm{Re}\left[\sqrt{1 - j\sigma/\omega\varepsilon_0\varepsilon_r}\right]}. \tag{2.14}$$

This wavelength shortening effect yields a propagation speed v slower than light in the human body as $v = \lambda/t$. Moreover, if $\sigma >> \omega\varepsilon_0\varepsilon_r$ which implies a good conductor, we have $k'' \approx (\omega\mu_0\sigma/2)^{1/2}$. It can be concluded from Equation 2.12 that the electric field thus attenuates exponentially with k'' inside the human body. The exponential attenuation characteristic differs from the wave propagation in free space. It is a main propagation mechanism in in-body communication.

In Equation 2.12, the distance where the field E_z is attenuated to $1/e$ (0.368 or −8.68 dB) of its initial value E_{z0} is called the skin depth or penetration depth, expressed by δ. This is defined as $k''\delta = 1$ or

$$\delta = \frac{1}{k''}. \tag{2.15}$$

Table 2.6 gives the penetration depth for some main tissues and organs at typical body area communication frequencies. The higher the frequency, the smaller the penetration depth. For the high water content tissue such as muscle, the penetration depth is of the order of 20 cm at 10 MHz and 1 cm at 5 GHz. However,for the low water content tissue such as fat, the penetration depth is of the order of 100 cm at 10 MHz and 5 cm at 5 GHz. The significant difference in the penetration depth must be considered especially in in-body communication.

Let us go back to Figure 2.11 to consider the path loss. We assume the lossy medium to be a typical body tissue of muscle. The path loss is defined as $20\log_{10}(E_{Z0}/E_z)$. From $e^{k''d}$ we obtain the path loss versus frequency at two typical distances of d (10 and 5 cm) from the internal digestive organs to the body surface (Figure 2.12). As can be seen from Figure 2.12, at 10 cm depth, to allow a 50 dB path

Table 2.6 Penetration depth (cm) for some main tissues and organs

Tissue	10 MHz	400 MHz	2.45 GHz	5 GHz
Skin dry	56.06	5.53	2.26	1.05
Skin wet	31.04	5.84	2.20	0.95
Fat	106.14	30.90	11.70	4.93
Muscle	21.88	5.26	2.23	0.93
Bone cancellous	53.17	11.00	2.87	1.20
Bone cortical	96.86	21.32	4.58	1.77
Bone marrow	235.38	43.49	12.88	5.11
Brain gray matter	39.32	5.66	2.07	0.88
Brain white matter	53.52	7.95	2.65	1.09
Vitreous humor	13.16	3.17	1.78	0.80
Blood	16.31	3.43	1.61	0.73
Heart	26.38	4.69	1.76	0.79
Kidney	27.22	4.19	1.61	0.76
Liver	34.24	6.03	2.09	0.88
Lung deflated	26.94	5.95	2.21	0.91
Lung inflated	38.98	7.30	3.02	1.36
Esophagus	19.60	4.56	1.91	0.79
Stomach	19.60	4.56	1.91	0.79
Colon	26.47	5.11	1.93	0.83
Duodenum	19.60	4.56	1.91	0.79
Small intestine	15.17	2.60	1.26	0.67

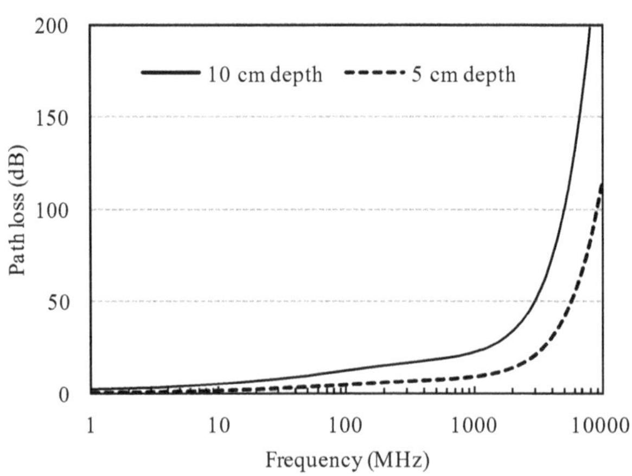

Figure 2.12 Frequency dependence of path loss for muscle tissue at a depth of 10 and 5 cm

loss for the in-body transmission, a frequency below 3 GHz should be chosen. If a 100 dB path loss is acceptable in the receiver, the frequency can be increased to 5 GHz. On the other hand, at 5 cm depth, the path loss may be lower than 50 dB up to 6 GHz. These results suggest that in-body communication is more appropriate at lower frequencies from the point of view of penetration depth. The penetration depth is therefore an important index for choosing an appropriate frequency band, especially in in-body communication.

2.6 In-Body Absorption Characteristic

The attenuation of transmitting power in the human body is due to absorption by body tissue. The absorption characteristic of body tissue is frequency-dependent. At low frequencies, the penetration depth is large and therefore the electromagnetic wave can go into the depths of the human body. At high frequencies, however, the penetration depth becomes shallow and therefore the electric field concentrates on the body surface. Although the conductivity σ is larger at higher frequencies than that at lower frequencies, the SAR in the depths of the human body is often smaller at higher frequencies because of the weaker electric field intensity penetration in the deep tissue.

This characteristic is easy to understand when we consider a plane wave incident to a semi-infinite slab of tissues. In this case an analytical expression is available which can provide a physical insight into the result. As a primary step for the physical insight, we consider a plane wave incident normal to the semi-infinite slab with a thickness l, as shown in Figure 2.13(a). The semi-infinite slab structure is assumed to be a layer of muscle, terminated by air. A transmission line model is given in Figure 2.13(b). Then we can apply the transmission line theory to obtain a SAR profile inside the slab structure.

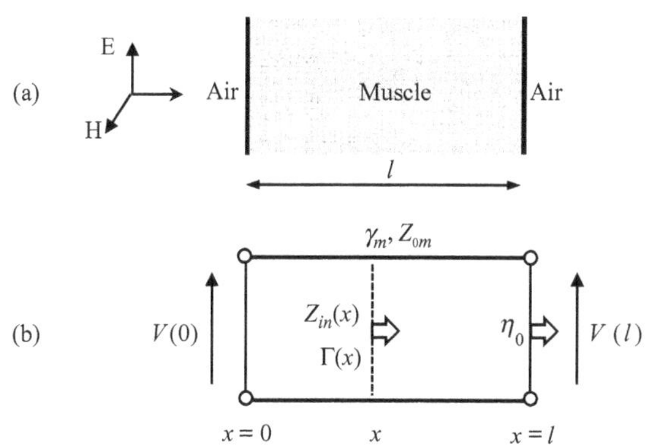

Figure 2.13 (a) Semi-infinite slab of tissue with a plane wave incident normal to it; (b) equivalent transmission line model

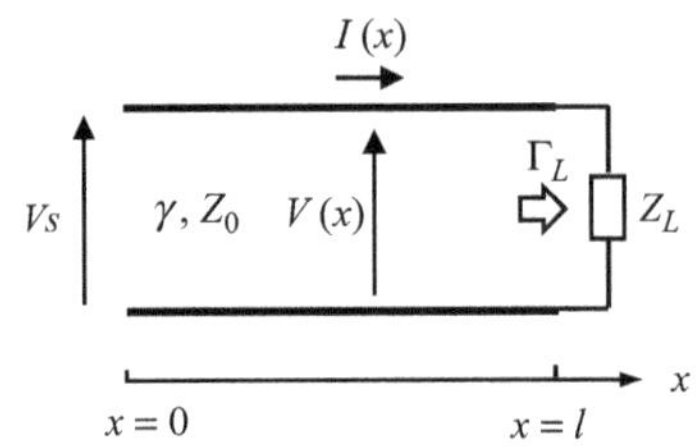

Figure 2.14 Basic transmission line model

Let us first consider the basic concept of a transmission line structure as in Figure 2.14. The transmission line is terminated at $x = l$ in a complex impedance Z_L, and the source is a phasor source with output voltage V_s. Let us define a complex voltage reflection coefficient $\Gamma(x)$ at a particular point x on the transmission line as the ratio of the phasor voltage of the backward and forward traveling waves. In terms of the reflection coefficient $\Gamma(x)$, the voltage and current expressions on the transmission line can be written as

$$V(x) = V^{+}e^{-\gamma x}[1 + \Gamma(x)] \tag{2.16a}$$

$$I(x) = \frac{V^{+}}{Z_0}e^{-\gamma x}[1 - \Gamma(x)] \tag{2.16b}$$

where V^{+} represents the forward traveling wave, and γ and Z_0 are the propagation constant and characteristic impedance of the transmission line, respectively. Then the input impedance at any point x on the transmission line is

$$Z_{in}(x) = \frac{V(x)}{I(x)} = Z_0\frac{1 + \Gamma(x)}{1 - \Gamma(x)}. \tag{2.17}$$

The reflection coefficient at the load is

$$\Gamma_L = \Gamma(l) = \frac{Z_L - Z_0}{Z_L + Z_0} \tag{2.18}$$

and the reflection coefficient at any x is related to the reflection coefficient at the load Z_L by

$$\Gamma(x) = \Gamma_L e^{-2\gamma(l-x)} = \frac{Z_L - Z_0}{Z_L + Z_0}e^{-2\gamma(l-x)}. \tag{2.19}$$

Submitting Equation 2.19 into Equation 2.16a we can get the voltage at any x as

$$V(x) = V^{+}e^{-\gamma x}\left[1 + \frac{Z_L - Z_0}{Z_L + Z_0}e^{-2\gamma(l-x)}\right]. \tag{2.20}$$

With the above basic knowledge of transmission line theory, now let us consider the equivalent transmission line model in Figure 2.13(b). Inside the muscle tissue, the propagation constant γ_m is given by

$$\gamma_m = j\omega\sqrt{\mu_0\dot{\varepsilon}_m} = j\omega\sqrt{\mu_0\varepsilon_0(\varepsilon_{rm} - j\sigma_m/\omega\varepsilon_0)} \tag{2.21}$$

and the characteristic impedance Z_{0m} is given by

$$Z_{0m} = \sqrt{\frac{\mu_0}{\dot{\varepsilon}_m}} = \sqrt{\frac{\mu_0}{\varepsilon_0(\varepsilon_{rm} - j\sigma_m/\omega\varepsilon_0)}}. \tag{2.22}$$

In addition, with Z_L equal to the air's intrinsic impedance η_0, we have the voltage at any x as

$$V(x) = V^+ e^{-\gamma_m x}\left[1 + \frac{\eta_0 - Z_{0m}}{\eta_0 + Z_{0m}} e^{-2\gamma_m(l-x)}\right]. \tag{2.23}$$

The power absorption in the human body is proportional to the squared electric field and the tissue conductivity. In the above transmission line model, therefore, the power absorption should be proportional to the squared voltage and the tissue conductivity. Assuming $l = 20$ cm, we show the power absorption as a function of the depth x in Figure 2.15 at frequencies of 10 MHz, 400 MHz and 4 GHz (nearly the center frequency of the UWB low band), respectively. The power absorptions are normalized to that at $x = 0$ for 10 MHz for comparison. As can be seen, the higher frequencies yield obviously significant power absorption at the tissue surface, which actually results in shallower penetration into the tissue. As the frequency decreases, however, the power concentration at the tissue surface becomes not so obvious so that the electromagnetic wave can go deeper into the human body.

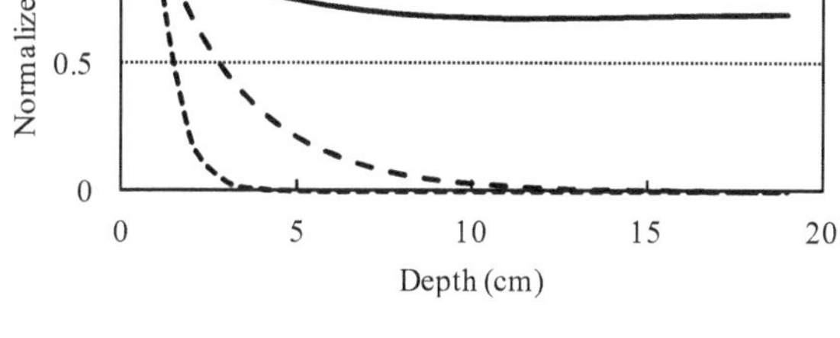

Figure 2.15 Normalized power absorption versus tissue depth at different frequencies

The quantization of power absorption usually uses the SAR. SAR is defined as power absorbed by unit mass

$$SAR = \frac{\sigma}{\rho} E^2 \tag{2.24}$$

in units of W/kg, where ρ is the mass density and E is the root mean squared electric field intensity in the human body. The SAR is a quantity often used for safety evaluation of electromagnetic interaction with the human body because it acts as a means to cause a temperature rise in the human body. In short, the larger the power absorbed by the human body, the larger the path loss. From the point of view of in-body communication, we have to pay attention to the path loss because it is needed in the system design such as the link budget evaluation. However, from the point of view of safety evaluation, we should pay attention to the SAR because it is related to a temperature rise in the human body where a rise exceeding 1 °C is considered to be unsafe.

When a human body in thermal equilibrium is exposed to electromagnetic fields, the resultant temperature rise can be calculated from a bio-heat equation which takes into account heat exchange mechanisms such as heat conduction, blood flow, and electromagnetic heating. The bio-heat equation is given by (Pennes, 1948)

$$\rho C_P \frac{\partial T}{\partial t} = \nabla(K \nabla T) + \rho SAR - b(T - T_b) \tag{2.25}$$

with the boundary condition

$$K \frac{\partial T}{\partial n} = -h(T - T_a) \tag{2.26}$$

where T is temperature, C_P is specific heat, K is thermal conductivity, b is a constant related to the blood flow, T_b is blood temperature, T_a is ambient temperature, h is the convective heat transfer coefficient, n is unit vector normal to the body surface, and SAR is the input electromagnetic heating source. It should be noted that there are two kinds of convective heat transfer coefficient h. One is h_a, the convective heat transfer coefficient from the human body surface to the ambient temperature, and the other is h_b, the convective heat transfer coefficient from the internal surface to the internal cavity, which is generally larger than h_a. These thermal parameters can be found in some physiological textbooks (Guyton and Hall, 1996), but a comprehensive database for the thermal parameters does not exist. By solving Equation 2.25 with the SAR as excitation, the temperature distribution inside the human body as a function of time can be obtained. As for the steady-state temperature rise, it can be calculated from the difference between the temperature T and T_0 where T_0 is the normal temperature in the unexposed body (with SAR = 0) at the thermal equilibrium state.

2.7 On-Body Propagation Mechanism

The on-body propagation mechanism is not as simple as the in-body propagation mechanism. The on-body propagation mechanism may depend on working frequency so that different propagation mechanisms may be involved.

Here we attempt to derive a general explanation for the propagation mechanism in various frequency bands based on electromagnetic theory. Let us first consider the electric field from a vertical dipole in free space (Figure 2.16). In a spherical coordinate system (r, ϕ, θ), the electric field at r is given by

$$E_\theta = j\frac{I\Delta z}{4\pi}\eta_0 k_0^2\left[\frac{1}{k_0 r} - j\frac{1}{(k_0 r)^2} - \frac{1}{(k_0 r)^3}\right]\sin\theta \cdot e^{-jk_0 r} \tag{2.27}$$

where I is the wire current in amperes, Δz is the dipole length in meters, $\eta_0 = (\mu_0/\varepsilon_0)^{1/2}$ is the intrinsic impedance of free space, and k_0 is the wavenumber of free space. Equation 2.27 contains terms in $1/r$, $1/r^2$, and $1/r^3$. In the near field, the $1/r^3$ term is dominant. This term is called the electrostatic field component. As the distance r increases, the $1/r^3$ and $1/r^2$ terms attenuate rapidly so that the far field is dominated by the $1/r$ term which is known as the radiation field. That is to say, the $1/r$, $1/r^2$, and $1/r^3$ terms correspond to the electric fields in the far-field, induction-field, and near-field region of the dipole, respectively. This basic concept is helpful to understand the propagation mechanism of on-body communication.

To derive the propagation mechanism from a theoretical approach, we simplify the human body to be a semi-infinitely large lossy dielectric medium with relative permittivity ε_r and conductivity σ. When a unit vertical dipole is placed on the

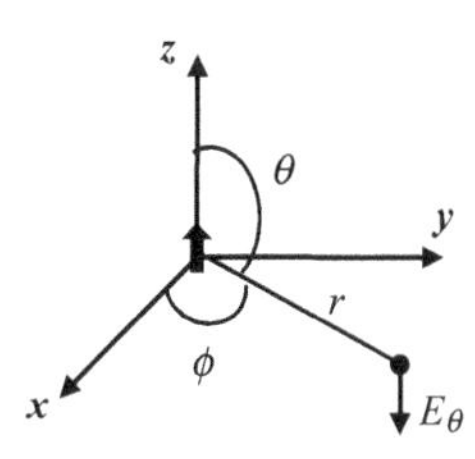

Figure 2.16 Electric field from a vertical dipole in free space in a spherical coordinate system

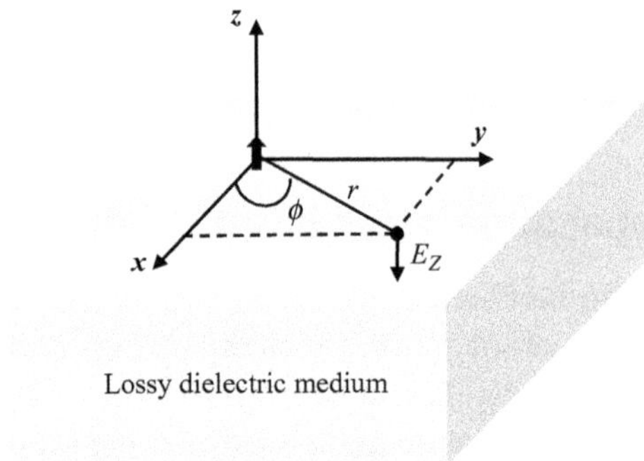

Figure 2.17 Electric field from a unit vertical dipole on a semi-infinite large lossy dielectric medium

boundary plane as shown in Figure 2.17, the electric field from the dipole is given by (Norton, 1937)

$$E = jk_0 \left[II + \frac{1}{k_0^2} \nabla\nabla \cdot II \right] \tag{2.28}$$

where II is the wave potential which has only a single component

$$II_z = \int_0^{\infty} \frac{2}{l + u^2 m} J_0(\lambda_0 r) \lambda_0 \, d\lambda_0 \tag{2.29}$$

where J_0 is the first kind Bessel function with order zero, and

$$l^2 = \lambda_0^2 - k_0^2 \tag{2.30}$$

$$m^2 = \lambda_0^2 - k^2 \tag{2.31}$$

$$k^2 = k_0^2 \left(\varepsilon_r - j \frac{\sigma}{\omega \varepsilon_0} \right) \tag{2.32}$$

$$u = k_0 / k. \tag{2.33}$$

Under the cylindrical coordinate system (r, ϕ, z), we have

$$E_z = jk_0 \left[II_z + \frac{1}{k_0^2} \frac{\partial^2}{\partial z^2} II_z \right] \tag{2.34}$$

$$E_r = \frac{j}{k_0} \frac{\partial^2}{\partial r \partial z} II_z \tag{2.35}$$

$$E_\phi = 0. \tag{2.36}$$

So the problem of determining the electric field intensity on the lossy dielectric plane can be reduced to calculate Π_z. After some complex mathematics operations, we may obtain the z-directed electric field on the lossy dielectric surface as (Bae *et al.*, 2012)

$$E_z = 2jk_0\left[G_S + j\frac{1}{k_0 r} - \frac{1}{(k_0 r)^2}\right]\frac{e^{-jk_0 r}}{r} \tag{2.37}$$

where

$$G_S = (1 - u^2 + u^4)F \tag{2.38}$$

$$F \equiv 1 + j\sqrt{\pi w}e^{-w}\mathrm{erfc}\left(-j\sqrt{w}\right) \tag{2.39}$$

$$w = \frac{1}{2}jk_0 r u^2(1 - u^2) \tag{2.40}$$

and

$$\mathrm{erfc}\left(-j\sqrt{w}\right) = \frac{2}{\sqrt{\pi}}\int_{-j\sqrt{w}}^{\infty} e^{-t^2}dt. \tag{2.41}$$

In Equation 2.37, the first term proportional to the inverse of the surface distance, that is, the $1/r$ term, contains an additional gain factor $G_S = (1 - u^2 + u^4)F$. Since G_S is related to the dielectric properties of the lossy dielectric medium, the first term in Equation 2.37 can be approximately regarded as a wave propagating along the surface of the medium. Strictly speaking, a surface wave is a signal that propagates along a boundary of two kinds of different media without radiation to the outside. In this sense the first term in Equation 2.37 is not a strict surface wave because it also radiates towards the outside of the boundary. However, in view of the fact that this field component decreases with an increase in the surface distance r, we refer to it as a "surface propagation component". In addition, the other two terms ($1/r^2$ and $1/r^3$ terms) in Equation 2.37 correspond to the induction and electrostatic field components of the dipole, respectively. The surface propagation component should predominate at larger transmission distances, whereas the transmission in shorter distances may be dominated only by the electrostatic field and the induction field.

The propagation mechanism of on-body communication can be therefore divided into three parts: the surface propagation of the $1/r$ term, the reactive induction of the $1/r^2$ term, and the electrostatic coupling of the $1/r^3$ term. Which term is dominant is not dependent on the actual propagation distance r but the distance normalized to the wavelength. That is to say, at a specified distance r, the size of contribution from the three different mechanisms depends on the frequency. At low frequencies, the wavelength is large and the normalized distance is consequently small. This makes the electrostatic coupling contribute more to the on-body propagation. On the other hand, as the frequency increases, the shorter wavelength yields a larger normalized distance. This makes the $1/r$ term or the surface propagation term begin to have a significant contribution whereas the electrostatic coupling term becomes negligible. It should be noted that, however, since the surface propagation term consists of a gain factor $G_S = (1 - u^2 + u^4)F$, the larger the gain factor, the larger the field component of surface propagation.

Let us see the frequency dependence of the gain factor G_S. We assume the semi-infinitely large lossy dielectric medium to be muscle. With Equations 2.33 and 2.38–2.41, we can obtain the gain factor G_S as a function of frequency. In the calculation of Equation 2.41 for the complementary error function with a complex argument, we employ the following Taylor series expression for complex argument z

$$\text{erfc}\,(z) = 1 - \text{erf}\,(z) = 1 - \frac{2}{\sqrt{\pi}}\sum_{n=0}^{\infty}\frac{(-1)^n z^{2n+1}}{n!(2n+1)}. \tag{2.42}$$

Figure 2.18 shows the calculation result at the unit propagation distance (1 m). Another representation is to plot it as a function of $k_0 r$ also at the unit propagation distance, as shown in Figure 2.19. As can be seen, the gain factor decreases with frequency or $k_0 r$, which means that the surface propagation component attenuates more rapidly at higher frequencies. When the frequency is below 100 MHz, the degradation in the gain factor of the surface propagation component is relatively smooth with frequency and is always larger than 0.85. Above 100 MHz, however, the gain factor degrades rapidly with frequency and reaches 0.27 at 5 GHz. Similarly, from Figure 2.19, the corresponding $k_0 r$ is about 2.0 for a gain factor of the surface propagation component larger than 0.85.

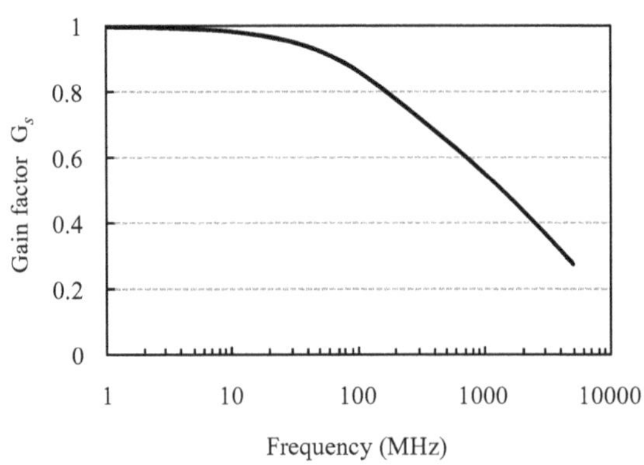

Figure 2.18 Gain factor G_S versus frequency at a unit propagation distance

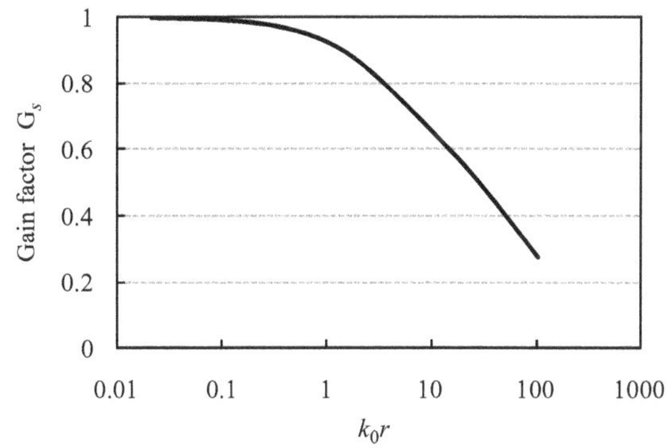

Figure 2.19 Gain factor G_S versus k_0r at a unit propagation distance

In order to distinguish the size of contribution from each mechanism to the received field component, the percentage of the surface propagation component, the induction field component, and the electrostatic field component, that is, the three components in Equation 2.37, are calculated. Figure 2.20 shows the percentage of contribution as a function of frequency, and Figure 2.21 shows it as a function of k_0r. From Figure 2.20, the three field components are found to be equal at 50 MHz. The corresponding k_0r value is 1.0 in Figure 2.21. Therefore, at frequencies below 50 MHz, the electrostatic field component is dominant and electrostatic coupling is the main on-body propagation mechanism. On the other hand, when the frequency exceeds 50 MHz, the surface propagation term becomes dominant. In particular, at frequencies larger than 100 MHz, the surface propagation component is larger than the sum of the other two components and acts as the main on-body propagation mechanism. The corresponding value of k_0r is about 2.0.

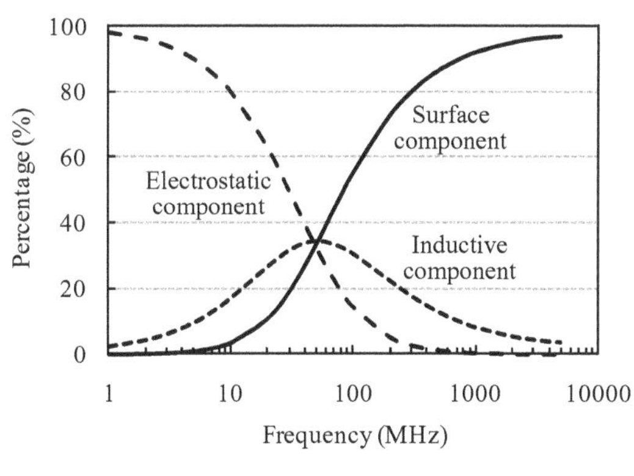

Figure 2.20 Percentage of contributions from different propagation mechanisms versus frequency at a unit propagation distance

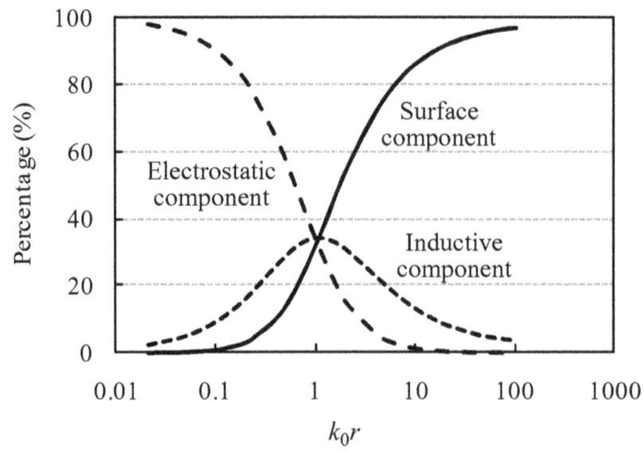

Figure 2.21 Percentage of contributions from different propagation mechanisms versus k_0r at a unit propagation distance

Based on the above observations, we now summarize the on-body propagation mechanism for typical candidate frequencies of body area communications, that is, 10 MHz, 400 MHz and the UWB band.

- At 10 MHz, almost 80% of the received field component is contributed by the electrostatic field term at a unit on-body communication distance. The signal transmission is thus realized mainly by electrostatic coupling in this frequency band.
- At 400 MHz, almost 80% of the received field component is contributed by the surface propagation term, which acts as a main on-body propagation mechanism.
- In the UWB band, more than 95% of the received field component is contributed by the surface propagation term. It completely dominates the on-body propagation.

Although the findings are derived from a semi-infinite large plane medium with the dielectric properties of muscle, they are useful in understanding the basic on-body propagation mechanism. Of course, the actual body tissue and shape may result in some deviations for the values in Figures 2.20 and 2.21.

In addition, we can rewrite Equation 2.37 as

$$E_Z = 2j\left[\frac{2\pi}{c}G_S f + j\frac{1}{r} - \frac{c}{2\pi}\frac{1}{r^2}\frac{1}{f}\right]\frac{e^{-jk_0 r}}{r} \tag{2.43}$$

with $k_0 = 2\pi f/c$ (where c is the speed of light) to derive an expression for the received electric field as a function of frequency. It is evident that the third term of electrostatic coupling at low frequencies exhibits a low-pass filter feature, whereas the first term of surface propagation at high frequencies exhibits a high-pass filter feature. In total the on-body propagation has a band-stop filter feature. The stop-band is dependent on the body tissue properties and propagation distance. This characteristic makes the choice of whether the lower HBC band or the higher

UWB band become more reasonable for on-body communication. Moreover, the electrostatic coupling is little affected by propagation distance, whereas the gain factor G_S decreases rapidly in the UWB band. This feature further suggests that the HBC band may be more suitable than the UWB band for on-body communication.

2.8 Diffraction Characteristic

Since the human body has a complicated surface shape, the actual propagation on it may differ from the propagation along a smooth plane. If the direct propagation path between the transmitter and receiver is obstructed by an obstacle, electromagnetic waves may travel into the shadow zone behind the obstacle. The apparent bending of waves around small obstacles is known as diffraction. Diffraction occurs when the dimensions of the obstacle are not greater than dozens of times of the wavelength. This is generally fulfilled in body area communications where the wavelengths are above several centimeters. In on-body communication the obstacle may be a part of the body or the curved body surface, whereas in in-body communication the obstacle may be the organs. Since the diffracted wave propagates on the body surface with the speed of light, it is also called a creeping wave. This wave can actually be considered as the surface propagation component as in the previous section. The difference is that it propagates along a curved surface. In this sense the diffraction may act as a propagation mechanism at higher frequencies on the curved part of the human body.

Let us consider an infinitely long circular cylinder with radius a as shown in Figure 2.22. A z-directed magnetic line source of unit strength is located at $(x, y) = (a, 0)$ as the transmitter and excites a transverse electric (TE) field. The corresponding magnetic field on the cylinder surface is z-directed and is given by (Paknys, 1993)

$$H_z(\phi) = \frac{j\omega\varepsilon_0}{2\pi ka}\int_{-\infty}^{\infty} \frac{H_\eta^{(2)}(ka)}{H_\eta^{(2)\prime}(ka)} e^{-j\eta\phi} d\eta \tag{2.44}$$

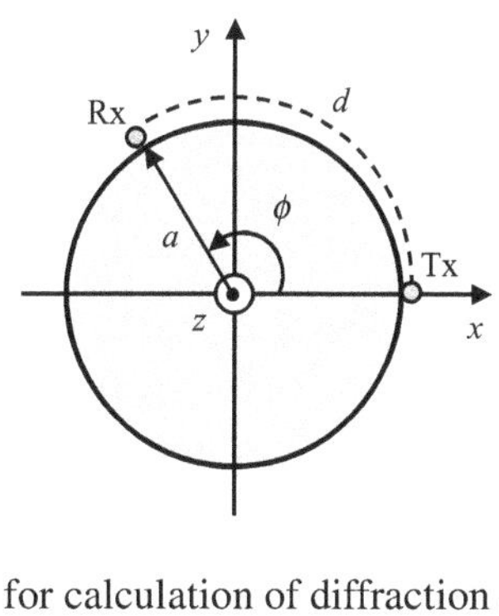

Figure 2.22 Schematic for calculation of diffraction around a circular cylinder

where k denotes the wavenumber, and $H_{\eta}^{(2)}(ka)$ and $H_{\eta}^{(2)\prime}(ka)$ are the second kind Hankel function and its derivative, respectively. In order to calculate the magnetic field $H(\phi)$ numerically, the Hankel functions in the numerator and denominator of Equation 2.44 are replaced with the Watson approximation. This yields

$$H_z(\phi) \approx -\frac{f\varepsilon_0\sqrt{\xi}}{\sqrt{2kd}} e^{-jkd} \int_{-\infty}^{\infty} \frac{w_2(\tau)}{w'_2(\tau)} e^{-j\xi\tau} d\tau \tag{2.45}$$

where

$$w_2(\tau) = \sqrt{\pi}[Bi(\tau) - jAi(\tau)] \tag{2.46}$$

is known as the Fock-type Airy function in which $Ai(\tau)$ is the Airy function and $Bi(\tau)$ is the second kind Airy function,

$$\xi = (ka/2)^{1/3}\phi \tag{2.47}$$

$$\tau = \frac{\eta - ka}{(ka/2)^{1/3}} \tag{2.48}$$

and d is the distance between the line-source-type transmitter and the receiving point on the cylinder surface. Hence the diffracted field for TE polarization can be calculated from Equation 2.45 as a function of the angle ϕ between the transmitter and the receiving point. In the calculation of the diffracted fields, MATLAB® provides a conventional tool to treat the Airy functions in Equation 2.45.

Figure 2.23 shows the diffracted fields calculated on the cylinder surface as a function of ϕ at 10 MHz and 5 GHz, respectively. The circular cylinder is assumed to be muscle tissue and the radius $a = 15$ cm. The diffracted field is normalized to that in the vicinity of the excitation source. As expected, the diffracted field attenuates with increasing angle. The attenuation is larger at higher frequencies. For $\phi = 180°$, which corresponds to the case where the transmitter and the receiving point are located on opposite sides of the cylinder, the diffracted field attenuates around 40 dB at 10 MHz and 50 dB at 5 GHz. The attenuation at 400 MHz band is between the two curves in Figure 2.23 but closer to that at 5 GHz. Such a level attenuation or path loss is of an acceptable order for on-body transmission. The diffraction phenomenon is thus a useful mechanism and provides the possibility for on-body communication along the curved body surface.

On the other hand, for the infinitely long circular cylinder in Figure 2.22, if we assume a ϕ-directed magnetic line source located at $(r, \phi) = (a, 0)$ in a cylindrical

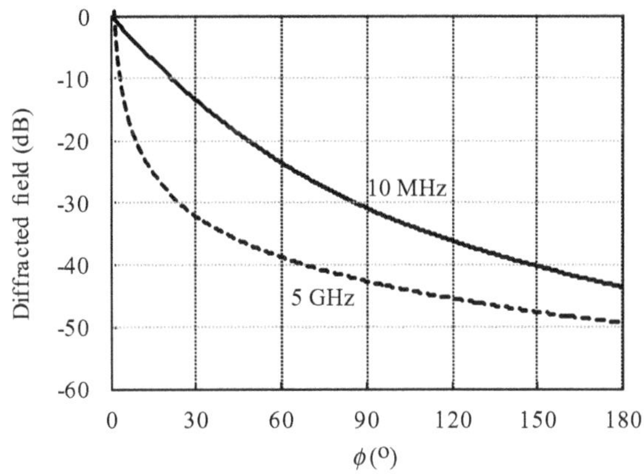

Figure 2.23 Diffracted field along a circular cylinder of muscle with radius of 15 cm (TE polarization)

coordinate system, it will excite a transverse magnetic (TM) field with both r component and ϕ component. That is to say, the excited magnetic field is given by

$$H = \hat{r}H_r + \hat{\phi}H_\phi. \tag{2.49}$$

However, at the body surface, that is, when $r = a$, H_r will vanish so that we only need to consider the H_ϕ component. H_ϕ is given by

$$H_\phi(\phi) = -j\frac{\sqrt{\varepsilon_0/\mu_0}}{2\pi a}\int_{-\infty}^{\infty}\frac{H_\eta^{(2)\prime}(ka)}{H_\eta^{(2)}(ka)}e^{-j\eta\phi}d\eta \tag{2.50}$$

for a unit strength line source. In order to derive an easy approximation solution for Equation 2.50, again, the Fock-type Airy function $w_2(\tau)$ in Equation 2.46 can be used. Then Equation 2.50 becomes

$$H_\phi(\phi) \approx \frac{\sqrt{\varepsilon_0/\mu_0}}{\sqrt{2k\pi}}\left(\frac{\xi}{d}\right)^{3/2} e^{-jkd}\int_{-\infty}^{\infty}\frac{w'_2(\tau)}{w_2(\tau)}e^{-j\xi\tau}d\tau \tag{2.51}$$

where ξ and τ can be found from Equations 2.47 and 2.48, respectively.

Figure 2.24 shows the diffracted fields for TM polarization, which are calculated using Equation 2.51 on the cylinder surface as a function of ϕ at 10 MHz and 5 GHz, respectively. The circular cylinder is assumed to be muscle tissue and the radius $a = 15$ cm. The diffracted fields are normalized to that in the vicinity of the excitation source. As can be seen, the diffracted field attenuates larger than that in TE polarization. It attenuates around 80 dB at 10 MHz and 90 dB at 5 GHz for $\phi = 180°$. Theses field characteristics reasonably suggest that

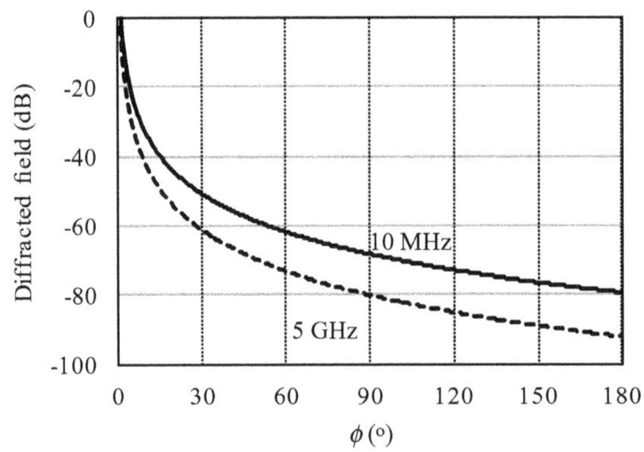

Figure 2.24 Diffracted field along a circular cylinder of muscle with radius of 15 cm (TM polarization)

a diffracted wave should propagate easier when it has an electric field component normal to the curved surface.

Although the diffraction characteristics are derived from a circular cylinder in this section, they are basically applicable to a human body. In view of the acceptable attenuation or path loss level especially in the TE polarization, the diffraction phenomenon should be considered as a valid mechanism in on-body transmission. It should be noted that, however, at low frequencies, the larger penetration depth may make the wave penetrate the human body so that a direct path exists between the transmitter and receiver in addition to the path of the diffracted wave. In this case, the received field signal should be the sum of the direct wave and the diffracted wave.

References

Altman, P.L. and Dittmer, D.S. (1974) *Biology Data Book: Blood and Other Body Fluids*, Federation of American Societies for Experimental Biology, Washington, DC.

Bae, J., Cho, H., Song, K., Lee, H., and Yoo, H.-J. (2012) The signal transmission mechanism on the surface of human body for body channel communication. *IEEE Transactions on Microwave Theory and Techniques*, **60**(3), 582–593.

Gabriel, C. (1996) Compilation of the dielectric properties of body tissues at RF and microwave frequencies. Brooks Air Force Technical Report, AL/OE-TR-1996-0037.

Guyton, A.C. and Hall, J.E. (1996) Body temperature, temperature regulation, and fever, in *Textbook of Medical Physiology*, W. B. Saunders, Philadelphia, PA, pp. 889–904.

IEE of Japan (1995) *Biological Effects of Electromagnetic Fields and Measurement*, Corona Publishing, Tokyo, pp. 51–54.

Lichtenecker, K. (1926) Die dielektrizitatskonstante naturlicher und kunstlicher mischkorper. *Physikalische Zeitschrift*, **27**, 115–158.

Misra, D. (1987) A quasi-static analysis of open-ended coaxial line. *IEEE Transactions on Microwave Theory and Techniques*, **35**, 925–928.

Norton, K.A. (1937) The propagation of radio waves over the surface of the earth and in the upper atmosphere Part II. *Proceedings of the IRE*, **25** (9), 1203–1236.

Paknys, R. (1993) Uniform asymptotic formulas for the creeping wave field on or off a cylinder. *IEEE Transactions on Antennas and Propagation*, **41** (8), 1099–1104.

Pennes, H.H. (1948) Analysis of tissue and arterial blood temperature in resting forearm. *Journal of Applied Physics*, **1**, 93–122.

Peyman, A., Rezazadeh, A.A., and Gabriel, C. (2001) Changes in the dielectric properties of rat tissue as a function of age at microwave frequencies. *Physics in Medicine and Biology*, **46** (6), 1617–1629.

Schwan, H.P. (1957) Electrical properties of tissue and cell suspensions, in *Advances in Biological and Medical Physics*, vol. **5**, Academic Press, New York, pp. 147–209.

Wang, J., Fujiwara, O., and Watanabe, S. (2006) Approximation of aging effect on dielectric tissue properties for SAR assessment of mobile telephones. *IEEE Transactions on Electromagnetic Compatibility*, **48** (2), 408–413.

3

Electromagnetic Analysis Methods

The difficulty in measurement of an actual human body, especially inside a human body, means that numerical electromagnetic analysis methods play an important role in body area communications. Numerical electromagnetic analysis methods are especially helpful to clarify propagation characteristics and establish channel models in or on the body area. Rapid progress with computers has enabled high-level numerical simulation with the aid of high-resolution human body models. In this chapter, we will introduce three representative numerical methods for electromagnetic analysis: the finite-difference time-domain (FDTD) method; the finite element method (FEM); and a hybrid method combining the method of moments (MoM) and the FDTD method. After introducing the basis and algorithm for the numerical methods, human body models for use in numerical analysis will also be described.

3.1 Finite-Difference Time-Domain Method

3.1.1 Formulation

The FDTD method (Taflove and Hagness, 2000) is currently the most widely accepted numerical electromagnetic analysis method for human bodies.

The FDTD formulations are derived from the following Maxwell time-domain equations

$$\begin{aligned} \nabla \times E &= -\mu \frac{\partial H}{\partial t} \\ \nabla \times H &= \varepsilon \frac{\partial E}{\partial t} + \sigma E \end{aligned} \tag{3.1}$$

where E is the electric field, H is the magnetic field, ε is the permittivity, μ is the permeability and σ is the conductivity. To derive the algorithm, we rewrite

Body Area Communications: Channel Modeling, Communication Systems, and EMC, First Edition.
Jianqing Wang and Qiong Wang.

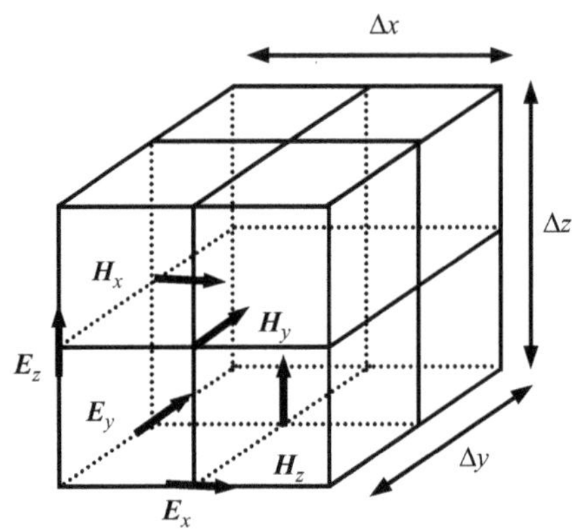

Figure 3.1 Yee cell in FDTD algorithm

Equation 3.1 as

$$\begin{aligned}\frac{\partial E}{\partial t} &= -\frac{\sigma}{\varepsilon}E + \frac{1}{\varepsilon}\nabla \times H \\ \frac{\partial H}{\partial t} &= -\frac{1}{\mu}\nabla \times E\end{aligned} \tag{3.2}$$

The FDTD discretization for the above equations is based on a Yee cell approach as shown in Figure 3.1. A special feature of the Yee cell is that the electric field E and magnetic field H components are staggered one half space cell apart. That is to say, the E field is generally assigned at the edges of the Yee cell, and the H field is assigned on the faces of the Yee cell, which facilitates the differencing scheme. If we denote the field F (E or H) at spatial location $(i\Delta x, j\Delta y, k\Delta z)$ and time step $n\Delta t$ by

$$F(i\Delta x, j\Delta y, k\Delta z, n\Delta t) = F^n(i, j, k) \tag{3.3}$$

using the central difference approximations such as

$$\begin{aligned}\frac{\partial F}{\partial x} &\approx \frac{F^n\left(i+\frac{1}{2}, j, k\right) - F^n\left(i-\frac{1}{2}, j, k\right)}{\Delta x} \\ \frac{\partial F}{\partial t} &\approx \frac{F^{n+\frac{1}{2}}(i, j, k) - F^{n-\frac{1}{2}}(i, j, k)}{\Delta t}\end{aligned} \tag{3.4}$$

we can discretize Equation 3.2 and consequently derive the electric and magnetic field components. In the Yee cell notation, the electric field E is assigned at $t=(n-1)\Delta t$, $n\Delta t$, $(n+1)\Delta t$, and the magnetic field H is assigned at $(n-1/2)\Delta t$, $(n+1/2)\Delta t$. Then the time differentiation in Equation 3.2 for E must be conducted at $t=(n-1/2)\Delta t$, and the time differentiation in Equation 3.4 for H must be conducted at $t=n\Delta t$. According to the notation in Equation 3.4, Equation 3.2 can be written as

$$\begin{aligned}\frac{E^n - E^{n-1}}{\partial t} &= -\frac{\sigma}{\varepsilon}E^{n-1/2} + \frac{1}{\varepsilon}\nabla \times H^{n-1/2} \\ \frac{H^{n+1/2} - H^{n-1/2}}{\partial t} &= -\frac{1}{\mu}\nabla \times E^n\end{aligned} \tag{3.5}$$

where

$$E^{n-1/2} = \frac{E^{n-1} + E^n}{2}. \tag{3.6}$$

Thus we have

$$\begin{aligned}E^n &= \frac{1 - \dfrac{\sigma\Delta t}{2\varepsilon}}{1 + \dfrac{\sigma\Delta t}{2\varepsilon}}E^{n-1} + \frac{\Delta t/\varepsilon}{1 + \dfrac{\sigma\Delta t}{2\varepsilon}}\nabla \times H^{n-1/2} \\ H^{n+1/2} &= H^{n-1/2} - \frac{\Delta t}{\mu}\nabla \times E^n\end{aligned} \tag{3.7}$$

That is to say, from the electric field E^{n-1} at $t=(n-1)\Delta t$ and the magnetic field $H^{n-1/2}$ at $t=(n-1/2)\Delta t$, we can get E^n at $t=n\Delta t$. And from the magnetic field $H^{n-1/2}$ at $t=(n-1/2)\Delta t$ and the electric field E^n at $t=n\Delta t$, we can get $H^{n+1/2}$ at $t=(n+1/2)\Delta t$.

In the Cartesian coordinates, Equation 3.2 can be written as

$$\frac{\partial E_x}{\partial t} = \frac{1}{\varepsilon}\left(\frac{\partial H_z}{\partial y} - \frac{\partial H_y}{\partial z} - \sigma E_x\right) \tag{3.8a}$$

$$\frac{\partial E_y}{\partial t} = \frac{1}{\varepsilon}\left(\frac{\partial H_x}{\partial z} - \frac{\partial H_z}{\partial x} - \sigma E_y\right) \tag{3.8b}$$

$$\frac{\partial E_z}{\partial t} = \frac{1}{\varepsilon}\left(\frac{\partial H_y}{\partial x} - \frac{\partial H_x}{\partial y} - \sigma E_z\right) \tag{3.8c}$$

$$\frac{\partial H_x}{\partial t} = \frac{1}{\mu}\left(\frac{\partial E_y}{\partial z} - \frac{\partial E_z}{\partial y}\right) \tag{3.8d}$$

$$\frac{\partial H_y}{\partial t} = \frac{1}{\mu}\left(\frac{\partial E_z}{\partial x} - \frac{\partial E_x}{\partial z}\right) \tag{3.8e}$$

$$\frac{\partial H_z}{\partial t} = \frac{1}{\mu}\left(\frac{\partial E_x}{\partial y} - \frac{\partial E_y}{\partial x}\right) \tag{3.8f}$$

Its z components, for example, are as follows according to Equation 3.7:

$$E_z^n\left(i,j,k+\frac{1}{2}\right) = \frac{1-\dfrac{\sigma(i,j,k+1/2)\Delta t}{2\varepsilon(i,j,k+1/2)}}{1+\dfrac{\sigma(i,j,k+1/2)\Delta t}{2\varepsilon(i,j,k+1/2)}} E_z^{n-1}\left(i,j,k+\frac{1}{2}\right) + \frac{\dfrac{\Delta t}{\varepsilon(i,j,k+1/2)}}{1+\dfrac{\sigma(i,j,k+1/2)\Delta t}{2\varepsilon(i,j,k+1/2)}}$$
$$\times\left[\frac{H_y^{n-1/2}\left(i+\frac{1}{2},j,k+\frac{1}{2}\right) - H_y^{n-1/2}\left(i-\frac{1}{2},j,k+\frac{1}{2}\right)}{\Delta x} - \frac{H_x^{n-1/2}\left(i,j+\frac{1}{2},k+\frac{1}{2}\right) - H_x^{n-1/2}\left(i,j-\frac{1}{2},k+\frac{1}{2}\right)}{\Delta y}\right] \tag{3.9}$$

$$H_z^{n+1/2}\left(i+\frac{1}{2},j+\frac{1}{2},k\right) = H_z^{n-1/2}\left(i+\frac{1}{2},j+\frac{1}{2},k\right) + \frac{\Delta t}{\mu(i+1/2,j+1/2,k)}$$
$$\times\left[\frac{E_y^n\left(i+1,j+\frac{1}{2},k\right) - E_y^n\left(i,j+\frac{1}{2},k\right)}{\Delta x} - \frac{E_x^n\left(i+\frac{1}{2},j+1,k\right) - E_x^n\left(i+\frac{1}{2},j,k\right)}{\Delta y}\right]. \tag{3.10}$$

Its x and y components are also in the same form. For applying the FDTD method to human bodies, the Yee cells correspond completely to the cells in human body models. By assigning each cell a corresponding permittivity and a conductivity, we can easily model the anatomical tissues and organs, and calculate the interior electric and magnetic fields.

A noticeable problem in the assignment of permittivity and conductivity is that, referring to Figure 3.1 and Equation 3.7, the permittivity and conductivity are required at the edges of the cell. In other words, the permittivity and conductivity

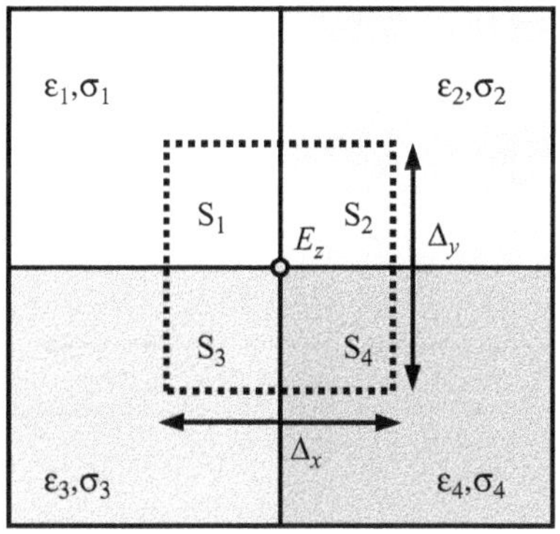

Figure 3.2 View of the *x-y* plane of four cells with different permittivities and conductivities

are required at the boundary of different tissues, because each cell is identified rigorously as belonging to one type of tissue in numerical human body models.

Let us consider the permittivity and conductivity at the boundary of four cells as shown in Figure 3.2. According to Ampere's law, along the closed curve C (denoted by dotted lines) with an area of S ($S=S_1+S_2+S_3+S_4$), we have

$$\begin{aligned} &\int_S \varepsilon \frac{\partial E}{\partial t} \cdot dS + \int_S \sigma E \cdot dS \\ &= \left(\frac{\varepsilon_1+\varepsilon_2+\varepsilon_3+\varepsilon_4}{4}\right)\frac{\partial E_z}{\partial t}\Delta x \Delta y + \left(\frac{\sigma_1+\sigma_2+\sigma_3+\sigma_4}{4}\right) E_z \Delta x \Delta y \\ &= \oint_C H \cdot dl \end{aligned} \tag{3.11}$$

which means the average permittivity and conductivity of the four neighboring cells should be used at the boundary.

3.1.2 Absorbing Boundary Conditions

The FDTD method requires discretization of the entire spatial domain over which the solution is to be calculated. However, it is impossible to discretize an infinite space, because of the finite memory capability of computers. The calculation domain, therefore, must be truncated to a finite size, even for an open-region problem. Once the infinite space of the open-region problem is truncated to a finite size, absorbing boundary conditions must be applied to the outside boundaries of the calculation domain in order to simulate the nonreflective nature of open space.

One of the most popular and effective absorbing boundary conditions is known as the perfectly matched layer (PML) (Berenger, 1994). The basic concept of PML is

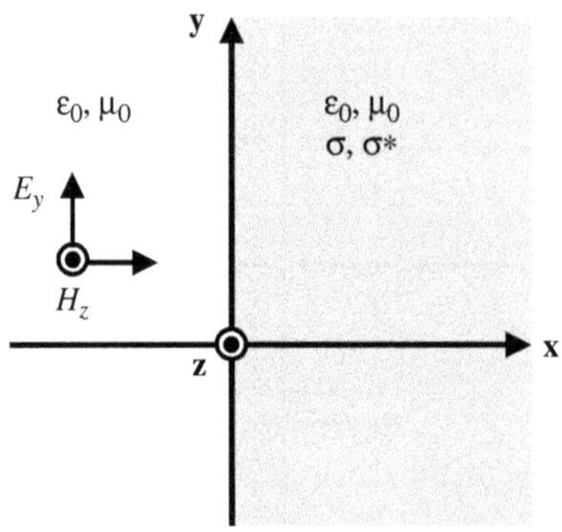

Figure 3.3 A plane wave traveling normal to a PML medium

based on impedance matching to minimize reflections. For a plane wave traveling normal to a PML medium as shown in Figure 3.3, the characteristic impedances in the free space and in the PML medium are

$$Z_0 = \sqrt{\frac{\mu_0}{\varepsilon_0}} \tag{3.12}$$

and

$$Z = \sqrt{\frac{\mu_0 + \frac{\sigma^*}{j\omega}}{\varepsilon_0 + \frac{\sigma}{j\omega}}} \tag{3.13}$$

respectively. Here σ^* is called the magnetic conductivity. From the impedance matching condition, that is, $Z = Z_0$, we have

$$\frac{\sigma}{\varepsilon_0} = \frac{\sigma^*}{\mu_0}. \tag{3.14}$$

That is to say, the reflection from the PML medium is zero for a normal incidence as long as Equation 3.14 is satisfied.

For incidence with an angle, however, zero reflection at the PML plane is also required. To do this, we first split each electric field and magnetic field component into two nonphysical subcomponents, respectively, that is,

$$\begin{aligned} E_x &= E_{xy} + E_{xz} \\ E_y &= E_{yx} + E_{yz} \\ E_z &= E_{zx} + E_{zy} \end{aligned} \tag{3.15}$$

$$\begin{aligned} H_x &= H_{xy} + H_{xz} \\ H_y &= H_{yx} + H_{yz} \\ H_z &= H_{zx} + H_{zy} \end{aligned} \tag{3.16}$$

where the second subscripts represent the traveling direction of the wave. In each traveling direction, we then assign a respective electric and magnetic conductivity $(\sigma_x, \sigma_y, \sigma_z, \sigma^{*}_{x}, \sigma^{*}_{y}, \sigma^{*}_{z})$ to characterize the PML medium. In such a way, Maxwell's equations in the PML medium become 12 equations in the Cartesian coordinates:

$$\begin{aligned} \varepsilon_0 \frac{\partial E_{xy}}{\partial t} + \sigma_y E_{xy} &= \frac{\partial H_z}{\partial y} \\ \varepsilon_0 \frac{\partial E_{xz}}{\partial t} + \sigma_z E_{xz} &= -\frac{\partial H_y}{\partial z} \end{aligned} \tag{3.17a}$$

$$\begin{aligned} \varepsilon_0 \frac{\partial E_{yz}}{\partial t} + \sigma_z E_{yz} &= \frac{\partial H_x}{\partial z} \\ \varepsilon_0 \frac{\partial E_{yx}}{\partial t} + \sigma_x E_{yx} &= -\frac{\partial H_z}{\partial x} \end{aligned} \tag{3.17b}$$

$$\begin{aligned} \varepsilon_0 \frac{\partial E_{zx}}{\partial t} + \sigma_x E_{zx} &= \frac{\partial H_y}{\partial x} \\ \varepsilon_0 \frac{\partial E_{zy}}{\partial t} + \sigma_y E_{zy} &= -\frac{\partial H_x}{\partial y} \end{aligned} \tag{3.17c}$$

$$\begin{aligned} \mu_0 \frac{\partial H_{xy}}{\partial t} + \sigma^{*}_{y} H_{xy} &= -\frac{\partial E_z}{\partial y} \\ \mu_0 \frac{\partial H_{xz}}{\partial t} + \sigma^{*}_{z} H_{xz} &= \frac{\partial E_y}{\partial z} \end{aligned} \tag{3.17d}$$

$$\begin{aligned} \mu_0 \frac{\partial H_{yz}}{\partial t} + \sigma^{*}_{z} H_{yz} &= -\frac{\partial E_x}{\partial z} \\ \mu_0 \frac{\partial H_{zx}}{\partial t} + \sigma^{*}_{x} H_{yx} &= \frac{\partial E_z}{\partial x} \end{aligned} \tag{3.17e}$$

$$\begin{aligned} \mu_0 \frac{\partial H_{zx}}{\partial t} + \sigma^{*}_{x} H_{zx} &= -\frac{\partial E_y}{\partial x} \\ \mu_0 \frac{\partial H_{zy}}{\partial t} + \sigma^{*}_{y} H_{zy} &= \frac{\partial E_x}{\partial y} . \end{aligned} \tag{3.17f}$$

Based on Equation 3.14, the matching conditions for PML are as follows.
At the PML planes normal to the x axis:

$$\begin{aligned} \frac{\sigma_x}{\varepsilon_0} &= \frac{\sigma^{*}_{x}}{\mu_0} \\ \sigma_y &= \sigma_z = 0. \end{aligned} \tag{3.18}$$

At the PML planes normal to the y axis:

$$\begin{aligned} \frac{\sigma_y}{\varepsilon_0} &= \frac{\sigma_y^*}{\mu_0} \\ \sigma_x &= \sigma_z = 0. \end{aligned} \tag{3.19}$$

At the PML planes normal to the z axis:

$$\begin{aligned} \frac{\sigma_z}{\varepsilon_0} &= \frac{\sigma_z^*}{\mu_0} \\ \sigma_y &= \sigma_z = 0. \end{aligned} \tag{3.20}$$

To incorporate PML into the FDTD procedure, the same discretization with the Yee cell algorithm for Equation 3.17 is necessary. For example, the discretization for E_{xy} in Equation 3.17a is

$$\begin{aligned} E_{xy}^{n}\left(i+\frac{1}{2},j,k\right) &= \frac{1-\dfrac{\sigma_y(j)\Delta t}{2\varepsilon_0}}{1+\dfrac{\sigma_y(j)\Delta t}{2\varepsilon_0}} E_{xy}^{n-1}\left(i+\frac{1}{2},j,k\right) + \frac{\Delta t/\varepsilon_0}{1+\dfrac{\sigma_y(j)\Delta t}{2\varepsilon_0}} \frac{1}{\Delta y} \\ &\times \left[H_z^{n-1/2}\left(i+\frac{1}{2},j+\frac{1}{2},k\right) - H_z^{n-1/2}\left(i+\frac{1}{2},j-\frac{1}{2},k\right)\right] \end{aligned} \tag{3.21}$$

and the discretization for H_{zx} in Equation 3.17f is

$$\begin{aligned} H_{zx}^{n+\frac{1}{2}}\left(i+\frac{1}{2},j+\frac{1}{2},k\right) &= \frac{1-\dfrac{\sigma_x\left(i+\frac{1}{2}\right)\Delta t}{2\varepsilon_0}}{1+\dfrac{\sigma_x\left(i+\frac{1}{2}\right)\Delta t}{2\varepsilon_0}} H_{zx}^{n-\frac{1}{2}}\left(i+\frac{1}{2},j+\frac{1}{2},k\right) \\ &- \frac{\Delta t/\mu_0}{1+\dfrac{\sigma_x\left(i+\frac{1}{2}\right)\Delta t}{2\varepsilon_0}} \frac{1}{\Delta x}\left[E_y^n\left(i+1,j+\frac{1}{2},k\right) - E_y^n\left(i,j+\frac{1}{2},k\right)\right] \end{aligned} \tag{3.22}$$

where σ_x changes only along the x direction, and σ_y changes only along the y direction.

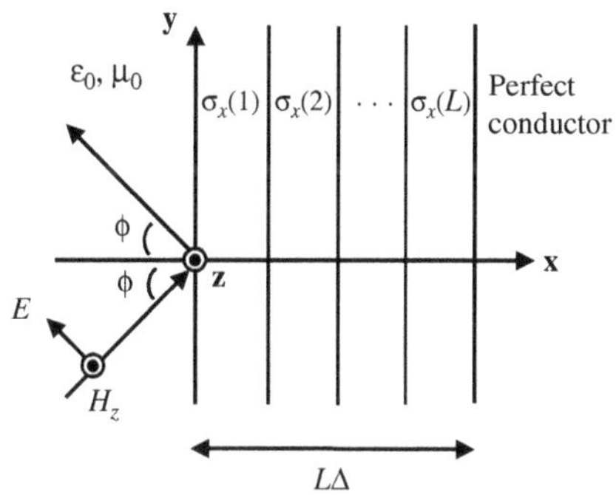

Figure 3.4 *L* layer PML for incidence with angle ϕ

Theoretically speaking, the PML provides a perfect absorption for traveling waves with any angle of incidence. However, in practice, the PML must be terminated, because of finite computer memory. Typically termination is accomplished using a perfect electric conductor, which introduces a reflection back into the calculation domain. The performance of PML is therefore characterized by three parameters: (a) thickness; (b) conductivity profile; and (c) the reflection coefficient at normal incidence.

Let us consider an L layer PML parallel to the w ($w = x$, y or z) plane for an incidence with angle ϕ, as shown in Figure 3.4. The impedance matching condition should be satisfied in all layers, that is, $Z_0 = Z_1 = \ldots = Z_L$, or

$$\frac{\sigma_w(1)}{\varepsilon_0} = \frac{\sigma_w^*(1)}{\mu_0}, \quad \frac{\sigma_w(2)}{\varepsilon_0} = \frac{\sigma_w^*(2)}{\mu_0}, \ldots, \frac{\sigma_w(L)}{\varepsilon_0} = \frac{\sigma_w^*(L)}{\mu_0}. \tag{3.23}$$

If we choose

$$\sigma_w = \begin{cases} \sigma_{\max}\left(\dfrac{L\Delta - w}{L\Delta}\right)^M & w < L\Delta \\ 0 & L\Delta < w < (N-L-1)\Delta \\ \sigma_{\max}\left(\dfrac{w-(N-L-1)\Delta}{L\Delta}\right)^M & w > (N-L-1)\Delta \end{cases} \tag{3.24}$$

where $\Delta = \Delta x$, Δy, or Δz, and N is the maximum cell number in the x, y or z direction, the reflection coefficient can be expressed as

$$|R(\phi)| \cong \exp\left[-\frac{2\sigma_{\max} L\Delta}{(M+1)\varepsilon_0 c}\cos(\phi)\right]. \tag{3.25}$$

Hence the required reflection coefficient is determined by the layer number L, the conductivity profile parameter M and $\sigma_{\max}$. In general, $M = 2 \sim 4$, $L = 4 \sim 16$, and $\sigma_{\max}$ is determined by the required reflection coefficient at $\phi = 0$, that is,

$$\sigma_{\max} = -\frac{(M+1)\varepsilon_0 c}{2L\Delta} \ln|R(0)|. \tag{3.26}$$

A typical choice for $R(0)$ may be -120 dB.

3.1.3 Field Excitation

In the FDTD calculation, an energy source must be coupled into the algorithm to excite the electric and magnetic fields. Typical energy sources in body area communications are transmitting antennas. For a transmitting antenna, a voltage source $v(t)$ is commonly applied at the feeding gap of the antenna and then is converted into the electric field via $E(t) = v(t)/\Delta$, where Δ is the cell size. The feeding current can be calculated as $i(t) = \oint_C H(t)dl$ by integrating the magnetic field along a closed curve C according to Ampere's law. Then the input impedance $Z_{in}(\omega)$ of the antenna can be obtained by

$$Z_{in}(\omega) = \frac{V(\omega)}{I(\omega)} = \frac{F\{v(t)\}}{F\{i(t)\}} \tag{3.27}$$

where the notation $F\{\,\}$ indicates Fourier transform.

For the voltage source $v(t)$, it may be a sinusoidal voltage with an angular frequency ω. It can also be a pulse voltage which is efficient to derive wide band characteristics. A Gaussian pulse is a typical excitation source in FDTD simulation in view of its smooth frequency spectrum and ease of formulation. The Gaussian pulse is expressed as

$$v(t) = \frac{A}{\sqrt{2\pi}\sigma_t} e^{-\frac{(t-t_0)^2}{2\sigma_t{}^2}} \tag{3.28}$$

where the coefficient σ_t is often set to be $\sigma_t = \tau_0/4\sqrt{2}$. Figure 3.5 shows its time waveform and frequency spectrum. The frequency spectrum contains a direct current (DC) component. If we do not need the DC component, we can use a derivative Gaussian pulse as the voltage source. A general expression of the derivative Gaussian pulse is

$$v^{(n)}(t) = -\frac{n-1}{\sigma_t{}^2} v^{(n-2)}(t) - \frac{t}{\sigma_t{}^2} v^{(n-1)}(t) \tag{3.29}$$

where n denotes the nth derivative.

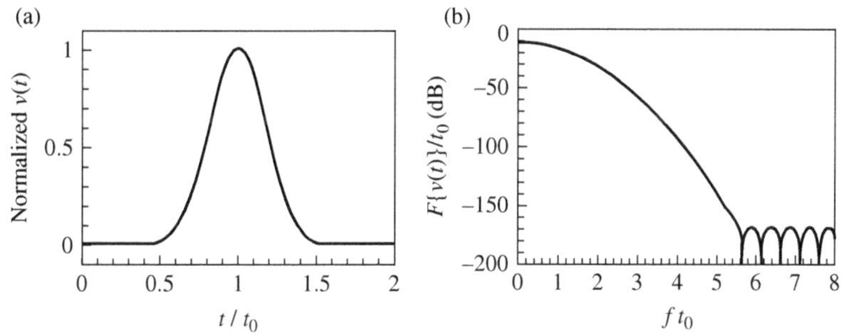

Figure 3.5 (a) Time domain Gaussian pulse waveform and (b) its frequency spectrum

3.1.4 FDTD Flow Chart and Code

Figure 3.6 shows a flow chart of FDTD calculation. First, the initialization for various field components and setting for various parameters of the analysis target are performed. In the 'Update E-field' subroutine, E^{n-1} and $H^{n-1/2}$ are the inputs and E^n is calculated as the output. In the 'Apply absorbing boundary of E-field' subroutine,

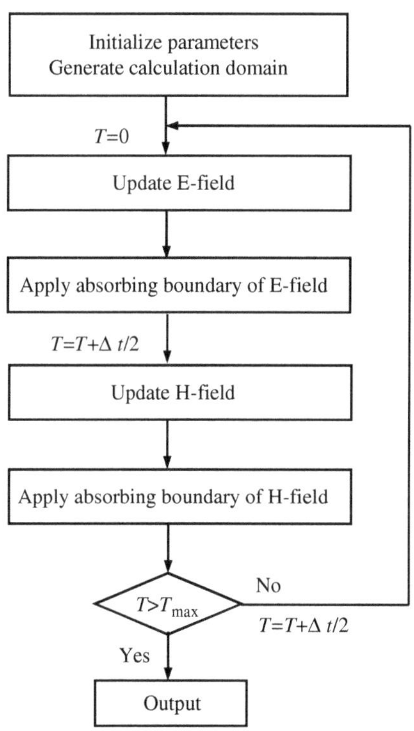

Figure 3.6 Flow chart of the FDTD method

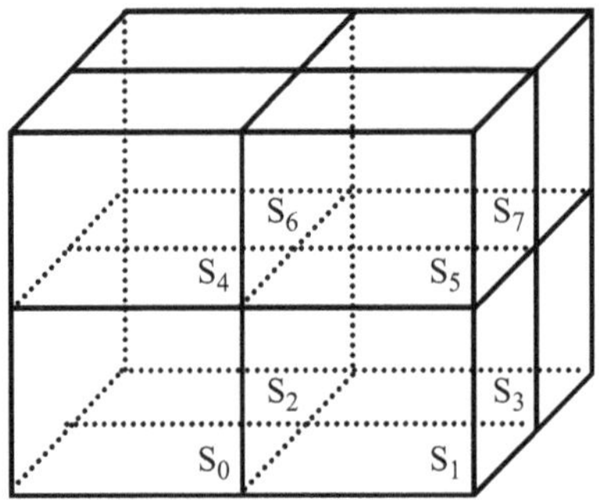

Figure 3.7 Division of FDTD space into eight subspaces. Each subspace is handled by one CPU

E^n at the boundary is determined from E^n and E^{n-1} inside the boundary based on the PML principle. Similarly, in the 'Update H-field' subroutine, E^n and $H^{n-1/2}$ are the inputs and $H^{n+1/2}$ is calculated as the output. In the 'Applying absorbing boundary of H-field' subroutine, $H^{n+1/2}$ at the boundary is determined from $H^{n+1/2}$ and $H^{n-1/2}$ inside the boundary based on the PML principle. This process is repeated until the exciting voltage signal travels through the entire analysis domain and reaches a steady state.

Two FDTD codes in Fortran are available to download from the website www.wiley.com/go/wang/bodyarea. The first one is tuned for a vector processor computer. The scenario is a short dipole excited with a Gaussian pulse in free space. The outputs are feeding voltage and current on the short dipole and electric and magnetic fields at some observation points.

The second one is written for use in parallel personal computers with a message passing interface (MPI) library (Wang *et al.*, 2004). The entire FDTD calculation domain is divided into eight subspaces, as shown in Figure 3.7. Each subspace is handled by one CPU, and all the CPUs execute the same program. To calculate the field in the boundary cells, we need to know the field in the cells belonging to the neighboring subspaces. This is realized by using the MPI library to transfer the data from one CPU to another CPU. For simplicity, let us consider a one-dimensional case as shown in Figure 3.8. We divide the whole FDTD space into two subspaces: in this case, S_0 and S_1. S_0 and S_1 are handled by CPU_0 and CPU_1, respectively. In the E-field calculation, we pass H_M in subspace S_0 to S_1, which is used as H_0 for obtaining E_1 in S_1. In the H-field calculation, we pass E_1 in subspace S_1 to S_0, which is used as E_{M+1} for obtaining H_M in subspace S_0. It is straightforward to extend this algorithm to a three-dimensional case to obtain a practical parallel FDTD code. Data transfer among the CPUs can be conducted using a gigabit network interface.

Figure 3.9 shows the flow chart of the MPI-based parallel FDTD code. The MPI communication subroutines are added after applying absorbing boundary conditions for the electric field and magnetic field, respectively. A performance evaluation for the

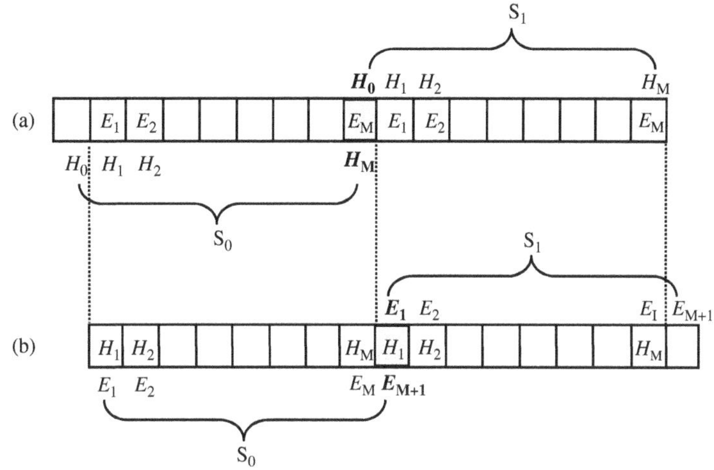

Figure 3.8 One-dimensional illustration of the MPI-based data communication between the subspaces used in the parallel FDTD code. (a) H_M in S_0 is passed to H_0 in S_1 for calculating E_1 in S_1. (b) E_1 in S_1 is passed to E_{M+1} in S_0 for calculating H_M in S_0 (Wang *et al.*, 2004). Reproduced with permission from Wang J., Fujiwara O., Watanabe S. and Yamanaka Y., "Computation with a parallel FDTD system of human-body effect on electromagnetic absorption for portable telephones," *IEEE Transactions on Microwave Theory and Techniques*, **52**, 1, 966–971, 2004. © 2004 IEEE

parallel code is given in Figure 3.10 for eight CPUs. The abscissa indicates the number of cells of the entire FDTD domain, and the left and right ordinates indicate the speed-up ratio and memory-down ratio, respectively. The speed-up ratio is defined as the ratio of the run time in one CPU case to that in eight CPUs, and the memory-down ratio is defined as the ratio of the memory required in each CPU when the calculation is conducted in one CPU to that in eight CPUs. For both the speed-up and memory-down ratios, the ideal value is eight because eight CPUs are used. As can be seen, the code exhibits an almost fixed speed-up ratio of six, which corresponds to a reasonable efficiency of 75%, and a memory-down ratio of approximately eight, which means that the overhead for the memory in the parallel code is almost negligible.

3.1.5 Frequency-Dependent FDTD Method

A pulse with a wide frequency band is often used as an excitation in the FDTD method, especially for getting wide band propagation characteristics. As described in the previous chapter, the permittivity and conductivity of biological tissue are frequency-dependent. A useful expression for the frequency-dependent tissue properties is known as the Debye approximation as shown in Section 2.3. With this approximation, it is possible to incorporate the frequency-dependent tissue properties into the FDTD method.

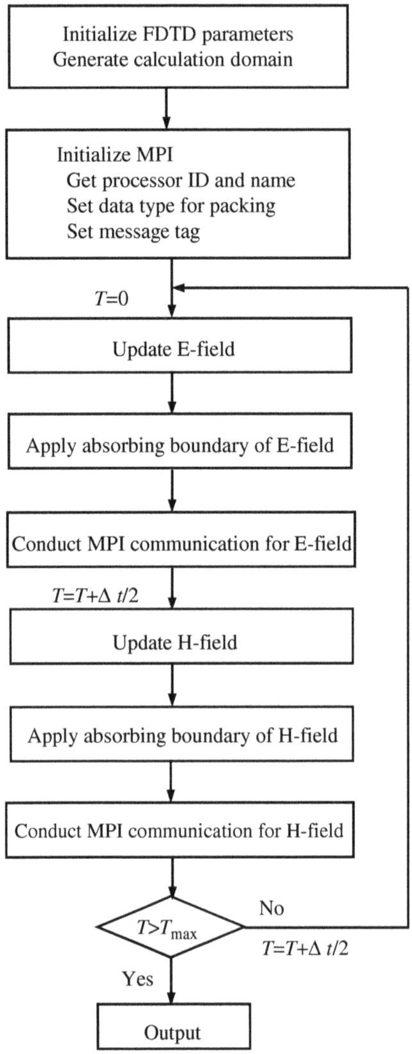

Figure 3.9 Flow chart of MPI-based parallel FDTD method

The Debye approximation for complex permittivity of biological tissue is expressed as

$$\dot{\varepsilon}_r(\omega) = \varepsilon_\infty + \chi(\omega) + \frac{\sigma_0}{j\omega\varepsilon_0} \tag{3.30}$$

where ε_∞ is the relative permittivity at infinite frequency, $\chi(\omega)$ is the frequency domain susceptibility, and σ_0 is the ionic conductivity at zero frequency. We can

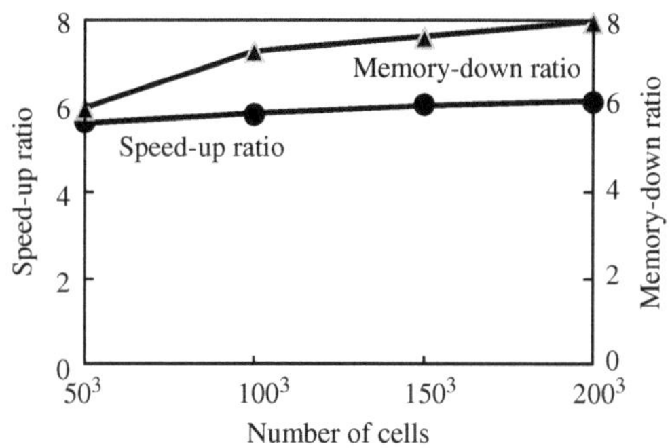

Figure 3.10 Performance of the parallel FDTD code (Wang *et al.*, 2004). Reproduced with permission from Wang J., Fujiwara O., Watanabe S. and Yamanaka Y., "Computation with a parallel FDTD system of human-body effect on electromagnetic absorption for portable telephones," *IEEE Transactions on Microwave Theory and Techniques*, **52**, 1, 966–971, 2004. © 2004 IEEE

relate the first and second terms in Equation 3.30 to the electric flux density $D(\omega)$ and the third term to the current density $J_0(\omega)$ because only the former two terms are due to the frequency dispersion of tissue. We then have

$$D(\omega) = \varepsilon_0[\varepsilon_\infty + \chi(\omega)]E(\omega) \tag{3.31}$$

$$J_0(\omega) = \sigma_0 E(\omega). \tag{3.32}$$

Since Maxwell's equations are to be solved iteratively in the time domain by using the FDTD method, we need to transfer Equations 3.31 and 3.32 into the time domain expressions. This can be realized by inverse Fourier transform. We thus have

$$\begin{aligned} D(t) &= \varepsilon_0\varepsilon_\infty E(t) + \varepsilon_0 \chi(t) * E(t) \\ &= \varepsilon_0\varepsilon_\infty E(t) + \varepsilon_0 \int_0^t \chi(\tau)E(t-\tau)d\tau \end{aligned} \tag{3.33}$$

$$J_0(t) = \sigma_0 E(t) \tag{3.34}$$

where the asterisk denotes the convolution. With the electric flux density $D(t)$, one of the Maxwell equations in Equation 3.1 can be expressed as

$$\nabla \times H(t) = \frac{\partial D(t)}{\partial t} + \sigma_0 E(t) \tag{3.35}$$

To incorporate Equations 3.33 and 3.34 into (3.35) for deriving a frequency-dependent FDTD formulation, we first discretize the time expression of

Equation 3.33 at $t = n\Delta t$ and $t = (n-1)\Delta t$. Then we have

$$
\begin{aligned}
D^n &= \varepsilon_0 \varepsilon_\infty E^n + \varepsilon_0 \sum_{m=0}^{n-1} E^{n-m} \chi^m \\
&= \varepsilon_0(\varepsilon_\infty + \chi^0) E^n + \varepsilon_0 \sum_{m=0}^{n-2} E^{n-1-m} \chi^{m+1}
\end{aligned}
\tag{3.36}
$$

$$
D^{n-1} = \varepsilon_0 \varepsilon_\infty E^{n-1} + \varepsilon_0 \sum_{m=0}^{n-2} E^{n-1-m} \chi^m \tag{3.37}
$$

and consequently

$$
\frac{D^n - D^{n-1}}{\Delta t} = \nabla \times H^{n-\frac{1}{2}} - \sigma_0 E^{n-\frac{1}{2}}. \tag{3.38}
$$

Substituting Equations 3.36 and 3.37 into Equation 3.38 we obtain the update equation for electric field E as

$$
\begin{aligned}
E^n = {} & \frac{2\varepsilon_\infty - \sigma_0 \Delta t/\varepsilon_0}{2\varepsilon_\infty + 2\chi^0 + \sigma_0 \Delta t/\varepsilon_0} E^{n-1} + \frac{2}{2\varepsilon_\infty + 2\chi^0 + \sigma_0 \Delta t/\varepsilon_0} \Phi^{n-1} \\
& + \frac{2\Delta t/\varepsilon_0}{2\varepsilon_\infty + 2\chi^0 + \sigma_0 \Delta t/\varepsilon_0} \nabla \times H^{n-\frac{1}{2}}
\end{aligned}
\tag{3.39}
$$

where

$$
\Phi^{n-1} = \sum_{m=0}^{n-2} E^{n-1-m} (\chi^m - \chi^{m-1}) \tag{3.40}
$$

and

$$
\chi^m = \int_{m\Delta t}^{(m+1)\Delta t} \chi(t) dt. \tag{3.41}
$$

The time-domain susceptibility $\chi(t)$ depends on the order of the Debye approximation. For the first-order Debye expression

$$
\chi(\omega) = \frac{\Delta\varepsilon}{1 + j\omega\tau} = \frac{\varepsilon_s - \varepsilon_\infty}{1 + j\omega\tau} \tag{3.42}
$$

we have

$$\chi(t) = \frac{\varepsilon_s - \varepsilon_\infty}{\tau} e^{-\frac{t}{\tau}} U(t) \tag{3.43}$$

where ε_s is the static permittivity at zero frequency, and τ is the relaxation time constant. Having found $\chi(t)$ and using the definition in Equation 3.41, we can obtain

$$\chi^m = (\varepsilon_s - \varepsilon_\infty) e^{-\frac{m\Delta t}{\tau}} \left[1 - e^{-\frac{\Delta t}{\tau}}\right] \tag{3.44}$$

and

$$\Delta\chi^m = \chi^m - \chi^{m-1} = (\varepsilon_s - \varepsilon_\infty) e^{-\frac{m\Delta t}{\tau}} \left[1 - e^{-\frac{\Delta t}{\tau}}\right]^2. \tag{3.45}$$

Substituting the above two equations into Equation 3.40 yields a simple update equation for Φ^{n-1} as

$$\Phi^{n-1} = E^{n-1} \Delta\chi^0 + e^{-\frac{\Delta t}{\tau}} \Phi^{n-2}. \tag{3.46}$$

With this update equation, the frequency-dependent permittivity and conductivity are well incorporated into the FDTD algorithm. Applying this approach to high-order Debye approximations is straightforward. Figure 3.11 shows the flow chart of the frequency-dependent FDTD method. This method is especially effective for a pulse excitation to obtain a wide band propagation characteristic over the human body.

3.2 MoM-FDTD Hybrid Method

The MoM is another popular method to solve electromagnetic problems, especially electromagnetic scattering problems. When electromagnetic waves are incident on a conducting body, it produces currents in the body surface. These currents radiate and produce the scattered field. Such a scattering problem is governed by Green's functions. Based on the Green's functions in the frequency domain, MoM constructs electric field or magnetic field integral equations for the currents which are on the surface of the conducting body. By modeling the surface with a number of small meshes and assigning a surface current distribution at each edge of the meshes, MoM reduces the Green's-function-governed operator equations to a system of linear equations which is written in a matrix form. By solving the system of linear equations, the surface current at each mesh is determined, and from all these surface currents the scattering field can be calculated.

An obvious advantage of MoM is its high accuracy in modeling a complicated surface structure. This feature is especially effective in solving the electromagnetic

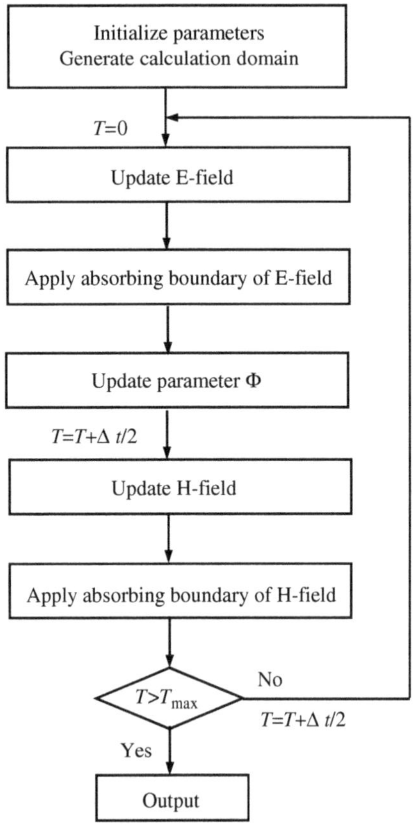

Figure 3.11 Flow chart of frequency-dependent FDTD method

problems with small-size and high-precision antennas. However, MoM is usually limited to homogeneous media because it requires calculating only boundary values on the surface, rather than values throughout the body. This yields a considerable restriction on its application to in-body communication where we need to know the received field signal inside the human body.

In this section, we introduce a hybrid technique which combines the MoM and FDTD method for analyzing body area propagations. The former is used to model the transmitting antenna outside a human body to calculate the scattered electric field from the antenna in the entire analysis region in the absence of the human body. By considering such an electric field as an incident field to the human body, the total field in or on the human body can be then calculated using a scattered field FDTD algorithm because the FDTD method is suited to model the human body's detailed anatomy. Such a hybrid technique combines the advantages of both the MoM and FDTD methods.

3.2.1 MoM Formulation

The boundary condition on a perfectly electric conductor (PEC) states that tangential electric fields on the surface must be zero, that is,

$$\boldsymbol{n} \times [E_i(r) + E_s(r)] = 0 \quad r \text{ on PEC} \tag{3.47}$$

where $\boldsymbol{n}$ is the normal direction vector pointing outwardly, E_i is the incident electric field, E_s is the scattered electric field which can be expressed in terms of the spatial convolution of electric current density J and the free-space Green's function in the frequency domain as

$$E_s(r) = -jk_0\eta_0 \int_S G(r, r') \left[J(r') + \frac{1}{k_0^2} \nabla' \nabla' \cdot J(r') \right] dr' \tag{3.48}$$

where k_0 is the wavenumber, η_0 is the intrinsic impedance of free space, and

$$G(r, r') = \frac{1}{4\pi} \frac{e^{-jk_0|r-r'|}}{|r - r'|} \tag{3.49}$$

represents the free-space Green's function. By substituting Equation 3.48 into 3.47, the electric field integral equation (EFIE) for PEC is obtained as follows

$$\boldsymbol{n} \times E_i(r) = jk_0\eta_0 \boldsymbol{n} \times \int_S \left[J(r') + \frac{1}{k_0^2} \nabla' \nabla' \cdot J(r') \right] G(r, r') dr'. \tag{3.50}$$

For a transmitting antenna, it can be considered as a PEC. In the MoM, the PEC surface is usually modeled either as a wire mesh – the so-called wire-grid model – or as a surface partitioned into smooth or piecewise-smooth patches – the so-called surface patch model. The wire-grid model makes all numerical integrals in the moment matrix one dimensional, and has been successfully applied in solving far-field radiation problems in particular. However, this approach is not well suited for calculating near-field and surface currents because of some existing problems such as the presence of fictitious loop currents in the solution, and difficulties in interpreting calculated wire currents and relating them to equivalent surface currents. Surface patch models can overcome these difficulties. For modeling arbitrarily shaped surfaces, planar triangular patch models are particularly appropriate. For example, triangular patches are capable of accurately conforming to any geometrical surface or boundary, the patch scheme is easily specified for computer input, and a varying patch density can be used according to the resolution required in the surface geometry or current.

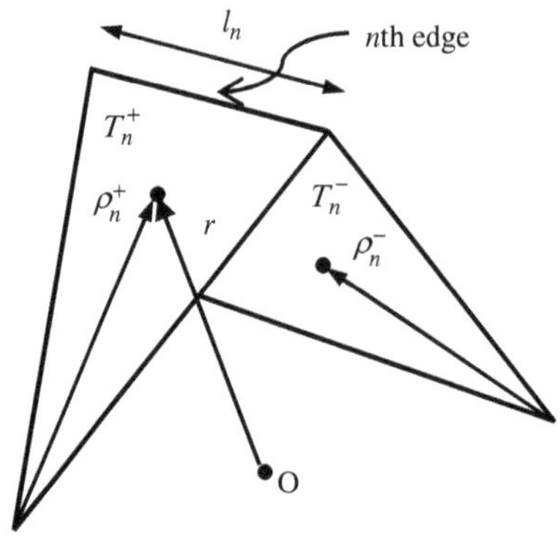

Figure 3.12 Two triangles and geometrical parameters associated with an edge in the triangular patch modeling

With the triangular patch modeling, the current distribution at each edge of the patches can be represented by using the sum of basis functions as

$$J(r) = \sum_{n=1}^{N} a_n f_n(r) \tag{3.51}$$

where N is the total number of edges, a_n is an unknown coefficient and $f_n(r)$ is a basis function associated with the nth edge to approximately represent the surface current. Figure 3.12 shows two triangles, T_n^+ and T_n^-, corresponding to the nth edge of a triangulated surface (Rao, Wilton, and Glisson, 1982). Points in T_n^+ can be designated by the position vector $\boldsymbol{\rho}_n^+$ defined with respect to the free vertex of T_n^+. Similar remarks apply to the position vector $\boldsymbol{\rho}_n^-$ except that it is directed toward the free vertex of T_n^-. The plus or minus designation of the triangles is determined by the choice of a positive current reference direction for the nth edge, the reference for which is assumed to be from T_n^+ to T_n^-. Referring to Figure 3.12, the basis function associated with the nth edge can be defined as

$$f_n(r) = \begin{cases} \dfrac{l_n}{2A_n^+}\boldsymbol{\rho}_n^+ & r \text{ in } T_n^+ \\ \dfrac{l_n}{2A_n^-}\boldsymbol{\rho}_n^- & r \text{ in } T_n^- \\ 0 & \text{otherwise} \end{cases} \tag{3.52}$$

where l_n is the length of the edge and $A_n^{\pm}$ is the area of triangle $T_n^{\pm}$.

By substituting Equation 3.51 into 3.50 and testing the EFIE with the curl of the basis functions, that is, $\hat{\boldsymbol{n}} \times f_m(r)$, the matrix equation for the EFIE is finally expressed as

$$[z_{mn}]_{N\times N}[a_n]_{N\times 1} = [v_m]_{N\times 1} \qquad m, n = 1, 2, \ldots, N \tag{3.53}$$

$$z_{mn} = \int_{f_m}\int_{f_n}\left[f_m(r)\cdot f_n(r') - \frac{1}{k_0^2}[\nabla\cdot f_m(r)][\nabla'\cdot f_n(r')]\right]G(r,r')dr'dr \tag{3.54}$$

$$v_m = -\frac{j}{k_0\eta_0}\int_{f_m} f_m(r)\cdot E_i(r)dr. \tag{3.55}$$

The matrix equation 3.53 can be solved using a direct solution of the linear system, typically via LU decomposition, or by making use of an iterative algorithm of linear system solutions. Once the currents are obtained, the induced electromagnetic field can be calculated in the post-processing.

3.2.2 Scattered Field FDTD Formulation

Total electromagnetic fields are the sum of the incident and scattered fields, that is,

$$E = E_i + E_s \tag{3.56}$$

$$H = H_i + H_s \tag{3.57}$$

where E, E_i and E_s are the total, incident and scattered electric fields, respectively, and H, H_i and H_s represent their magnetic counterparts. All quantities in Equations 3.56 and 3.57 are functions of both time and space. Maxwell's equations in the form of separate fields are generally expressed as

$$\nabla\times E_s = -u\frac{\partial H_s}{\partial t} - (\mu-\mu_0)\frac{\partial H_i}{\partial t} \tag{3.58}$$

$$\nabla\times H_s = \varepsilon\frac{\partial E_s}{\partial t} + \sigma E_s + \left[(\varepsilon-\varepsilon_0)\frac{\partial E_i}{\partial t} + \sigma E_i\right]. \tag{3.59}$$

Since the human body is a lossy dielectric medium with electrical properties ε, σ and μ_0, Equation 3.58 reduces to

$$\nabla\times E_s = -u_0\frac{\partial H_s}{\partial t}. \tag{3.60}$$

Thus, there is no need to obtain information on the incident magnetic field. As long as we know the incident electric field, for example, by using the MoM as described in Section 3.2.1, we can calculate the scattered field due to the human body and then the total field from Equations 3.59 and 3.60.

After some manipulation of discretization for Equations 3.59 and 3.60, the scattered field FDTD formulation is derived straightforwardly as follows

$$E_s^n = \frac{1-\dfrac{\sigma\Delta t}{2\varepsilon}}{1+\dfrac{\sigma\Delta t}{2\varepsilon}} E_s^{n-1} + \frac{\dfrac{\Delta t}{\varepsilon}}{1+\dfrac{\sigma\Delta t}{2\varepsilon}} \nabla \times H_s^{n-\frac{1}{2}} - \frac{\dfrac{\sigma\Delta t}{\varepsilon}}{1+\dfrac{\sigma\Delta t}{2\varepsilon}} E_i^{n-\frac{1}{2}} - \frac{(\varepsilon-\varepsilon_0)\dfrac{\Delta t}{\varepsilon}}{1+\dfrac{\sigma\Delta t}{2\varepsilon}} \frac{\partial E_i^{n-\frac{1}{2}}}{\partial t} \tag{3.61}$$

$$H_s^{n+\frac{1}{2}} = H_s^{n-\frac{1}{2}} - \frac{\Delta t}{\mu_0} \nabla \times E_s^n \tag{3.62}$$

where Δt is the time step, and n is the temporal step index.

3.2.3 Hybridization of MoM and FDTD Method

The concept of the hybrid MoM-FDTD method is depicted in Figure 3.13. There are two analysis regions: one is the MoM region including the transmitting antenna structure, and the other is the FDTD region including the human body to be irradiated by the electromagnetic field from the antenna. J_a denotes the electric current density on the antenna. J_s and M_s denote the electric and magnetic current densities on a closed surface surrounding the human body, respectively. The closed surface can be chosen arbitrarily as long as it totally encompasses the human body. The following describes the calculation procedure of the hybrid MoM-FDTD method (Chakarothai *et al.*, 2012).

1. Construct the antenna structure with many small meshes and apply a voltage at the edge of the mesh corresponding to the antenna feeding point. The voltage at

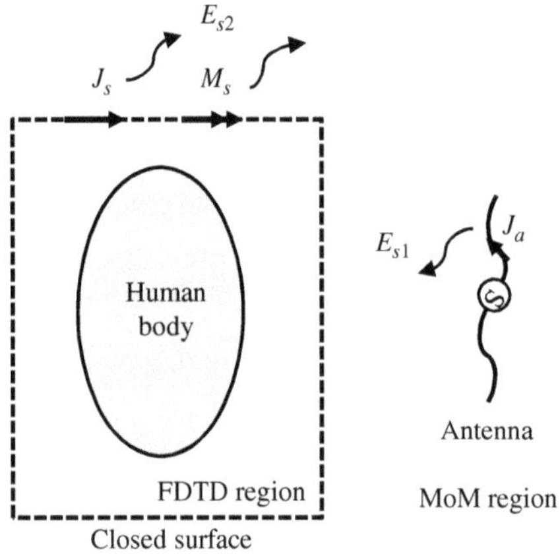

Figure 3.13 Concept of hybrid MoM-FDTD method

the feeding edge is given based on the delta gap feeding method as

$$v_n^{feed} = -\frac{j}{k_0\eta_0} V^{in} l_n \tag{3.63}$$

where V^{in} is the input voltage of the antenna and l_n is the length of the feeding edge. According to the MoM algorithm, we can derive the matrix equation 3.53 and solve the matrix equation via LU decomposition to obtain the antenna current density J_a. In this step, the human body is excluded from the calculation.

2. Once J_a is determined, the corresponding electric field, which is radiated from the antenna and denoted by E_{s1} in Figure 3.13, can be numerically calculated via Equation 3.48 at an arbitrary observation point in the analysis region. Since the calculated electric field E_{s1} is in the frequency domain, we then transform it into the time domain by using a phasor expression. The time-domain E_{s1} is then used as an incident field in the scattered field FDTD formulation of Equation 3.61, that is,

$$E_i(r,t)\big|^{t=n\Delta t} = |E_{s1}(r)|\cos[\omega n\Delta t + \angle E_{s1}(r)] \tag{3.64}$$

$$\left.\frac{\partial E_i(r,t)}{\partial t}\right|^{t=n\Delta t} = -\omega|E_{s1}(r)|\sin[\omega n\Delta t + \angle E_{s1}(r)] \tag{3.65}$$

where ω denotes the angular frequency, and $|E_{s1}(r)|$ and $\angle E_{s1}(r)$ are the amplitude and phase of the scattered electric field, respectively.

3. Calculate the scattered electric field E_s and scattered magnetic field H_s in the human body with the scattered field FDTD method. In this step the antenna is excluded from the FDTD region and the absorbing condition is applied at the truncated boundary of the FDTD region.
4. Calculate the electric and magnetic current densities J_s and M_s on the closed surface surrounding the human body from the equivalence principle of electromagnetic fields, that is,

$$J_s(r) = \boldsymbol{n}(r) \times H_s(r) \tag{3.66}$$

$$M_s(r) = -\boldsymbol{n}(r) \times E_s(r) \tag{3.67}$$

where E_s and H_s on the closed surface are also calculated by using the scattered field FDTD method. Since the FDTD cells are as small as in the order of one-tenth of a wavelength, we can consider $J_s(r)$ and $M_s(r)$ as infinitesimal electric and magnetic dipole moments, respectively, on the face of the FDTD cell which corresponds to point r. The infinitesimal dipole moments at the ith face on the closed surface can be expressed as

$$\boldsymbol{j}_i = J_s(r_i)\Delta s \tag{3.68}$$

$$\boldsymbol{m}_i = M_s(r_i)\Delta s \tag{3.69}$$

where r_i is defined at the center of the face with an area Δs. Then the electric field scattered from the human body to an arbitrary point outside the closed surface can be analytically calculated from the following expression

$$\begin{aligned} E_{s2}(r) &= jk_0 \sum_{i=1}^{N_d} \left[\eta_0(\boldsymbol{p}_i - \boldsymbol{j}_i) + \frac{3\boldsymbol{p}_i - \boldsymbol{j}_i}{j\omega\varepsilon_0} D + (R_i \times \boldsymbol{m}_i) D \right] \frac{e^{-jk_0 R_i}}{4\pi R_i} \\ \boldsymbol{p}_i &= \frac{(R_i \cdot \boldsymbol{j}_i) R_i}{R_i^2}, \quad D = \frac{1}{R_i}\left(1 + \frac{1}{jk_0 R_i}\right), \quad R_i = r - r_i \end{aligned} \tag{3.70}$$

where N_d is the total number of electric and magnetic dipole moments. Such an analytical calculation does not need to discretize the free space volume and can greatly save computational resources.

5. Substitute the scattered field E_{s2} in Equation 3.70 into 3.55 to obtain a new induced voltage at each edge element in the MoM region by

$$v_m = -\frac{j}{k_0 \eta_0} \int_{f_m} f_m(r) \cdot E_{s2}(r) dr. \tag{3.71}$$

Since each vector element v_m is treated as an additional source in the MoM, the current density J_a obtained in this step must be added to that obtained in the previous turn of iterations. Steps 1–4 are thus repeated until a convergence of required quantity, such as the electric field, is reached.

The hybrid MoM-FDTD method provides a useful alternative when the FDTD method is difficult in modeling a transmitting antenna with very small size or very complicated structure. The convergence is usually reached after several iterations.

3.3 Finite Element Method

For a time-harmonic field, that is, the field quantities are harmonically oscillating functions with a single angular frequency ω, Maxwell's equations can be expressed in a simplified form with complex phasor notation as

$$\begin{aligned} \nabla \times E + j\omega B = 0 \\ \nabla \times H - j\omega D = J. \end{aligned} \tag{3.72}$$

To solve Equation 3.72, we may first convert the first-order differential equations involving both E and H field quantities into second-order differential equations involving only E or H field. We can then obtain the differential equation for E by eliminating H with the aid of the constitutive relations $D = \varepsilon E$, $B = \mu H$, and $J = \sigma E$. That is

$$\nabla \times \left(\frac{1}{\mu} \nabla \times E \right) - \omega^2 \dot{\varepsilon} E = -j\omega J_s \tag{3.73}$$

where J_s is an impressed source current. Similarly, we can eliminate E to derive the equation for H as

$$\nabla \times \left(\frac{1}{\dot{\varepsilon}} \nabla \times H \right) - \omega^2 \mu H = \nabla \times \left(\frac{1}{\dot{\varepsilon}} J_s \right). \tag{3.74}$$

FEM (Jin, 1993) is a numerical technique used to solve the above partial differential equations by transforming them into a system of equations. The principle of the method is to replace entire analysis domain by a number of subdomains in which the unknown function is represented by simple interpolation functions with unknown coefficients. That is to say, the solution of the entire system is approximated by a finite number of unknown coefficients. Then a system of equations is obtained, and the solution is achieved by solving the system of equations. Since FEM is not as popular as the FDTD method in conjunction with numerical human body models described in the next section, here we only give a simple introduction to its fundamental concept.

In general, a partial differential equation can be symbolically expressed with a differential operator L so that

$$L(u) = s \tag{3.75}$$

where s is an excitation, and u is the unknown quantity. To solve this equation, FEM employs a variational method in which the problem to be solved is formulated in terms of a variational expression, referred to as functional. The functional, denoted as F, is defined as a mapping that assigns a number to a function. This makes the functional a logical extension of the concept of the function, which assigns one number to another number. For instance, if v is a known function, then $F[v]$ will be a unique number assigned to the function v. The minimum of the functional corresponds to the differential equation under given boundary conditions. Consequently the approximation solution of Equation 3.75 can be obtained by minimizing the functional with respect to its variables.

A finite element analysis usually includes four steps: (a) discretization of the analysis domain; (b) selection of interpolation functions; (c) formulation of a system of equations; and (d) solution of the system of equations.

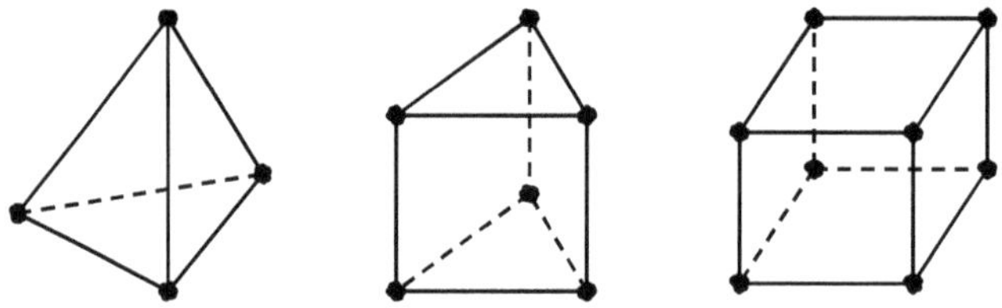

Figure 3.14 Three-dimensional element shapes: tetrahedron, triangular prism, and rectangular brick

In the first step, that is, the discretization of the analysis domain Ω, the domain Ω is subdivided into a number of small subdomains. The subdomains are referred to as elements from $m=1$ to M. The elements may have the shape of a tetrahedron, triangular prism, or rectangular brick, as shown in Figure 3.14. The tetrahedrons are the best suited for arbitrary-shaped volumes and the rectangular bricks are more straightforward for applying to the usual numerical human body models. In the finite element solutions, the problem is formulated in terms of the unknown function u at nodes associated with the elements. A tetrahedron has four nodes and a rectangular brick has eight nodes. A complete description of a node contains its coordinate values, local number, and global number. The local number and global number indicate the node's position in the element and in the entire system, respectively. Figure 3.15 illustrates the concept of local number and global number.

The second step is to select an interpolation function which approximates the unknown solution within an element. The interpolation is usually selected to be a first-order polynomial for simplicity. The unknown solution in the element m can be then expressed as

$$\tilde{u}^m = \sum_{l=1}^{N_m} w_l^m u_l^m = [u^m]^T [w^m] \tag{3.76}$$

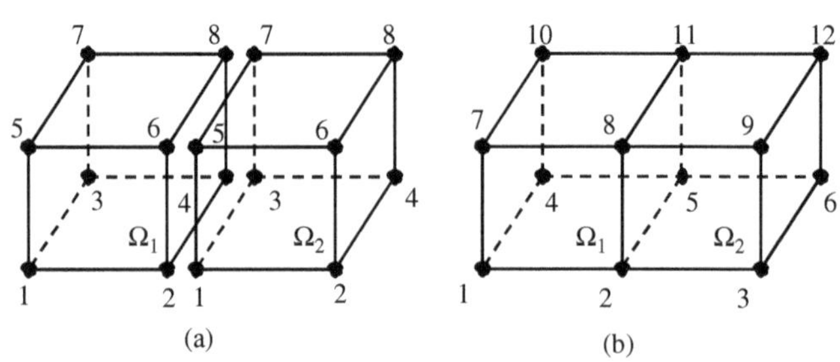

Figure 3.15 Concept of (a) local numbering and (b) global numbering

where N_m is the number of nodes in the element m, u_l^m is the value of u at node n in the element m, and w_l^m is the interpolation function which is nonzero only within element m. In this equation, the coefficients to be determined are u_l^m.

To proceed to the third step for obtaining a system of equations, let us first define the inner product of two functions u and v, denoted by angular brackets, as

$$<u, v> = \int_\Omega u \cdot v \, d\Omega \tag{3.77}$$

where Ω is the domain of u and v. With this definition, if the differential operator L satisfies

$$<Lu, v> = <u, Lv> \tag{3.78}$$

the following functional

$$F(\tilde{u}) = \frac{1}{2} <\tilde{u}, L(\tilde{u})> - <s, \tilde{u}> \tag{3.79}$$

should give the solution to Equation 3.75 at its minimum with respect to $\tilde{u}$. In other words, Equation 3.75 is equivalent, in the mathematical sense, to minimizing the functional F. For the functional F to have a minimum, its first-order derivative with respect to $\tilde{u}$ should vanish.

In the entire analysis domain comprised of M elements, the functional F can be expressed as

$$F(\tilde{u}) = \sum_{m=1}^{M} F^m(\tilde{u}^m) \tag{3.80}$$

and

$$F^m(\tilde{u}^m) = \frac{1}{2} <\tilde{u}^m, L(\tilde{u}^m)> - <s, \tilde{u}^m> \tag{3.81}$$

where the domain of $\tilde{u}^m$ is Ω_{m}. Substituting Equation 3.76 into 3.81, we have

$$F^m = \frac{1}{2} [u^m]^T [K^m]^T [u^m] - [u^m]^T [b^m] \tag{3.82}$$

where $[K^m]$ is an $N_m \times N_m$ matrix and $[b^m]$ is an $N_m \times 1$ column vector with

$$K_{nl}^m = <w_n^m, L(w_l^m)> \tag{3.83}$$

and

$$b_n^m = <s, w_n^m> . \tag{3.84}$$

Substituting F^m into Equation 3.80 and adopting the global node numbers, we obtain

$$F = \frac{1}{2}[u]^T[K][u] - [u]^T[b] \tag{3.85}$$

where $[K]$ is an $N \times N$ matrix with N being the total number of unknowns or nodes, $[u]$ is an $N \times 1$ unknown coefficient vector, and $[b]$ is an $N \times 1$ known vector. To minimize F we let its partial derivatives with respect to u_n ($n = 1,2, \ldots, N$) to zero. This yields a system of linear algebraic equations

$$[K][u] = [b] \tag{3.86}$$

in matrix form.

The fourth or final step in the finite element analysis is to solve the system of equations 3.86 for $[u]$. Although the dimension of the system or the number of unknowns may be as large as several hundred thousands, there are many well-developed black-box program packages available for solving it. In such a way, FEM can be used to solve the vector wave equations 3.73 and 3.74 for propagation characteristic analysis in body area communications.

In body area propagation analysis the radiation may take place in an open region. Covering the entire domain with finite elements is impossible. So the analysis domain has to be truncated to a finite size by using a domain termination boundary. Over this terminal boundary, a radiation condition or an absorbing boundary condition must be enforced such that no reflection takes place as the wave impinges on the boundary from the inside of the domain. The simplest boundary condition is to set a circular boundary at ρ, and enforce the Sommerfeld radiation condition

$$\frac{\partial u}{\partial n} = -jku(\rho) \tag{3.87}$$

where n is the normal to the circular boundary and k is the wavenumber. The Sommerfeld radiation condition assumes the radiating waves are purely traveling, and therefore the terminal boundary should be located at a considerable distance from the analysis object. To overcome this problem, the PML as described in the FDTD method can also be used as an absorbing boundary condition in FEM. This implementation amounts to introducing an additional layer into the finite element domain which works to attenuate the field as it penetrates into the layer.

3.4 Numerical Human Body Model

An anatomically based human body model is essential for numerical simulation in body area communications. Such a numerical human body model is developed commonly from magnetic resonance imaging (MRI) or computed tomography (CT) scans. MRI or CT provides gray-scale image data as many transverse slices, at a designated spacing, from the head to feet of the human body. The resolution in each slice is in the order of several millimeters. In general, it is easy to identify bone from CT images, but it is more difficult to segment interior organs. MRI data are generally superior to CT data in identifying interior tissues, because of their high contrast. Consequently MRI data are used more often in the development of numerical models. The drawback of MRI is that the image acquisition is time-consuming and automatic identification of various organs is difficult.

In order to develop a human body model for numerical electromagnetic analysis, original gray-scale data must be interpreted into tissue types, which is known as a process of segmentation. This is because the permittivity and conductivity are required for each tissue in solving the Maxwell equations. However, the gray scales in magnetic resonance images do not correspond to tissue types directly. Different tissues can have the same gray-scale values, which makes it impossible to automatically identify the original images into corresponding tissues or organs with sufficient accuracy using presently available image-processing techniques. All tissue- and organ-identification processing is therefore performed manually or semi-automatically, to a large extent. Guidance from medical specialists is also necessary for accurate segmentation.

The gray-scale images are first rescaled to produce appropriate cells. Each cell in the images is then identified rigorously as belonging to one type of tissue by, for example, assigning each cell a red-green-blue code that identifies the discrete tissue type of that particular cell. This process can be performed with commercial software such as Adobe Photoshop. All identified transverse images are then combined to obtain a three-dimensional numerical model. A fine adjustment is generally required to smoothly connect each slice in the three orthogonal planes (axial, sagittal, and coronal).

Figure 3.16 shows an example of a numerical human body model (Nagaoka *et al.*, 2004). The human body model was developed from MRI data using the method just described. It consists of 51 tissue types such as skin, fat, muscle, bone, white matter, gray matter, heart, liver, lung, and so on. The model is comprised of a huge number of cubic cells of size $2 \times 2 \times 2$ mm, with a very high complexity for numerical electromagnetic analysis. It can be obtained from the National Institute of Information and Communications Technology (NICT), Japan, for academic use.

However, it is evident that the above-described approach is cumbersome and time-consuming when developing a numerical human body model. To overcome this drawback, a promising approach is to develop an anthropomorphic numerical model, which is known as the dielectric anatomical model (Mazzurana *et al.*, 2003). In this

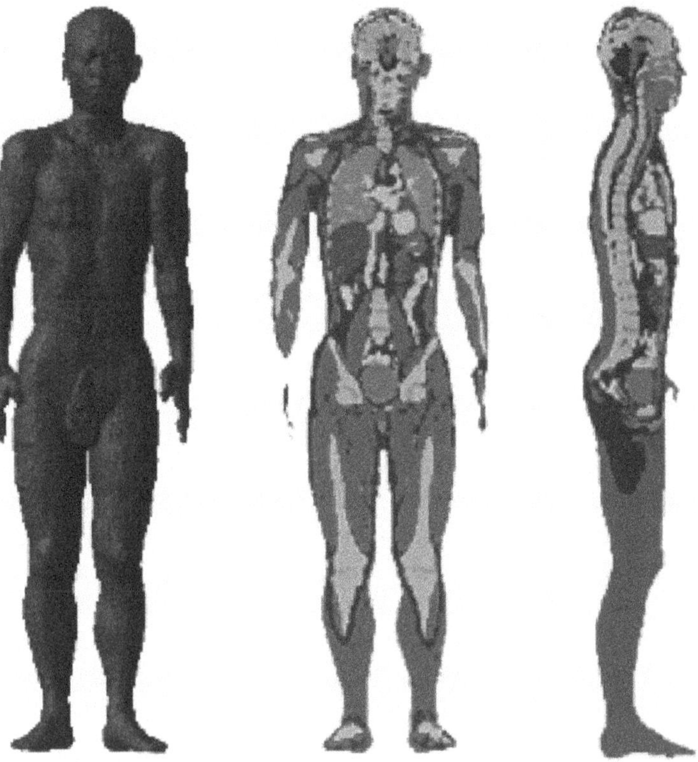

Figure 3.16 Numerical human body model developed from MRI data (Nagaoka *et al.*, 2004)

approach, the gray scales in magnetic resonance images are correlated to the complex permittivity by some continuous transfer functions. This approach makes it possible to semi-automatically obtain a numerical human body model from the raw MRI data.

According to this approach, we first need to subsequently acquire two turbo spin-echo sequences with an echo time T_E and two different repetition times T_R (short and long) for the MRI data. The spin-echo signal intensity of each pixel in the magnetic resonance images is related to the spin–lattice relaxation time T_1 which characterizes the human tissues. By dividing the images with short repetition time, pixel by pixel, by the images with long repetition time, we can obtain the ratio image intensity

$$I_{ratio}(r) = k_{ratio} \frac{1 - 2e^{-\frac{T_{R,short} - T_E/2}{T_1(r)}} + e^{-\frac{T_{R,short}}{T_1(r)}}}{1 - 2e^{-\frac{T_{R,long} - T_E/2}{T_1(r)}} + e^{-\frac{T_{R,long}}{T_1(r)}}} \tag{3.88}$$

where the gray levels in MRI data depend only on T_1 and on the factor $k_{ratio} = k_{short}/k_{long}$. The factor k, which depends on the type of sequence and on the acquisition parameters, is independent of the position r. Since the gray level of each pixel of the short image is lower than that of the long image, I_{ratio} takes values between 0 and 1.

Next, in place of the tissue segmentation and subsequent assignment of complex permittivity, we can determine the complex permittivity values automatically from T_1 in the ratio images. This is because that there is a relationship to relate T_1 to water content and another one to relate water content to complex permittivity. T_1 is usually influenced by the frequency and the amount of water in a human body. In particular, in high water content tissues, water may exist in two states: free water, associated with a longer relaxation time; and bound water, associated with a shorter relaxation time. The measured spin–lattice relaxation time T_1 is actually a weighted average of the free and bound water times. The larger the water content is, the longer the spin-lattice relaxation time T_1. Based on this monotonic relationship we can derive an equation to link the water fraction of various human tissues to the relaxation time T_1 by

$$\frac{1}{f_w} = A + \frac{B}{T_1} \tag{3.89}$$

where A and B are two factors which depend on the magnetic field strength.

On the other hand, water content is known to be related to the complex permittivity of human tissue. Since T_1 determines the gray level in the ratio image, and also is related to the water content, some continuous transfer functions such as $\varepsilon_r(I_{ratio})$ and $\sigma(I_{ratio})$ can be derived based on empirical data for some representative tissues. Figure 3.17 shows some examples for the transfer functions from the MRI ratio intensity to the relative permittivity and conductivity at frequencies of 100 MHz and 1 GHz. Symbols indicate Gabriel's data and vertical segments the literature variability (Foster and Schwan, 1996). Typical values for Equations 3.88 and 3.89 are reported as $A = 0.915$, $B = 293$ S, and $k_{ratio} = 1.04$.

In such a way, we can easily obtain an anthropomorphic numerical model of dielectric anatomy from the raw MRI data for numerical electromagnetic field analysis. The dielectric anatomical model also takes into account the complex permittivity variations in every tissue.

When the frequency is sufficiently high so that the skin depth is around several millimeters, the tissue and organs inside the human body become not so important. This unimportance in internal organs is also true for on- or off-body communication scenarios. In these cases, the body externals and posture play a more important role from the point of view of electromagnetic propagation characteristics. Hence we can employ a homogeneous body model in place of the anatomical one, and set the model to have skin's permittivity and conductivity in the GHz band and muscle's

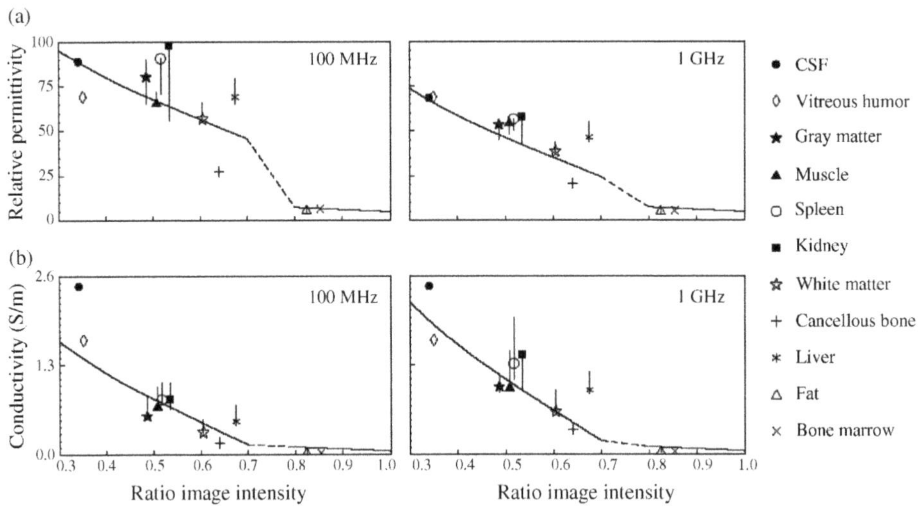

Figure 3.17 Transfer functions from the MRI ratio intensity to (a) the relative permittivity and (b) conductivity at frequencies of 100 MHz and 1 GHz. Symbols indicate Gabriel's data and vertical segments the literature variability (Mazzurana *et al.*, 2003). Reproduced with permission from Mazzurana M., Sandrini L., Vaccari A., Malacarne C., Cristoforetti L. and Pontalti R., "A semi-automatic method for developing an anthropomorphic numerical model of dielectric anatomy by MRI," *Physics in Medicine and Biology*, **48**, 3157–3170, 2003. © 2003 IOP Publishing

permittivity and conductivity below the GHz band, in view of the possible penetration depth. In addition, at lower frequencies (around MHz), the wave can completely penetrate into the human body. In this case, two-thirds of muscle's permittivity and conductivity are known to be a good approximation for average tissue properties of the human body.

The creation of a homogeneous human body model is much simpler than the anatomical one. It is also easier to modify the body model to have various postures, whereas this is almost impossible for an anatomical model. The body model with various postures is essential for on-body propagation characterization.

A simple method for making a homogeneous human body model with postures is to use the Poser software (http://poser.smithmicro.com). It prepares some basic three-dimensional poly-mesh human figures and with it the user can simply pose the human figures to have a required posture. By saving the three-dimensional human figure to a format readable in the FDTD code, a posed human body model with homogeneous tissue can be obtained.

In conjunction with adequate human body models, the numerical electromagnetic analysis method provides an effective means for body area propagation characterization and channel modeling.

References

Berenger, J.P. (1994) A perfectly matched layer for the absorption of electromagnetic waves. *Journal of Computational Physics*, **114**, 185–200.

Chakarothai, J., Wang, J., Fujiwara, O. *et al.* (2012) Implementation and validation of hybrid MoM-FDTD method for analysis of composite structure including inhomogeneous objects. IEICE Technical Report AP2012-7.

Foster, K.R. and Schwan, H.P. (1996) Dielectric properties of tissues, in *Handbook of Biological Effects of Electromagnetic Fields* (eds C. Polk and E. Postow), CRC Press, New York, pp. 25–102.

Jin F. J. (1993) *The Finite Element Method in Electromagnetics*, John Wiley & Sons, Ltd, New York.

Mazzurana, M., Sandrini, L., Vaccari, A. *et al.* (2003) A semi-automatic method for developing an anthropomorphic numerical model of dielectric anatomy by MRI. *Physics in Medicine and Biology*, **48**, 3157–3170.

Nagaoka, T., Watanabe, S., Saurai, K. *et al.* (2004) Development of realistic high-resolution whole-body voxel models of Japanese adult males and females of average height and weight, and application of models to radio-frequency electromagnetic-field dosimetry. *Physics in Medicine and Biology*, **49**, 1–15.

Rao, S.M., Wilton, D.R., and Glisson, A.W. (1982) Electromagnetic scattering by surfaces of arbitrary shape. *IEEE Transactions on Antennas and Propagation*, **30** (3), 409–418.

Taflove, A. and Hagness, S.C. (2000) *Computational Electrodynamics: The Finite-Difference Time-Domain Method*, Artech House Inc., Boston, MA.

Wang, J., Fujiwara, O., Watanabe, S., and Yamanaka, Y. (2004) Computation with a parallel FDTD system of human-body effect on electromagnetic absorption for portable telephones. *IEEE Transactions on Microwave Theory and Techniques*, **52** (1), 966–971.

4

Body Area Channel Modeling

4.1 Introduction

Body area communications differs from traditional radio communications due to human body proximity effects. Human body tissue is a complicated frequency-dependent dielectric material with relatively high permittivity and certain conductivity. Radio signals propagating in the body area are significantly affected by human body tissue (Zasowski *et al.*, 2006). Propagation mechanisms in the body area are also frequency-dependent as introduced in Chapter 2. The resulting received signal is subject to the transmission channel, the distance between the transmitter and the receiver, transmitter and receiver antenna positions, tissue dielectric properties along the transmission channel, body curvature and so on. Channel modeling is the initial step to explore and investigate the body area communications. An adequate propagation channel model is essential for the design of a body area communication system. The ultimate performance limits of body area communication systems, as well as the performance of practical systems, are determined by the channel they operate in.

Body area channels can be classified into wearable channels and implant channels according to the locations of the transmitter and receiver devices on or in or off the body. Wearable channels have all of the devices on body, while implant channels have some devices in body which communicate with on- or off-body devices. Figure 4.1 illustrates several typical body area channels (IEEE P802.15, 2009), in which channel A is an in-body channel, B is an in- to on-body channel, C is an on-body channel and D is an on- to off-body channel. On-body channel C is the wearable channel and can be characterized as a line-of-sight (LOS) or non-line-of-sight (NLOS) channel depending on the way the electromagnetic waves propagate between the two devices. The scenario with one device on the front of the body and the other device on the back of the body is characterized as a NLOS propagation, in

Body Area Communications: Channel Modeling, Communication Systems, and EMC, First Edition.
Jianqing Wang and Qiong Wang.

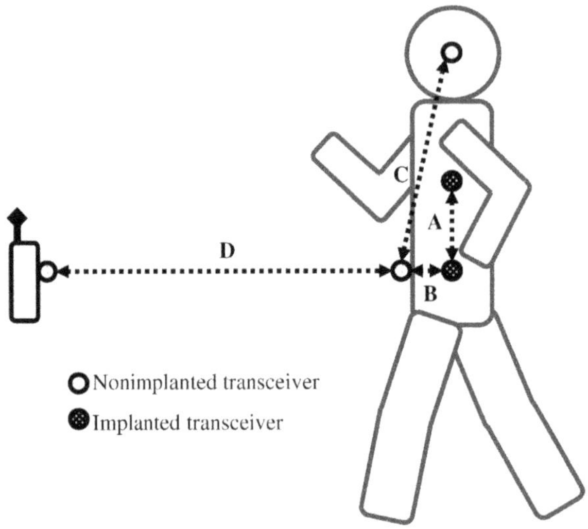

Figure 4.1 Typical communication channels for body area network

which the electromagnetic waves may experience loss due to diffraction around the body. Channels A and B are typical implanted channels. Wireless capsule endoscope technology will make use of channel B where the ingestible capsule can take pictures during its course through the digestive tract after being swallowed and transmit the real-time biological data from inside the body to on-body medical devices. Alternatively, the medical receiver may be located off the body at a certain short distance and therefore a channel consisting of B and D can be generated. An implanted pacemaker in the chest, communicating with an on-/off-body device, also brings about an implant channel resembling B or B plus D.

The wave propagation characteristics in the human body area are very complicated. There are absorption phenomena in the human body due to the lossy dielectric properties of body tissues as well as scattering due to the heterogeneous nature of body tissues. The transmitted signal is therefore largely attenuated. Moreover, diffractions and creeping waves exist along the body surface, where shadow fading arises due to the diffraction in the shadowed regions of the body. In addition, since the human body might take various postures, or one or more body parts may move during the communication period, multiple paths are created between the transmitter and receiver so that a transmitted signal can traverse. Consequently a received signal may end up being the superimposition of several attenuated, delayed, time-varying and eventually distorted replicas of a transmitted signal. For different body area channels, the dominant channel characteristics may vary. An on-body channel may suffer from multipath fading and shadowing due to body postures and movements more than an in-body channel, while an in-body channel undergoes severe signal decay during transmission inside the body more than an on-body channel. For different frequency candidates, like UWB, MICS and HBC, the channel characteristics will also differ due to the frequency-dependent dielectric properties of body tissues. Channel modeling needs to incorporate the wave propagation characteristics in the body area in the context of different channel types as well as different frequency bands.

In this chapter, we will generalize the channel model into a path loss model or a multipath channel model. The path loss model mainly describes the channel loss, including propagation loss, absorption loss as well as diffraction loss, while the multipath channel model focuses on the time-domain characteristics of the body area channel, particularly the multipath propagation characteristic. We will first address the path loss models for UWB, MICS and HBC bands in both on-body and in-body transmission channels. Then we will investigate the multipath channel model especially for on-body UWB. The methodology of characterizations and parameterizations of the UWB multipath channel will be described in detail.

4.2 Path Loss Model

4.2.1 Free-Space Path Loss

Path loss is the reduction/attenuation in power (density) of an electromagnetic wave as it propagates in a specific environment. In general, the path loss is due to many effects, such as free-space loss, refraction, diffraction, reflection, aperture-medium coupling loss, and absorption. In free space, path loss is influenced by terrain contours, environment, propagation medium, the distance between the transmitter and the receiver, and the height and location of antennas. Path loss involves propagation losses caused by the natural expansion of the radio wave front in free space, absorption losses/penetration losses when the signal passes through lossy media, diffraction losses when part of the radio wave propagation is obstructed by an obstacle, and losses caused by other phenomena.

In wireless communications, path loss is usually expressed in decibels and can be represented by the path loss exponent in the following simple formula

$$PL_{dB} = 10n\log_{10}(d) + C \tag{4.1}$$

where PL_{dB} is the path loss in decibels, n is the path loss exponent, d is the distance between the transmitter and the receiver, usually measured in meters, and C is a constant which accounts for system losses. This equation is referred to as the log-distance path loss model. The value of the path loss exponent is normally in the range of 2–6. The loss exponent of 2 is for propagation in free space. Larger exponent values are for lossy environments. In buildings, stadiums and other indoor environments, the path loss exponent can reach values in the range of 4–6.

The Friis transmission equation is usually used for the path loss in free space. It gives the power received by one antenna under idealized conditions when another antenna at a certain distance away is transmitting a known amount of power. The basic form of the Friis transmission equation is as follows:

$$\frac{P_r}{P_t} = G_t G_r \left(\frac{\lambda}{4\pi d}\right)^2 \tag{4.2}$$

which shows the ratio of power available at the output of the receiving antenna P_r to power input to the transmitting antenna P_t, where G_t and G_r are the gains of the transmitting and receiving antennas, respectively, λ is the wavelength, and d is the distance. Assuming two isotropic antennas, the free-space path loss in decibels can be expressed by a simplified formula:

$$PL_{dB} = 20\log_{10}\left(\frac{4\pi d}{\lambda}\right) \tag{4.3}$$

where the wavelength and distance have the same units. This simplified form applies only under ideal conditions: (1) no feed line or miscellaneous losses; (2) no obstructions or no multipath propagation; (3) no atmospheric absorption; (4) correctly aligned and polarized transmitter and receiver antennas; and (5) single frequency is assumed. It can be seen that the path loss exponent in free space is 2 according to Equation 4.3. In addition, it must be kept in mind that this equation is only accurate in the far field where spherical spreading can be assumed; it does not hold close to the transmitter, that is, in the near field region.

4.2.2 *On-Body UWB Path Loss*

In order to investigate the on-body UWB path loss characteristics, we need to assign the transmitter and receiver on the human body. In a practical scenario, many low power transmitters in the sensor mode send data from various spots on the body to one receiver in the master mode. A wireless network with multiple transmitters and one receiver on the body is therefore constructed. However, in the derivation of the path loss model using a numerical approach, multiple transmitter antennas on the body interact with each other and consequently result in an inexact electric field value in terms of on-body path loss evaluation. On the other hand, a set-up of one transmitter at a fixed position and multiple receivers at various different positions results in a low computation efficiency since a statistical analysis over the whole body is needed. Therefore, in order to enhance the computation efficiency, one transmitter antenna and multiple receiving points are constructed on an anatomical human body model. Although it is not consistent with the realistic usage scenario, this does not affect the analysis or the conclusions.

Figure 4.2 shows the numerical simulation set-up for the on-body path loss on a human body with a Hertzian dipole fixed on the left chest as the transmitter and

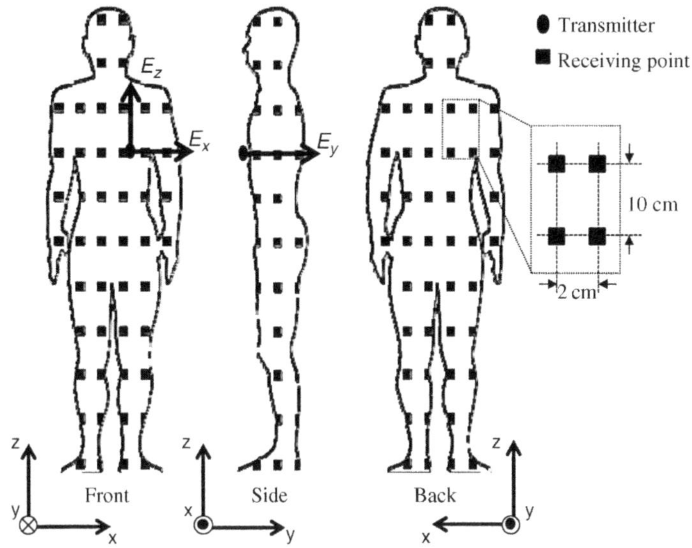

Figure 4.2 Configuration of transmitter and receiving points on the body

multiple receiving points distributed along the whole body surface. The transmission distance is calculated around the perimeter of the body since it has been shown that waves diffract around the body in the UWB band (Fort *et al.*, 2006; Zasowski *et al.*, 2006). The receiving points are separated by 10 cm in the longitudinal direction, while the separation between receiving points in the horizontal is 2 cm. Using the FDTD method, a total of 869 receiving points are calculated which consists of 250 points on the front, 345 points on the side and 274 points on the back of the body. Both the transmitter and the receiving points have a spacing distance of 2 mm above the body surface.

Since the UWB frequency band ranges from 3.1 to 10.6 GHz, the practically usable skin depth of the biological tissue in this frequency band is only several millimeters. This suggests that the surface tissue layers should dominate the propagation characteristics of the body. In fact, based on the computational comparison between the received voltages for a homogeneous body model comprising only skin and a heterogeneous body model comprising 51 different tissue types, it is found that the differences in the received voltages under these two situations never exceed 5%. Therefore we can assume a homogeneous human body of skin and employ the one-relaxation Debye equation to simulate its dielectric properties (Gabriel, 1996).

The choice of the transmitted pulse is crucial since it affects the energy spectral density (ESD) of the transmitted signal. The widely used UWB signal is based on a Gaussian pulse which can be generated the most easily by a pulse generator

(Benedetto and Giancola, 2004). A Gaussian pulse $v(t)$ can be represented by the following expression

$$v(t) = \frac{A}{\sqrt{2\pi\sigma^2}} e^{-\frac{t^2}{2\sigma^2}} \tag{4.4}$$

where A is the amplitude, σ^2 is the variance and $\alpha = (4\pi\sigma^2)^{1/2}$ is a shape factor.

Gaussian derivatives have a zero DC offset which facilitates the pulses to be radiated in an efficient way. Actually, Gaussian derivatives have been widely adopted to shape the UWB signal since a Gaussian pulse can have infinite differentiations. The nth-derivative Gaussian pulse is given by

$$v^{(n)}(t) = -\frac{n-1}{\sigma^2} v^{(n-2)}(t) - \frac{t}{\sigma^2} v^{(n-1)}(t) \tag{4.5}$$

and the corresponding Fourier transform is

$$X_n(f) = A(j2\pi f)^n e^{-\frac{(2\pi f\sigma)^2}{2}} \tag{4.6}$$

One of the most efficient and popular pulse derivatives is the second derivative of a Gaussian function (Win and Scholtz, 2000), described by

$$v^{(2)}(t) = -\frac{A}{(\sqrt{2\pi}\sigma^3)} \left(1 - \frac{t^2}{\sigma^2}\right) e^{-\frac{t^2}{2\sigma^2}} \tag{4.7}$$

Other pulse shapes have also been proposed such as the Laplacian (Conroy, LoCicero, and Ucci, 1999), compositions of Gaussian pulses and derivatives (Hämäläinen *et al.*, 2001), and Hermite pulses (Ghavami *et al.*, 2002).

The pulse shape factor and differentiation order are the two most significant parameters affecting the pulse waveform and ESD. Pulse duration is directly proportional to the shape factor α. Increasing α enlarges the pulse duration and conversely will decrease the pulse bandwidth. Therefore, adjusting the pulse duration can create an appropriate bandwidth for various UWB applications. Differentiation of the Gaussian pulse waveform can change the pulse shape without changing the pulse duration. However, differentiation influences both peak frequency and bandwidth of the pulse; increasing differentiation order can result in higher peak frequencies of the transmitted pulse. Differentiation is therefore an effective means of putting transmitted energy to higher frequency bands.

The ESD of a signal describes how the energy of a signal is distributed with frequency. If $v_s(t)$ is a finite-energy signal, the ESD $E(\omega)$ of the signal can be defined as the square of the magnitude of the continuous Fourier transform of the signal where

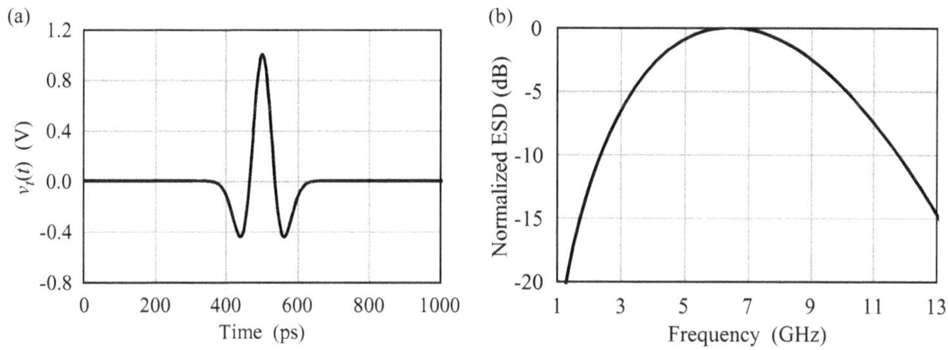

Figure 4.3 (a) Transmitted second derivative of Gaussian pulse and (b) corresponding ESD characteristics

the energy is taken as the integral of the square of the signal. The $E(\omega)$ can be expressed as follows

$$E(\omega) = \left| \frac{1}{\sqrt{2\pi}} \int_{-\infty}^{\infty} v_s(t) e^{-j\omega t} dt \right|^2 = \frac{V_s(\omega) V_s^*(\omega)}{2\pi} \tag{4.8}$$

where ω is the angular frequency and $V_s(\omega)$ is the continuous Fourier transform of $v_s(t)$, and $V_s^*(\omega)$ is its complex conjugate.

We now use the typical second-derivative Gaussian pulse as shown in Figure 4.3(a) as the transmitted UWB pulse. As described above, pulse width is directly related to the shape factor α. A pulse width of 280 ps is used in order to have most of the energy between 3.1 GHz and 10.6 GHz. Figure 4.3(b) shows the corresponding ESD characteristics where the –10 dB bandwidth covers the whole UWB band with a central frequency around 6.5 GHz. With different pulse width, the 10 dB bandwidth and the central frequency will change accordingly.

The path loss can be obtained from the following equation

$$PL_{dB}(d) = 10 \log_{10} \left[\frac{E_t(d)}{E_r(d)} \right] + G_{t,dB} + G_{r,dB} \tag{4.9}$$

where $E_t(d)$ and $E_r(d)$ are the amounts of transmitted and received energy, which can be calculated from the transmitted and the received pulse voltage waveforms. As shown in Figure 4.2, the transmitting Hertzian dipole can locate along three orthogonal directions x, y and z. According to the radiation characteristics of the Hertzian dipole, the orientation of the transmitting dipole and received field components perpendicular to the human body are advantageous for wave diffraction around the body. In addition, the antenna gains are also needed as a function between the transmitter and receiver for removing the antenna effect, although the

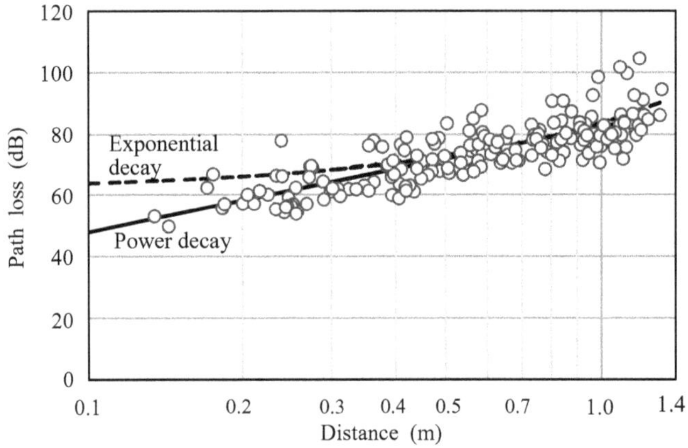

Figure 4.4 Path loss fitting versus distance along the human body under the condition that both the transmitter and receiver antennas are y-directed

antenna effect is usually difficult to remove completely because the body area propagation is in a near-field region. Figure 4.4 shows the calculated path loss with both y-directed transmitting and receiving components versus distance on the human body model. It is clear that the path loss increases with distance as expected, and that there is a large variance around the mean path loss. Keep in mind, in Figure 4.4, the horizontal axis is not the straight-line distance but the one traveled by the wave around the perimeter of the body between the transmitter and the receiving points, because the major components reaching the receivers are due to diffraction around the body.

There are two potential path loss models for possible fitting to the calculated results. One is according to the empirical power decay law as

$$PL_{dB} = PL_{0,dB} + 10n \log_{10}\left[\frac{d}{d_0}\right] \tag{4.10}$$

where $PL_{0,dB}$ is the path loss at distance d_0 and n is the path loss exponent. The other is according to an exponential fitting as

$$PL = Ae^{\alpha d} \tag{4.11}$$

where A is the excitation coefficient and α is the attenuation coefficient.

The fitting results for the calculated path loss data based on the two equations in the least-square-error sense are also shown in Figure 4.4 by the solid line and the dashed line, respectively. As can be seen, comparing with Equation 4.11, Equation 4.10 gives a better fitting, which suggests that the power decay law is more

Table 4.1 Estimated PL_0 and n in the log-distance path loss model for various directed transmitter antenna and receiving components

Transmitting antenna	Received components	Receiving antenna					
		Front		Side		Back	
		PL_0 [dB]	n	PL_0 [dB]	n	PL_0[dB]	n
X	X	50.2	4.05	48.9	4.61	81.7	3.35
	Y	54.1	3.69	48.5	4.42	83.8	3.43
	Z	56.0	2.46	51.4	3.61	81.5	2.62
Y	X	44.8	3.87	45.8	3.92	73.0	2.46
	Y	43.0	3.75	44.1	3.85	73.0	2.98
	Z	45.8	2.55	47.6	2.90	74.6	1.73
Z	X	47.7	3.81	53.2	3.82	82.3	2.60
	Y	50.1	3.62	52.3	3.66	81.9	3.21
	Z	49.6	2.60	55.9	2.88	83.0	1.95

appropriate for the UWB path loss modeling on the human body. This finding is identical to the results in Fort *et al.* (2006).

In fact, the path loss model in Equation 4.10 is essentially similar to the log-distance path loss model in Equation 4.1. Table 4.1 shows the estimated $PL_{0,dB}$ and n for all three directed transmitting dipoles and receiving components. $PL_{0,dB}$ is obtained at $d_0 = 0.1$ m. It is found that the path loss at the first 0.1 m is around 43–83 dB, and the exponent n is around 2–5. Furthermore, for the transmitter, the y-directed dipole yields a smaller reference path loss $PL_{0,dB}$. On the other hand, at the receiving points, the z-directed field components always have the smallest path loss exponent n. Taking the two parameters into consideration, the y-directed transmitting dipole and z-directed receiving component may be the best option from all of the results.

As noticed from Figure 4.4 there is a large fluctuation in the simulated path loss values around the fitted mean path loss. Actually, the fluctuation of the simulated path loss is mainly due to the shadowing effect of the body. The shadowing is induced by the diffraction in the shadowed regions of the body. The multipath-induced fading from body postures is not taken into account here since a static human body model is assumed. Shadowing can directly result in the variations of the received signal at the receiver front-end. The amplitude variation caused by shadowing is often defined as the difference between the calculated path loss values and the mean path loss. It can be modeled using a log-normal distribution with a standard deviation according to the log-distance path loss model in Equation 4.10 which represents the mean path loss. Combination of both the statistical variance of path loss and the fitted mean path loss can result in a complete path loss model.

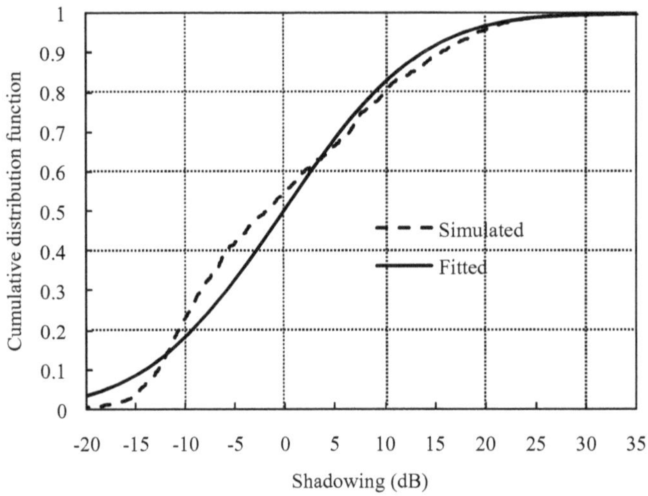

Figure 4.5 Cumulative distribution function of shadow fading around the human body

A comprehensive path loss model can be therefore rewritten as follows

$$PL_{dB} = PL_{0,dB} + 10n\log_{10}\frac{d}{d_0} + S_{dB} \tag{4.12}$$

where S_{dB} is the difference of the calculated path loss values in decibels and the mean path loss in decibels derived from Equation 4.10 at the same distance. Such a model is called as a static shadow fading channel afterwards. S_{dB} represents the shadow fading in decibels with normal distribution since the logarithm of a log-normal distribution is normally distributed. It has a mean value 0 dB and standard deviation σ_{dB}. σ_{dB} reflects how concentrated the path loss is around its mean value as well as the degree of the shadow fading strength. The parameter σ_{dB} of the shadow fading can be determined based on statistical fitting. Figure 4.5 shows the fitted cumulative distribution function (CDF) of normal distribution for the shadow fading with y-directed transmitter and the total receiving component. The standard deviation is fitted as $\sigma_{dB} = 10.8$ dB, which illustrates the large variance of the path loss values due to the strong shadowing effect around the human body.

4.2.3 In-Body UWB Path Loss

As shown in Chapter 2, the interaction of electromagnetic waves and the human body differs with different frequencies. For different communication scenarios in the body area, frequency selection should be designed to facilitate the wave propagation for specific communication purposes. The recent research results for the on-body UWB propagation (Fort *et al.*, 2006; Zhao *et al.*, 2006; Taparugssanagorn *et al.*, 2008; Wang *et al.*, 2009) and the in-body UWB propagation (Khaleghi and

Balasingham, 2009; Wang, Masami, and Wang, 2011) has concluded that requirements on frequencies are different due to different operating environments. For UWB, wearable body area networks may cover the full band from 3.1 to 10.6 GHz, while implant body area networks have to be limited in the low band, for example from 3.4 to 4.8 GHz. This is because the skin depth or the penetration depth of human tissue becomes very small with increasing frequency.

We introduce here two in-body UWB channels in the context of two implant communication services. One is wireless capsule endoscope for non-invasive gastrointestinal tract imaging, in which the exterior receiver is assumed to be on the torso surface. This scenario builds an in-body to on-body channel around the abdomen district. The other service is the radio communication from an implanted cardiac pacemaker or other electronic implants inside the chest area with an off-body medical device. This communication scenario forms an in-body to off-body channel in the chest area.

4.2.3.1 In-Body to On-Body UWB Path Loss

Differing from the on-body UWB path loss, the in-body channel investigation should employ a heterogeneous human body model so as to accurately explore the wave propagation in the in-body channel.

Wireless capsule endoscope was first introduced in 2001 (Qureshi, 2004). The ingestible capsule can take pictures during its course through the digestive tract after being swallowed. It can therefore transmit real-time biological data from inside the body to exterior medical instruments, which aids the non-invasive diagnosis. A UWB capsule endoscope has also been proposed to ensure real-time video monitoring of the internal parts of a patient's body. Some research work has been done about radio signal propagation inside the human body within the FCC allocated UWB frequency band (Chavez-Santiago *et al.*, 2009; Shi and Wang, 2010). Highly frequency-dependent signal attenuation has been concluded within the UWB band, which creates a big challenge for UWB capsule system design.

Figure 4.6 shows an anatomically based human body model with an on-body elliptic disk dipole antenna to receive the capsule endoscope data. The elliptic disk dipole antenna is optimized to operate on the body surface. The dimensions of major axis radius a and minor axis radius b are adjusted to be $a = 7$ mm and $b = 5$ mm, which yields a voltage standing wave ratio (VSWR) smaller than 2.0 from 3.4 to 4.8 GHz. The VSWR is defined as $\mathrm{VSWR} = (1 + |\Gamma|)/\ (1 - |\Gamma|)$, where Γ is the antenna reflection coefficient which can be calculated from the antenna input impedance. A value of VSWR smaller than 2.0 is usually acceptable in practical applications.

The receiving antenna is taken to be at five locations on the front of the body on the abdomen, as shown in Figure 4.6. On the other hand, the transmitting antenna of the capsule endoscope is taken to be a 4-mm-long dipole. It moves

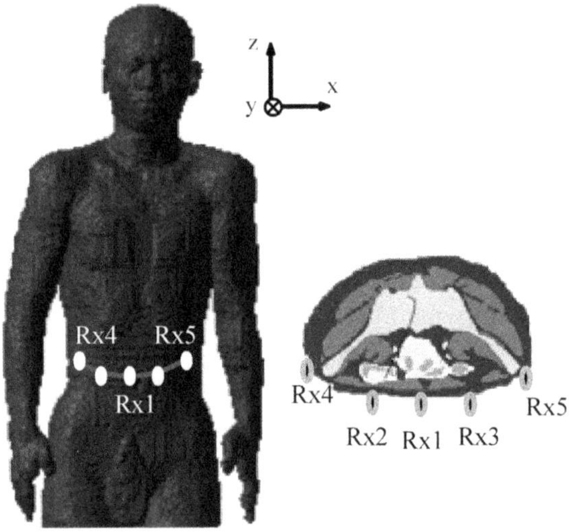

Figure 4.6 Human body model and receiving antenna locations

mainly along the small intestine at 33 locations, and at each location it has three different directivities (x, y, and z) respectively. As a result, for each on-body receiver location, there are 99 sets of data for extracting the path loss and shadow fading characteristics.

In the characterization of the in-body to on-body channel for the capsule endoscope application, a sine-modulated Gaussian pulse with a center frequency of 4.1 GHz and a pulse width of 2.1 ns is used as the transmitted UWB signal in order to have most of the energy in the UWB low band. The signals are received by the on-body receivers. Then the path loss is obtained from

$$PL_{dB} = 10\log_{10}\frac{\int_{f_L}^{f_H}|F[v_t(t)]|^2 df}{\int_{f_L}^{f_H}|F[v_r(t)]|^2 df} \tag{4.13}$$

where $v_t(t)$ is the transmitted signal voltage, $v_r(t)$ is the received signal voltage calculated by FDTD simulation, $F[\ ]$ denotes the Fourier transform, $f_L = 3.4$ GHz and $f_H = 4.8$ GHz.

According to the propagation mechanism as discussed in Chapter 2, the signal exponentially attenuates inside a lossy dielectric medium. However, the finite thickness of the human body makes the log-distance path loss model in Equation 4.12 applicable to approximate the exponential attenuation behavior over a short distance such as the body cross-section thickness. In the log-distance path loss model, the mean path loss can be fitted based on the reference path loss $PL_{0,dB}$ at a distance

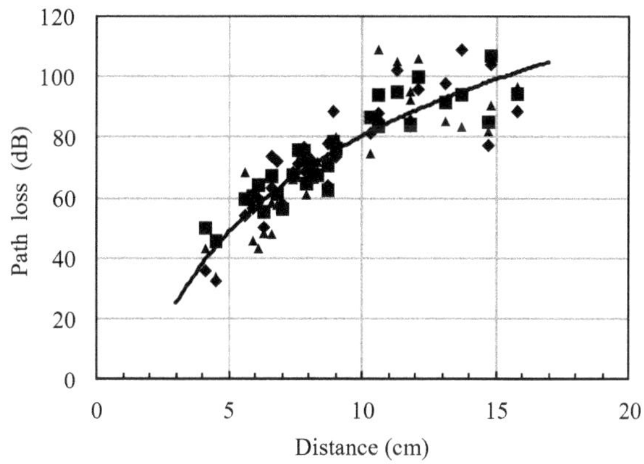

Figure 4.7 Path loss versus distance at a receiver location (Rx1) on the front of the abdomen

$d_0 = 0.05$ m. Figure 4.7 shows the FDTD-calculated path losses (symbols) and the fitted mean path losses (curve). The results for the receiver located in front of the abdomen (Rx1) are demonstrated. Figure 4.8 gives the CDF of shadow fading S_{dB} for Rx1. It is found that the shadow fading S_{dB} follows the normal distribution. The mean value is of course zero, and the standard deviation is $\sigma_{dB} = 8.2$ dB. The derived fitting parameters for all five receiver locations in terms of the least-square-errors are shown in Table 4.2. The fitting parameters are similar at different receiver locations. Moreover, due to the rapid attenuation inside the human body at UWB frequencies, the path loss exponent n looks large. The standard variance of the shadow fading is in the range of 7.1–8.5 dB.

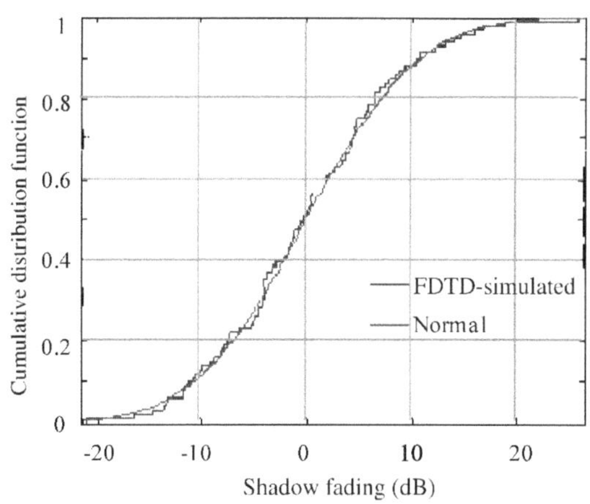

Figure 4.8 Cumulative distribution function of shadow fading at a receiver location (Rx1) on the front of the abdomen

Table 4.2 Fitted parameters for path loss expression

Rx location	$PL_{0,dB}$	n	σ (dB)
Rx1	48.6	10.6	8.2
Rx2	48.5	8.0	7.7
Rx3	46.7	12.6	8.5
Rx4	44.5	12.8	7.1
Rx5	52.4	6.2	7.8

4.2.3.2 In-Body to Off-Body UWB Path Loss

Figure 4.9 shows an anatomical human body model with an elliptic disk dipole antenna implanted in the chest for cardiac pacemaker application. The body model is the upper part of the anatomically based whole body human model. The elliptic disk dipole antenna is implanted at a depth of about 1.5 cm, just between the fat and muscle layer, from the left-chest surface. To cover the UWB low band from 3.4 to 4.8 GHz, the antenna is designed to have a major axis radius $a = 12$ mm and a minor axis radius $b = 10$ mm, which yields an acceptable VSWR over the frequency band of 3.4–4.8 GHz both inside and outside the human body. Although the antenna is not

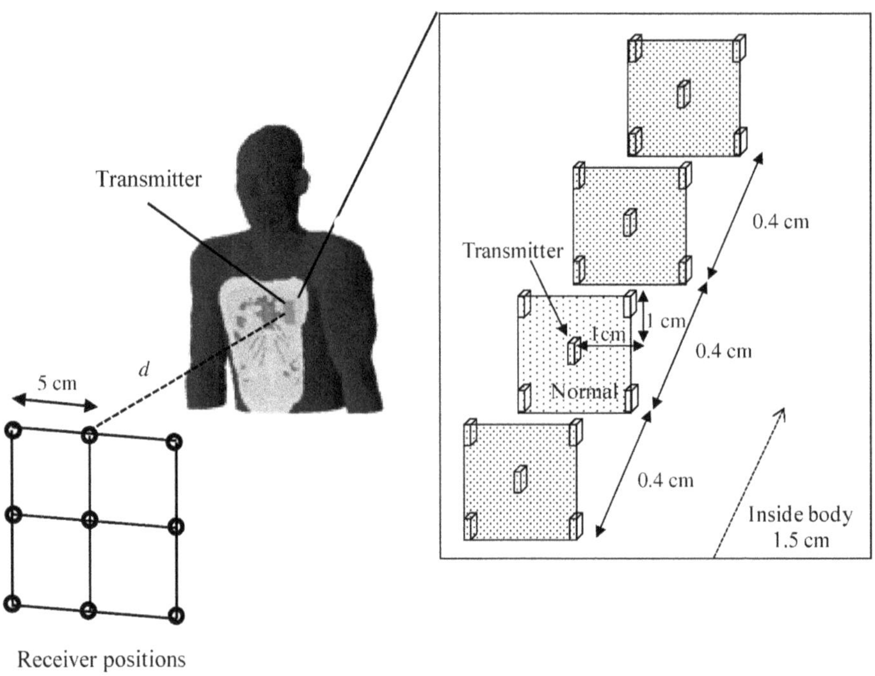

Figure 4.9 Location configuration of implanted transmitter and external receiver

small enough for practical implant applications, its flat VSWR characteristic with frequency is suitable to derive a general-purpose channel model. Using two of the above-described elliptic disk dipole antennas, one as the transmitting antenna inside the body and the other as the receiving antenna outside the body, the in-body to off-body channel can be characterized based on the FDTD numerical calculation. The UWB signal to be transmitted is also a sine-modulated Gaussian pulse with a center frequency of 4.1 GHz and a pulse width of 2.1 as in the previous in-body to on-body UWB path loss case.

As for the location configurations of the implanted transmitter and external receiver, as shown in Figure 4.9, the transmitting antenna is embedded inside the left chest. It covers a total of 20 locations ranging from 1.1 to 2.3 cm from the chest surface along the transverse plane and 2 cm in the coronal plane. This is to take into account the randomness of the implanted location in actual usage. Meanwhile, location variation can be used to imitate the implanted position change due to human respiration. The receiver is assumed to be an external coordinator to collect the medical data transmitted from the implanted antenna. It is located in a 10 cm square area in the front of the left chest at a distance of up to 40 cm from the chest surface. For each square area, nine receiver locations are considered to cover the randomness in receiver location setting. As a result, 180 combinations of the transmitter and receiver are simulated in total.

Figure 4.10 shows the average path loss versus the distance from the implant transmitter location. It is found that the path loss at the body surface, that is, at a distance $d = 1.5$ cm from the transmitter, is about 54 dB. At the off-body receiver location at a distance up to 40 cm from the body surface, the path loss seems to decay in an exponential manner. Also in Figure 4.10, an approximate curve

$$PL = Ae^{\alpha(d-d_0)} \tag{4.14}$$

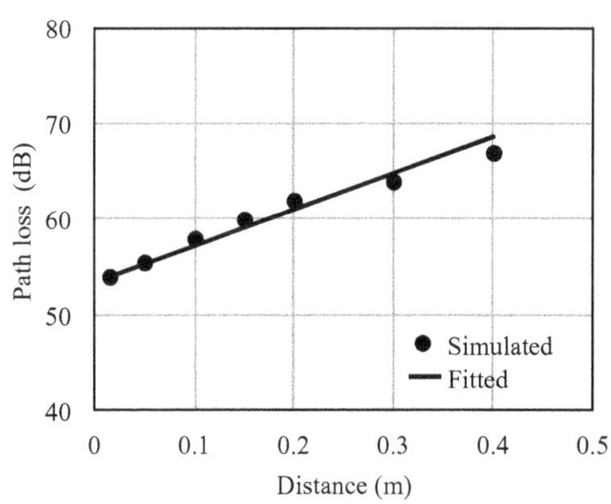

Figure 4.10 Path loss versus distance of the off-body receiver from the body surface

Table 4.3 Fitted parameters for path loss expression

d_0 (m)	A_{dB}	α (m^{-1})
0.015	54	8.8

is given to fit the calculated average path loss in the least-square-error sense, where d_0 is the implant depth with a typical value of 1.5 cm. Equation 4.14 exhibits fair agreement with the FDTD calculated result when $A = 251188.6$ or 54 dB and $\alpha = 8.8\,m^{-1}$. The exponential decay of power versus distance in the body vicinity is identical to the basic wave feature at the boundary of air and a dielectric medium. Table 4.3 lists the fitted parameters.

4.2.4 In-Body MICS Band Path Loss

MICS as a standard medical implant communication band between 402 MHz and 405 MHz provides a low bit rate communication service compared with the UWB low band. Since it utilizes narrow band signals, the path loss characteristics in the MICS band are derived based on the ratio of the received average power and transmitted average power, rather than the ratio of energy as in the UWB case. The wireless capsule endoscope scenario is employed again here to address the in-body to on-body MICS band path loss characteristics.

Similar to the in-body to on-body UWB capsule endoscope case, the transmitting antenna is set to be a 4-mm dipole located at 30 different locations inside the human digestive system, including four locations in the esophagus, four locations in the stomach, nine locations in the small intestine and 13 locations in the large intestine, respectively. The receiving antenna is a 20-mm dipole which locates around the body torso. Figure 4.11 shows the configuration of the seven receiver locations:

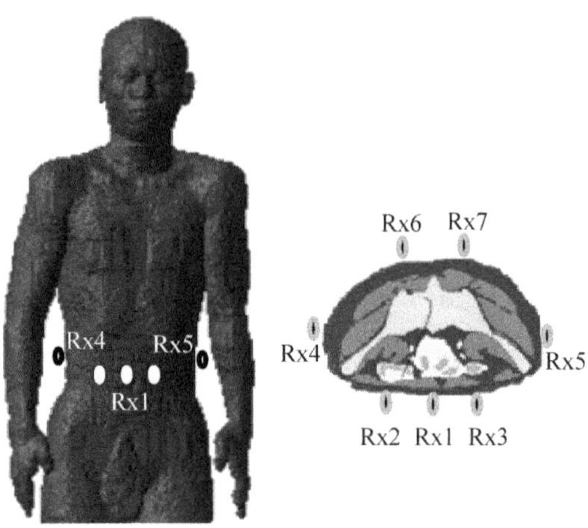

Figure 4.11 Receiver locations around the human body for MICS band capsule endoscope

Rx1, Rx2, and Rx3 at the belly region, Rx4 and Rx5 at the torso side, and Rx6 and Rx7 at the back. The receiving dipoles are along the body height direction at a spacing distance of 4 mm from the body surface.

The derivation of the path loss is based on the following equation

$$PL_{dB} = 10\log_{10}\frac{P_t}{P_r} + G_t + G_r \tag{4.15}$$

where P_t and P_r are the transmitted and received average power, respectively.

As for the UWB path loss, the MICS path loss can also be fitted to the log-distance path loss model in Equation 4.12. Figure 4.12 shows the FDTD-calculated path loss results as well as the fitted log-distance path loss model for the receiver location Rx1. Based on the log-distance path loss model, $PL_{0,dB}$ is 49.7 dB at a reference distance $d_0 = 0.05$ m and the path loss exponent n is 5.5. Figure 4.13 shows the CDF of the shadow fading S_{dB} for the receiver at location Rx1. The normal distribution is found to provide the superior fitting. The mean value is zero and the standard deviation is $\sigma_{dB} = 8.1$ dB. Table 4.4 lists the fitted parameters of the log-distance path loss model for all seven receiver locations around the body.

A more detailed path loss model for capsule endoscope in the MICS band is developed to also take into account the angle θ between the transmitting and receiving antennas (Aoyagi *et al.*, 2010). The model is based on a combination of FDTD simulation and phantom experiment. The derived path loss formula is as follows

$$PL_{dB}(d,\theta) = a \cdot d + b + P_{dB}(\theta) + S_{dB} \tag{4.16}$$

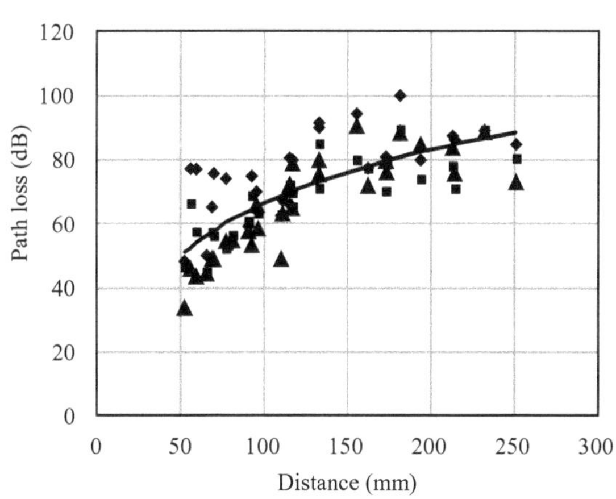

Figure 4.12 Path loss versus distance for receiver Rx1

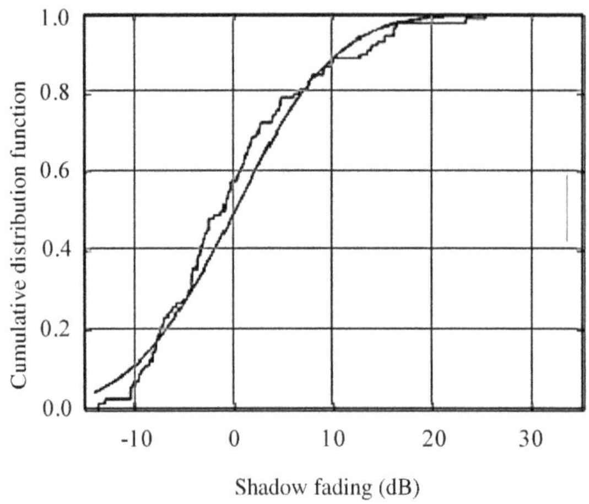

Figure 4.13 Cumulative distribution function of shadow fading for receiver Rx1

Table 4.4 Fitted parameters for path loss expression

Rx location	$PL_{0,dB}$	n	σ (dB)
Rx1	49.7	5.5	8.1
Rx2	38.0	6.7	7.4
Rx3	37.8	6.0	8.0
Rx4	40.4	6.4	7.9
Rx5	41.2	5.8	8.5
Rx6[a]	44.2	13.8	8.1
Rx7[a]	51.0	11.0	6.8

$d_0 = 0.05$ m; [a]$d_0 = 0.1$ m.

Table 4.5 Fitted parameters for path loss expression

a (dB/cm)	b (dB)	x_c	σ (dB)
1.92	39.85	0.145	6.59

where a is the gradient coefficient (dB/cm), b is the intercept coefficient (dB),

$$P_{dB}(\theta) = -20\log_{10}(|\cos(\theta)| \cdot (1 - x_c) + x_c) \quad (4.17)$$

with x_c as a coefficient of difference between the main and crossed electric field direction, and S_{dB} is the stochastic fluctuation or shadowing due to the organ structure. Table 4.5 gives the corresponding parameters in Equation 4.16, in which σ is the standard deviation for the normal distributed S_{dB}.

4.2.5 HBC Band Path Loss and Equivalent Circuit Expression

HBC uses the human body as a positive communication route to transmit data. It is superior to wireless transmission from the point of view of information security and electromagnetic compatibility. This is because (1) the signal is transmitted along the surface of the human body, and (2) radiation to outside the human body is extremely small. To meet these requirements, the tens of megahertz or less frequency band is favorable in HBC. In this frequency band the human body acts as a conducting medium so that we can model the electromagnetic coupling between the human body and the transceiver with capacitors. This structure forms a current loop which consists of the transmitter, body, receiver, and capacitive return path through the external ground. An equivalent circuit expression is useful to understand the transmission characteristics and is helpful in system design. In this section, we first introduce an alternative explanation for the HBC propagation based on surface wave theory. Then we give both a path loss formula and an equivalent circuit expression in the HBC band.

4.2.5.1 Basic Characteristics

HBC is considered mainly to take advantage of electrostatic coupling propagation. However, a surface wave approximation may also be effective to explore its basic characteristic (Wang, Nishikawa, and Shibata, 2009). Let us consider an infinitely long lossy dielectric cylinder in the cylinder coordinate system in Figure 4.14. The electric field will have only E_x and E_r components and the magnetic field will

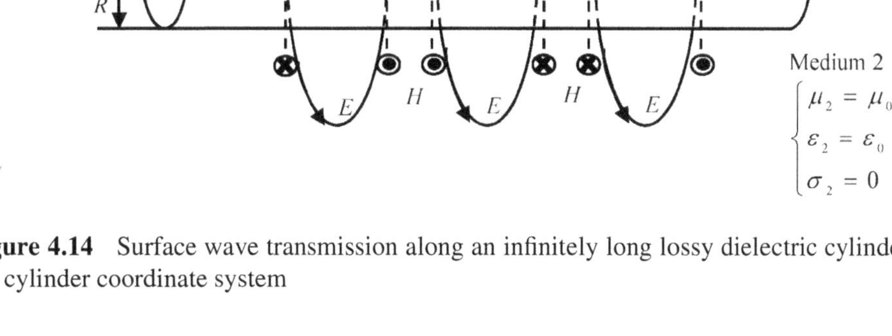

Figure 4.14 Surface wave transmission along an infinitely long lossy dielectric cylinder in the cylinder coordinate system

have only an H_ϕ component. In this case, after applying the usual sinusoidal time variation, the Maxwell equations become

$$\begin{aligned}
&\frac{\partial E_r}{\partial x} - \frac{\partial E_x}{\partial r} = -j\omega\mu_0 H_\phi \\
&-\frac{\partial H_\phi}{\partial x} = (\sigma + j\omega\varepsilon)E_r \\
&\frac{1}{r}H_\phi + \frac{\partial H_\phi}{\partial r} = (\sigma + j\omega\varepsilon)E_x.
\end{aligned} \tag{4.18}$$

For a surface wave, the field should decay exponentially with increasing distance along the surface. It is therefore reasonable to assume the electric field $E_x = A_x F(r) e^{-\gamma_H x}$ where $\gamma_H = \alpha_H + j\beta_H$ (with α_H the attenuation constant and β_H the phase constant) is the propagation constant along the cylinder surface in the x direction. Then we have (Barlow and Brown, 1962)

$$\frac{\partial^2 F(r)}{\partial r^2} + \frac{1}{r}\frac{\partial F(r)}{\partial r} - \gamma_V^2 F(r) = 0 \tag{4.19}$$

where γ_V is the propagation constant normal to the cylinder surface, that is, in the r direction, and

$$\gamma_{V1}^2 = -\left[\gamma_H^2 - j\omega\mu_0(\sigma + j\omega\varepsilon)\right] \qquad (r \leq R) \tag{4.20}$$

$$\gamma_{V2}^2 = -\left[\gamma_H^2 + \omega^2\mu_0\varepsilon_0\right] \qquad (r > R). \tag{4.21}$$

Equation 4.19 can be transformed into the standard form of Bessel's equation

$$\frac{\partial^2 F(r)}{\partial p^2} + \frac{1}{p}\frac{\partial F(r)}{\partial p} + F(r) = 0 \tag{4.22}$$

by putting $p = jur$. The appropriate solution to Equation 4.22 outside the cylinder surface is

$$F(r) = BH_0^{(1)}(p) = BH_0^{(1)}(j\gamma_{V2} r) \tag{4.23}$$

so that

$$E_{x2} = Be^{j\omega t} e^{-\gamma_H x} H_0^{(1)}(j\gamma_{V2} r) \tag{4.24}$$

$$E_{r2} = B\left(\frac{\gamma_H}{j\gamma_{V2}}\right) e^{j\omega t} e^{-\gamma_H x} H_1^{(1)}(j\gamma_{V2} r) \tag{4.25}$$

$$H_{\phi 2} = B\left(\frac{\omega\varepsilon_0}{\gamma_{V2}}\right) e^{j\omega t} e^{-\gamma_H x} H_1^{(1)}(j\gamma_{V2} r) \tag{4.26}$$

where B is a coefficient, and $H_0^{(1)}$ and $H_1^{(1)}$ are Hankel functions of the first kind. On the other hand, inside the cylinder surface, instead of Equation 4.23 we have

$$F(r) = B'J_0(p) = B'J_0(j\gamma_{V1}r). \tag{4.27}$$

Thus

$$E_{x1} = B'e^{j\omega t}e^{-\gamma_H x}J_0(j\gamma_{V1}r) \tag{4.28}$$

$$E_{r1} = B'\left(\frac{\gamma_H}{j\gamma_{V1}}\right)e^{j\omega t}e^{-\gamma_H x}J_1(j\gamma_{V1}r) \tag{4.29}$$

$$H_{\phi 1} = B'\left(\frac{\sigma + j\omega\varepsilon}{j\gamma_{V1}}\right)e^{j\omega t}e^{-\gamma_H x}J_1(j\gamma_{V1}r) \tag{4.30}$$

where B' is a coefficient, and J_0 and J_1 are Bessel functions of the first kind.

According to the boundary condition that the tangent components of the fields are continuous at the cylinder surface ($r = R$), we must have $E_{x1} = E_{x2}$ so that

$$BH_0^{(1)}(j\gamma_{V2}R) = B'J_0(j\gamma_{V1}R) \tag{4.31}$$

and $H_{\phi 1} = H_{\phi 2}$ so that

$$B\left(\frac{\omega\varepsilon_0}{\gamma_{V2}}\right)H_1^{(1)}(j\gamma_{V2}R) = B'\left(\frac{\sigma + j\omega\varepsilon}{j\gamma_{V1}}\right)J_1(j\gamma_{V1}R). \tag{4.32}$$

From Equations 4.31 and 4.32 we have

$$\begin{aligned}\frac{\gamma_V}{\omega\varepsilon_0}\frac{H_0^{(1)}(j\gamma_V R)}{H_1^{(1)}(j\gamma_V R)} &= \frac{j\sqrt{\gamma_V^2 + \omega^2\mu_0\varepsilon_0 + j\omega\mu_0(\sigma + j\omega\varepsilon)}}{\sigma + j\omega\varepsilon} \\ &\quad\times \frac{J_0(j\sqrt{\gamma_V^2 + \omega^2\mu_0\varepsilon_0 + j\omega\mu_0(\sigma + j\omega\varepsilon)}R)}{J_1(j\sqrt{\gamma_V^2 + \omega^2\mu_0\varepsilon_0 + j\omega\mu_0(\sigma + j\omega\varepsilon)}R)}.\end{aligned} \tag{4.33}$$

In Equation 4.33, $\gamma_V = \gamma_{V2}$, the propagation constant normal to the surface, is the only unknown quantity. Solving the nonlinear equation we can obtain γ_V, and substituting it into Equation 4.21 we can obtain γ_H, the propagation constant along the cylinder surface. Although Equation 4.33 is a nonlinear equation including special functions, it is possible to use Newton's method to find an approximation solution.

To compare the transmission characteristic on a human body with the assumed surface wave, we consider a circular cylinder as shown in Figure 4.15. The cylinder is a homogeneous medium of 3 cm in radius and 1 m in length. Its dielectric

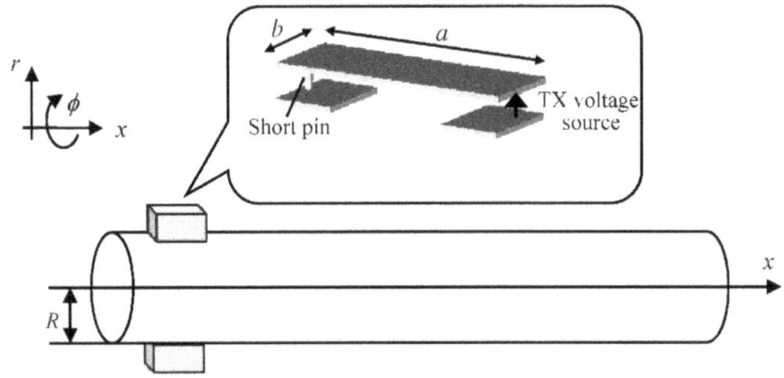

Figure 4.15 Cylinder model of arm

properties are adopted to have two-thirds the values of muscle to simulate a human arm. We make the signal excitation with two pairs of metal electrodes that are shown as the shaded regions at the left end of the cylinder. Each pair of metal electrodes has one signal electrode and two ground electrodes as also shown in Figure 4.15. The signal electrode and the left ground electrode are shorted, and the transmit signal is excited between the signal electrode and the right ground electrode by a voltage source. Such an electrode structure is based on two considerations. First, the major component of the electric field is induced normal to the cylinder surface in order to propagate along the cylinder as an approximated surface wave. Secondly, shorting the signal electrode and the left ground electrode reduces the stray capacitance between them and consequently makes impedance matching with the transmitter easy. An almost symmetric electromagnetic field is made at the top and bottom of the cylinder, and comparison with the theoretical solution of the surface wave becomes possible although a strictly symmetric wave is not induced.

The electromagnetic field analysis in the vicinity of the cylinder surface is possible by using the FDTD method. Then we can extract the electric fields along the cylinder surface and normal to the cylinder surface. Due to the limited length of the cylinder, reflection occurs at the cylinder end. The comparison with the theoretical solution to the surface wave is made in the region which is far enough from the cylinder end so as to reduce the influence of the reflection as much as possible.

Figure 4.16 shows the FDTD-calculated relative electric field distributions in a vertical cross-section. A tendency of the surface wave is seen because the E_r component transmits especially along the cylinder, and it attenuates slowly in the x direction and quickly in the r direction compared with the case of only air. Also, the E_ϕ component is small. Since the E_r and E_x components and the H_ϕ component are more predominant than other components, the transmission on the cylinder surface can be approximately considered as a surface wave.

Although the field distribution in the vicinity of the electrodes is quite complex and is influenced by the electrode structure, it is still possible to extract the propagation characteristics from the field distribution when the observation point is

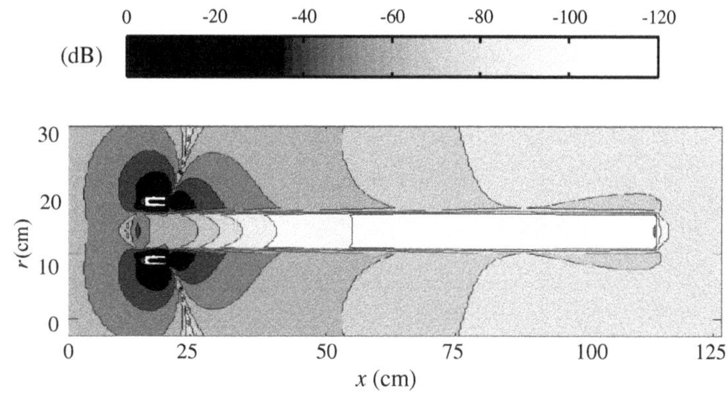

Figure 4.16 Relative electric field distribution of E_r component in a vertical cross-section (Wang, Nishikawa, and Shibata, 2009). Reproduced with permission from Wang J., Nishikawa Y. and Shibata T., "Analysis of on-body transmission mechanism and characteristic based on an electromagnetic field approach," *IEEE Transactions on Microwave Theory and Techniques*, **57**, 10, 2464--2470, 2009. © 2009 IEEE

somewhat far from the electrodes. From the FDTD-calculated field distributions, the principal field component E_r indeed decays exponentially with the distance from the transmitter electrodes. We therefore assume $E_r \propto e^{-\alpha_H x}$ at the cylinder surface in the x direction, and $E_r \propto e^{-\alpha_V r}$ in the middle of the cylinder in the r direction. We use this relationship to fit the data in Figure 4.17 based on the least square approximation. The data in Figure 4.17 are first divided into two areas in both the x direction and r direction, and the fitting is then done to the data in the latter area with the curve fitting tool in MATLAB®. In either of the two directions, two straight lines with different inclination rates are used to approximate the exponential field attenuation characteristic.

Next, we pay attenuation to E_r along the cylinder surface in the x direction because it attenuates quickly in the r direction normal to the cylinder surface. We change the exciting frequency from 10 to 150 MHz, and derive the attenuation constant α_H in the x direction along the cylinder surface as described above. On the other hand, we also obtain the theoretical α_H for an ideal infinitely long cylinder using the surface wave theory. Figure 4.18 shows the FDTD-derived and theoretical α_H values from 10 to 150 MHz. The FDTD result shows the same tendency as the theoretical result of the surface wave, and they are in fair agreement. This finding suggests that the signal transmission on the surface of a finitely long cylinder, which has a dielectric constant of the living body, is approximately expressible by the theoretical solution to the surface wave.

Moreover, α_H increase as the frequency rises between 10 MHz and 150 MHz. This is because that the permittivity of the living body decreases with the frequency, which makes the surface wave difficult to hold. A higher dielectric constant means

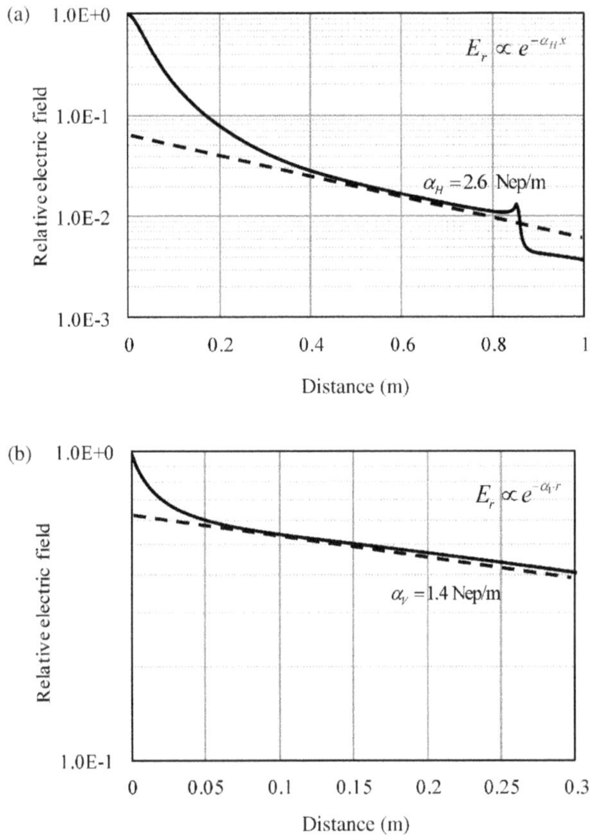

Figure 4.17 Exponential approximation of the dominant electric field component: (a) horizontal; (b) vertical

less loss for the surface wave transmission. From Figure 4.18, it is clear that lower frequency is more appropriate for propagation along the human body.

4.2.5.2 Path Loss

Based on a numerical and experimental approach, we derive a path loss formula for a typical HBC application. Figure 4.19 shows the human body model. It is homogeneous and is an average adult male size. The posture simulates a situation of communication by touching a receiver electrode. The transmitter electrode is arranged at 50 locations in total on the human body surface such as the shoulder, waist, belly, chest, and arm. On the other hand, the receiver is set up on the tip of a finger of the left hand. Then, using the FDTD method, we can obtain the received voltage at the receiver electrode, and the path loss from the ratio of the received voltage to the transmitted voltage. It should be noted that there are two kinds of distances between

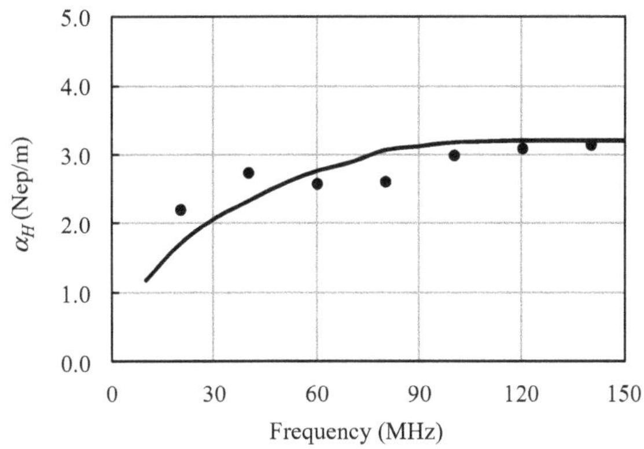

Figure 4.18 Comparison of the surface wave theory (line) and the FDTD-derived attenuation constant (symbols) (Wang, Nishikawa, and Shibata, 2009). Reproduced with permission from Wang J., Nishikawa Y. and Shibata T., "Analysis of on-body transmission mechanism and characteristic based on an electromagnetic field approach," *IEEE Transactions on Microwave Theory and Techniques*, **57**, 10, 2464–2470, 2009. © 2009 IEEE

the transmitter and the receiver. One is the straight line distance, and the other is the surface distance along the human body. According to the propagation mechanism, the surface distance should be used for the path loss formulation.

Now we formulate the path loss for the HBC channel. Since the propagation is in a near-field region very close to the excitation source, a log-distance path loss model is

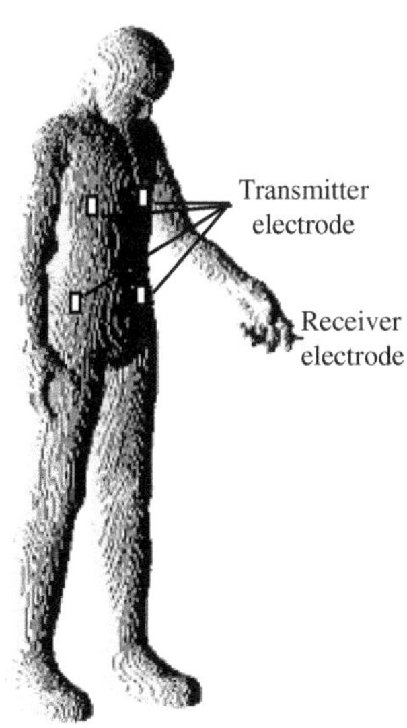

Figure 4.19 Human body model and representative transmitter locations

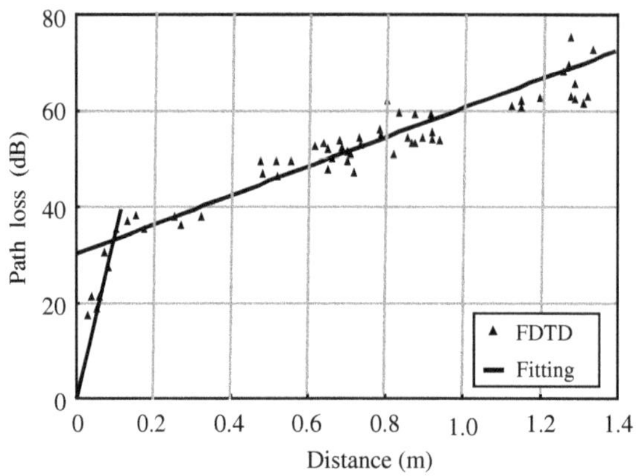

Figure 4.20 HBC path loss characteristic (Wang, Nishikawa, and Shibata, 2009). Reproduced with permission from Wang J., Nishikawa Y. and Shibata T., "Analysis of on-body transmission mechanism and characteristic based on an electromagnetic field approach," *IEEE Transactions on Microwave Theory and Techniques*, **57**, 10, 2464–2470, 2009. © 2009 IEEE

no longer applicable. Here we propose the following expression. That is

$$PL_{dB} = \begin{cases} \alpha_0 d & d < 0.1\text{ m} \\ PL_{0,dB} + \alpha_1(d - 0.1) & d > 0.1\text{ m} \end{cases} \tag{4.34}$$

where α_0 and α_1 ($=20\alpha_H \log_{10} e$) are the losses per unit distance in units of dB/m on the human body, d is the distance between the transmitter and the receiver, and $PL_{0,dB}$ is the path loss at the boundary ($d = 0.1$ m) of the two regions. Figure 4.20 shows the fitting result of the above expression and the calculated results. The parameters are given in Table 4.6.

4.2.5.3 Equivalent Circuit Expression

An equivalent circuit expression is also useful to explore the path loss in the HBC band. The simplest equivalent circuit is based on the assumption of the human body as a perfect conductor. This assumption makes the human body a single node. The single node is connected to the transceiver and external ground through lumped equivalent capacitors. Such a capacitive approach is shown in Figure 4.21. However,

Table 4.6 Fitted parameters for path loss expression

α_0 (dB/m)	α_1 (dB/m)	$PL_{0,dB}$
371.2	30.4	35.4

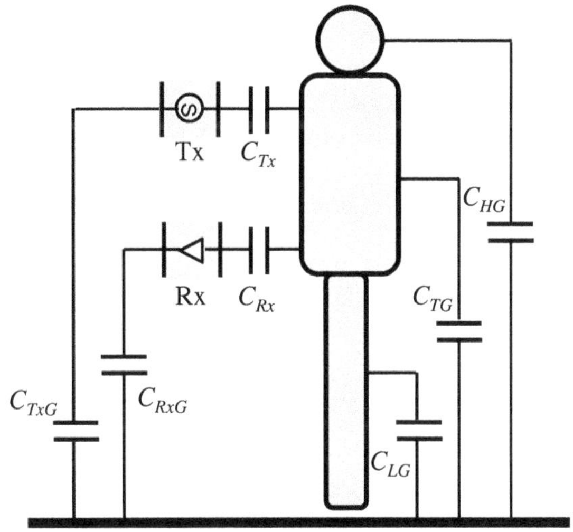

Figure 4.21 Capacitive approach

at tens of MHz or more, the body impedance cannot be ignored in reality. So we have to consider both the resistance component and capacitance component of the body itself, in addition to the capacitive components to the external ground in the equivalent circuit expression. These components will cause signal loss between the transmitter and receiver. Figure 4.22 shows a unit block with a RC parallel network and a shunt capacitor (Cho *et al.*, 2007). The unit block represents one part of the human body and the corresponding electrical coupling. The entire human body can be considered as a cascade of many unit blocks.

When considering a HBC scenario as in Figure 4.19, we can segment the human body into many unit blocks of 10 cm along the body length, and then determine the parameters of the equivalent circuit for each unit block, that is, the impedance of the parallel RC network and the coupling capacitance to the external ground.

The derivation of the impedance of parallel RC network is based on the dielectric properties of the human body. The dielectric properties of the human body are highly dependent on the frequency and tissue types. Due to the significant penetration depth

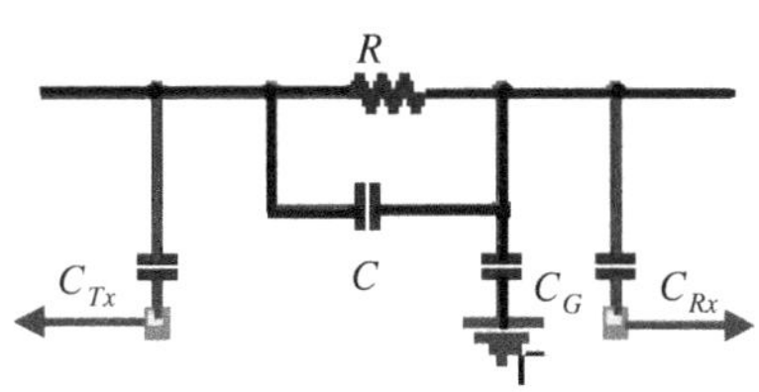

Figure 4.22 Unit block of equivalent circuit

in the HBC band, the human body can be approximated by a homogeneous medium with dielectric properties similar to two-thirds the value of muscle. The conductivity σ is within 0.4–0.5 S/m and the relative permittivity ε_r varies from 44 to 106 in the frequency band of 10–100 MHz. Therefore, the resistance R can be obtained from

$$R = L/\sigma S \tag{4.35}$$

and the capacitance C can be obtained from

$$C = \varepsilon_0 \varepsilon_r S/L \tag{4.36}$$

where L and S are the length and cross-sectional area of a unit block, respectively.

On the other hand, the coupling capacitance C_G to the external ground can be obtained by approximating the unit block as a conductive sphere in free space. For each segment of the body, we have a volume V. If we use a sphere to approximate this volume, the equivalent sphere's radius is

$$a = \left(\frac{3}{4\pi}V\right)^{1/3}. \tag{4.37}$$

Then the coupling capacitance C_G can be calculated from

$$C_G = 4\pi\varepsilon_0 a\left(1 + \frac{a}{2d} + \frac{(a/2d)^2}{1-(a/2d)^2} + \cdots\right) \tag{4.38}$$

where d is the distance of the center of the sphere to the ground.

It should be noted that the capacitance determined in the above way assumes that the human body is located in an open space. If any large conductive object is nearby, the capacitance will increase due to the existence of an additional coupling path. Moreover, the coupling capacitances between the transceiver grounds and the unit block are highly affected by the body configurations. In most cases, their values are smaller than 1 pF, with a smaller influence on the channel characteristic. An approach for accurately extracting the coupling capacitance is to use a numerical electromagnetic analysis tool to calculate quasi-static field components. Such an approach is especially effective for considering a complicated body shape and structure, but it suffers from a large computation burden.

By cascading these RC unit blocks, we can obtain a complete circuit model of the human body. Table 4.7 gives some typical values for the parameters in the parallel RC unit block, which are calculated using Equations 4.35–4.38. To consider the channel responses at various locations on the body, we can place the transceiver models at corresponding nodes of the equivalent circuit. In the equivalent circuit,

Table 4.7 Calculated parameters of equivalent circuit per 10 cm

Frequency (MHz)		10	40	60	80
ε_r		106.6	54.4	48.7	45.8
σ (S/m)		0.43	0.46	0.47	0.48
Head	R_H (Ω)	4.7	4.4	4.3	4.2
	C_H (pF)	463.3	237.9	213.0	200.3
	C_{HG} (pF)	12.2	12.2	12.2	12.2
Torso	R_T (Ω)	2.4	2.2	2.2	2.1
	C_T (pF)	917.4	468.2	419.1	394.2
	C_{TG} (pF)	15.6	15.6	15.6	15.6
Leg	R_L (Ω)	3.5	3.3	3.2	3.2
	C_L (pF)	622.9	317.9	284.6	267.6
	C_{LG} (pF)	14.8	14.8	14.8	14.8

the major return path is formed by the electrical coupling between the transceiver grounds and the external ground. A large ground plane or a special electrode for the return path is therefore advantageous to enhance the signal-to-noise ratio (SNR) of the received signal.

Figure 4.23 shows an example of an equivalent circuit with the transmitter and receiver at the torso. The head, torso and leg are modeled using three, five and nine unit blocks, respectively. Figure 4.24 shows simulation results for the path loss as a function of transmission distance by using the equivalent circuit. The simulation is conducted in the circuit simulator SPICE. The path loss is the ratio in decibels of the received voltage at the receiver electrode and the transmitted voltage at the transmitter electrode. Also shown in Figure 4.24 are the measured results. The transmitter electrode is excited by a crystal oscillator, and is fixed at the chest. The receiver is shifted along the body surface to detect the received voltage. The obtained path losses for five people are averaged and indicated by diamond symbols in the figure.

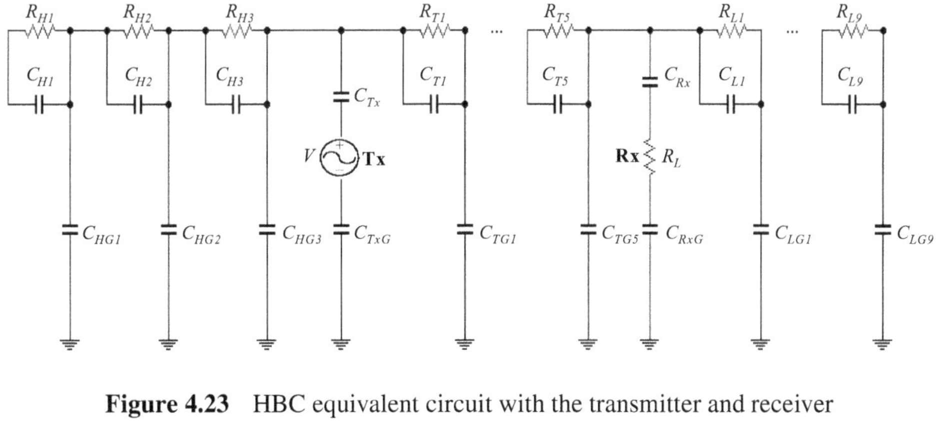

Figure 4.23 HBC equivalent circuit with the transmitter and receiver

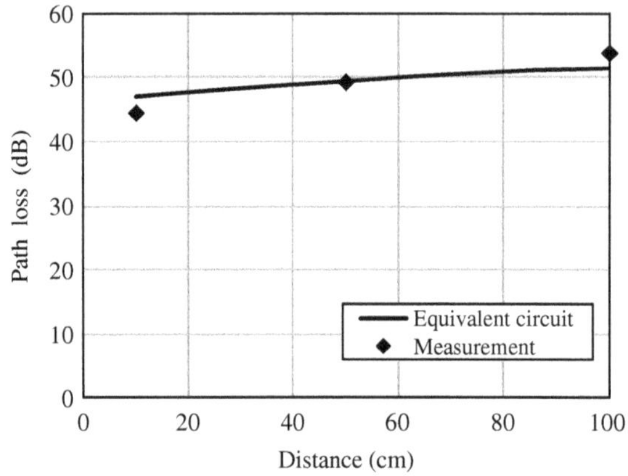

Figure 4.24 Path loss obtained by using the equivalent circuit and measurement

It can be seen that the simulated path loss has fair agreement with the measured results. It is also on the same order as the FDTD-calculated version in Figure 4.20. It is worth noting that the path loss does not exhibit significant frequency dependence between 10 MHz and 100 MHz. The distance dependence is also weak compared with other frequency bands. This feature suggests that HBC is a good choice for efficient on-body transmission.

Although this approach only gives a rough estimation for the path loss, it is much simpler and easier to use compared with electromagnetic field numerical simulation or measurement.

4.3 Multipath Channel Model

In wireless communications, the multipath propagation phenomenon is when the transmitted signal travels along many different paths to reach the receiver. The presence of multiple paths between transmitter and receiver introduces complexity in the channel modeling. The complexity depends on the distribution of the multipath intensity, relative propagation time of the waves and bandwidth of the transmitted signal. Therefore, the time-varying properties of the channel must be taken into account in the channel model.

Multipath channel models have already been widely investigated in indoor and outdoor wireless communications. Several models are available for both narrowband and wideband transmissions (Rice, 1959; Saleh and Valenzuela, 1987; Hashemi, 1993); the Saleh–Valenzuela channel model, sometimes abbreviated to the S–V model, has provided a comprehensive and standardized formalization on statistical modeling for indoor multipath propagation. For the body area multipath channel model, some effort has also been made in recent years based on experimental and numerical approaches (Fort *et al.*, 2006; Tang *et al.*, 2006; Zhao *et al.*, 2006; Wang *et al.*, 2009). A generalization of the derived channel model, however, has to take into account the statistical characteristics of body postures and movements.

In this section, we will present a body area UWB multipath channel model based on the classical Saleh–Valenzuela model. The Saleh–Valenzuela model is based on indoor multipath propagation measurements using radar-like pulses, which essentially covers a wideband propagation characteristic. And more importantly, it appears to be extendable to a body area UWB multipath channel model via appropriate modifications. We will first introduce the classical Saleh–Valenzuela model and then investigate in detail the body area UWB multipath channel model with a few modifications to the Saleh–Valenzuela model.

4.3.1 Saleh–Valenzuela Impulse Response Model

It is known that the impulse response model is a convenient model to characterize the multipath channel and any deterministic impulse response can be represented by a discrete tapped delay line model as long as the system is band-limited. The Saleh–Valenzuela model is expressed by a discrete impulse response model as follows

$$h(t) = \sum_k \beta_k e^{j\theta_k} \delta(t - \tau_k) \tag{4.39}$$

where β_k is the multipath power gain, θ_k is the associated phase shift, τ_k is the propagation delay, k is the path index and $\delta(\cdot)$ is the Dirac delta function. The parameters β_k, θ_k and τ_k are treated as virtually time-invariant random variables since the motion rate of the indoor subjects is very slow compared with the potential signaling rates. In this indoor channel model, the multipaths are modeled based on the observation that multipath contributions generated by the same pulse usually arrive in clusters. The clusters are formed by the building structure, while the individual paths are formed by objects in the vicinities of the transmitter and the receiver. The arrival time of the clusters is modeled as a Poisson process. Meanwhile, the multipaths within each cluster also arrive according to a Poisson process with a different rate. The multipath gain β_k is a statistically independent Rayleigh distributed random variable. The θ_k is an independent uniform variable over $[0,2\pi)$.

4.3.2 On-Body UWB Channel Model

Different from the path loss characteristics which can be extracted based on the whole body average, the characteristics of the on-body multipath channels can be expected to differ considerably for different propagation links due to the variability of the link geometry. Moreover, differences in body parts as well as in body curvatures in the vicinities of the transmitter and receiver result in distinct multipath characteristics. Therefore, we need to differentiate different transmission links and assign the transmitter and receiver points on the human model in the context of

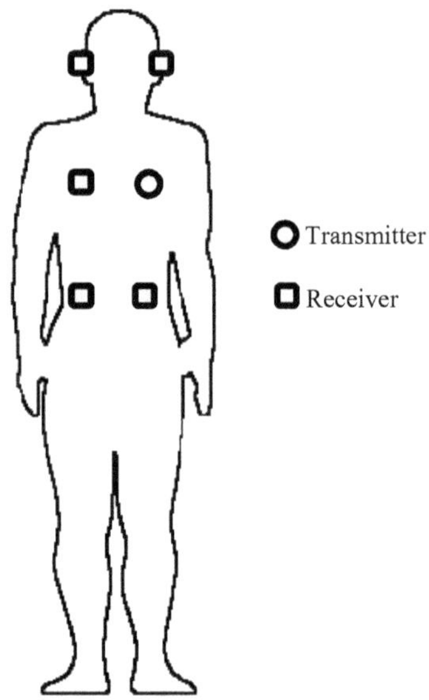

Figure 4.25 Representative transmitter and receiver locations on the human body

multipath channel characterization. As described in the path loss modeling, in order to improve the computation efficiency in the FDTD method, one transmitting antenna and multiple receiving points are constructed on the body model.

As shown in Figure 4.25, a Hertzian dipole transmitter is fixed on the left chest and five receiving points are assigned on the right chest, the left and right waists, and both ears, respectively. This assignment results in five transmission links: chest-to-right-chest link, chest-to-left-ear link, chest-to-right-ear link, chest-to-left-waist link and chest-to-right-waist link, respectively. The five transmission links can be considered as typical transmission links for medical and healthcare applications.

Since the on-body multipaths are mainly caused by the body movement, the effects of various body postures are investigated by simulating 35 different postures, as seen in Figure 4.26. The details of the postures are as follows:

Standing: 9 postures
Walking: 10 postures
Running: 10 postures
Sitting: 6 postures.

Simulating various postures is indispensible in order to obtain a statistical characterization of the transmission channels, since the posture of the user is generally not fixed in medical and healthcare applications.

Since the impulse response of the Saleh–Valenzuela model can be used after appropriate modifications (Molisch *et al.*, 2006) for UWB applications, here we

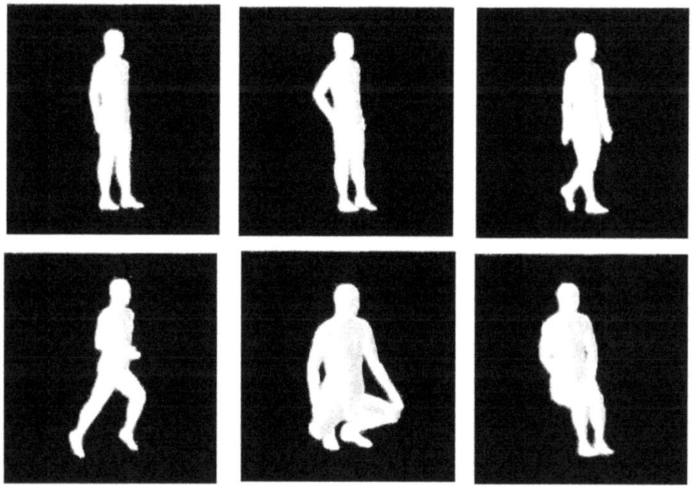

Figure 4.26 Some representative body postures

apply and modify the Saleh–Valenzuela model in order to characterize the on-body UWB multipath channel and to obtain the parameters required for the implementation. The modified Saleh–Valenzuela model in the time domain deserves practical implementation in system design and simulation.

4.3.2.1 Power Delay Profile

The power delay profile (PDP), $p(\tau)$, is a statistical expression of the transmission channel characteristics. It can be derived from the impulse response $h(t)$, that is,

$$h(t) = F^{-1}\{H(f)\} = F^{-1}\left\{\frac{F[v_r(t)]}{F[v_t(t)]}\right\} \tag{4.40}$$

and

$$p(\tau) = \langle h(\tau) \cdot h^*(\tau) \rangle \tag{4.41}$$

where $v_t(t)$ and $v_r(t)$ are the transmitted and the received pulse voltages respectively, $H(f)$ is the frequency-domain transfer function, and $F\{\cdot\}$ and $F^{-1}\{\cdot\}$ denote Fourier transform and inverse Fourier transform, respectively. The PDP characterizes the arrival time of the different multipath contributions versus the received mean power.

In the derivation of the impulse response $h(t)$, frequency-domain windowing has to be applied in the spectral analysis. Frequency-domain windowing can reduce the time side lobes in the impulse response using weighting in the frequency domain. Different windowing functions have different effects on the time-domain derivation.

This is mainly because different windows in the frequency domain will result in different time side lobes/spreads as well as different time resolution. Therefore, time side lobe and time resolution are the two most important factors in the windowing function selection. The rectangular window has excellent resolution characteristics, but its first side lobe is only 13.3 dB down. The Hanning window, also called the raised cosine window, can have a first side lobe 32 dB down while its main lobe has been widened compared with the rectangular window with twice the main lobe width of the rectangular window. The wide main lobe represents a degraded time-domain resolution due to a corresponding time spread. The Hamming window is a kind of optimized raised cosine window with simpler coefficients, which is optimized to minimize the maximum side lobe, giving it a peak side lobe level approximately 42 dB down. The Blackman window will result in a much higher peak side lobe compared with the Hanning and Hamming windows while the main lobe width is four times that of the rectangular window. Moderate windows should be selected in between the resolution extreme and side lobe extreme, such as Hanning and Hamming windows. The Hamming window has the same main lobe width as the Hanning window with a much lower time side lobe and hence could be a very good window candidate in the time-domain impulse response derivation.

The Hamming window is therefore employed in $F[v_t(t)]$ and $F[v_r(t)]$ with a frequency bandwidth of 14 GHz in view of the dominating frequency components of the employed second-derivative Gaussian pulses. The time resolution of the inverse Fourier transform for $h(t)$ can be approximated as the reciprocal of the bandwidth (1/14 GHz = 0.07 ns) multiplied by the additional window function bandwidth. Since the coefficient of the Hamming window is 2, this results in a 0.14.ns time resolution for the impulse response and the PDP.

Based on the data from all 35 postures, the average power delay profiles (APDPs) for five typical transmission links are shown in Figure 4.27. It can be noted that one cluster is sufficient for describing the PDP in all five representative transmission links. Each peak in the cluster may be attributed to a multipath, which results from the diffraction from the body surface or the reflection from a body part. We should note that no clusters corresponding to the surrounding environment are included and the channel model extraction is limited solely to the human body. From Figure 4.27, as expected, the APDP decays exponentially with the arrival time, that is, it can be approximately expressed as

$$p(\tau) = \Omega_0 e^{-\frac{\tau-\tau_0}{\gamma}} \tag{4.42}$$

where Ω_0 and τ_0 are the mean power gain and the arrival time of the first path, respectively. With the exponential fittings to the data in Figure 4.27, it is found that the decay time constants γ are 0.21, 0.26, 0.38, 0.30 and 0.5 ns, respectively for chest-to-right-chest link, chest-to-left-ear link, chest-to-right-ear link, chest-to-left-waist link and chest-to-right-waist link, respectively.

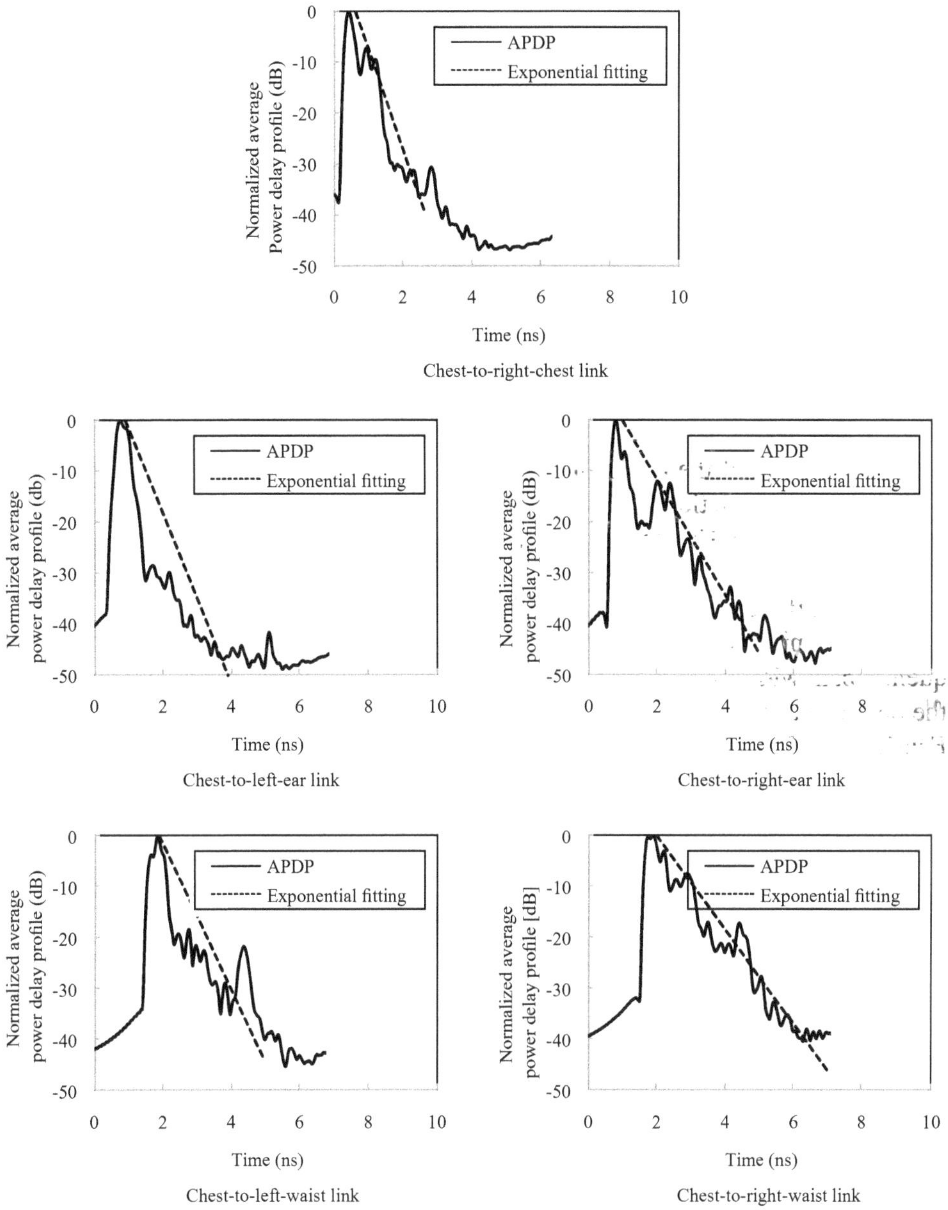

Figure 4.27 Average power delay profiles for five typical transmission links based on 35 postures

4.3.2.2 Power Gain Distribution

The log-normal distribution is reported in Fort *et al.* (2006) and Zhao *et al.* (2006) as an excellent fit to the power for all receiver locations. Here, we assume that all of the obvious peaks in the PDP correspond to the multipaths. The multipaths are characterized by identifying the corresponding peaks, and then obtaining the arrival time and the power of the individual paths. The peaks that have power gains up to 30 dB lower than the maximum peak value are taken into account in order to extract the channel parameters, while the other peaks are small enough to be ignored. The power distribution in the first and the second paths are also studied. Some possible candidates, such as the log-normal distribution, Rayleigh distribution, Rice distribution and Weibull distribution, are considered for the power model.

In order to pick the optimal distribution, the classical second-order Akaike information criterion ($\mathrm{AIC_c}$) (Akaike, 1973) is used to rank the fitting results from best to worst. The AIC is a measure of the relative goodness of fit of a statistical model. It is grounded in the concept of information entropy, in effect offering a relative measure of the information lost when a given model is used to describe reality.

The second-order $\mathrm{AIC_c}$ is defined as follows (Burnham and Anderson, 2002)

$$\mathrm{AIC_c} = -2\log_e(\ell(\hat{\theta}|\mathrm{data})) + 2K + \frac{2K(K+1)}{(n-K-1)} \tag{4.43}$$

where $\log_e(\ell(\hat{\theta}|\mathrm{data}))$ is the value of the maximized log-likelihood over the unknown parameters (θ), given the data and the model, K is the number of parameters estimated in that model, and n is the sample size. This equation is straightforward to compute since the log-likelihood is readily available from the maximum likelihood estimation. Intuitively, the first term indicates that better models have a lower $\mathrm{AIC_c}$ because the log-likelihood reflects the overall fit of the model to the data. The second part of the equation penalizes additional parameters so that the selected model best fits the data with the least number of parameters. In this way, the model with the lowest $\mathrm{AIC_c}$ approximates the "true" distribution with the minimum loss of information.

In practice, the value of the $\mathrm{AIC_c}$ by itself has no meaning. However, the relative values of $\mathrm{AIC_c}$ among the models can be used to rank the models from best to worst and to provide strength of evidence that one model is better than another. To facilitate this, the two following related metrics are normally reported

$$\Delta_\mathrm{i} = \mathrm{AIC_{c,i}} - \mathrm{Min}(\mathrm{AIC_c}) \tag{4.44}$$

$$w_i = \frac{\exp\left(-\frac{\Delta_i}{2}\right)}{\sum_{r=1}^{R}\exp\left(-\frac{\Delta_r}{2}\right)} \tag{4.45}$$

where $\mathrm{AIC_{c,i}}$ is the AIC value for the model index, and R is the number of models. Clearly, the best model among the set of models has a ΔAIC of 0. As a rule-of-thumb, $\Delta_i < 2$ suggests substantial evidence for the model, values between 3 and 7

Table 4.8 Comparison of fitting models of power distribution for the chest-to-right-waist link

Model	Δ	w
Lognormal	0	0.88
Rayleigh	173.75	0.0
Rice	14	0.0
Weibull	4	0.12

indicate that the model has considerably less support, while values >10 indicate that the model is very unlikely. The Akaike weights (w_i) provide a more precise measure of the strength of evidence and can be interpreted as the probability that the model is the best among the whole set of candidates. In addition, the ratio of two AIC weights indicates how much more likely one model is better compared with the other. Clearly, these metrics are more informative than a simple hypothesis test that can only pass or fail a model based on an arbitrary significance level without providing any strength of evidence or ranking. These advantages will become more apparent as we apply the metrics to our data in the following subsections.

As seen from Figure 4.28, the log-normal distribution provides a superior fit to the power distribution in the multipaths. Furthermore, Table 4.8 gives a comparison of the power distribution fitting models for the chest-to-right-waist link and the AIC parameters show further that the log-normal distribution gives the best fitting result. For the chest-to-right-waist link, the average standard deviation of the power variation is 7.87 dB, which will be used in the channel modeling. For other representative links, the conclusion that the log-normal distribution provides a superior fit to the powers still holds, although obviously the standard deviations are different. The physical meaning of the power variation following a log-normal distribution is easily understood, since the wave can be considered to propagate along the body area with magnitudes affected by statistically varying reflection and diffraction coefficients in the form of multiplication.

4.3.2.3 Arrival Time of the First Path

For various transmission links, the arrival time of the first multipath varies, and is determined mainly by the direct transmission distance. For all five transmission links, Gamma distribution is fitted to the arrival time data of the first path, in accordance with the AIC. The Gamma fittings of the arrival time of the first path for the five links are shown in Figure 4.29. For example, for the chest-to-right-waist link, the mean value and the standard deviation of the Gamma distribution are found to be 2.0 and 0.03 ns, respectively. The extremely small standard deviation suggests that fixing the arrival time of the first path at the mean value is reasonable.

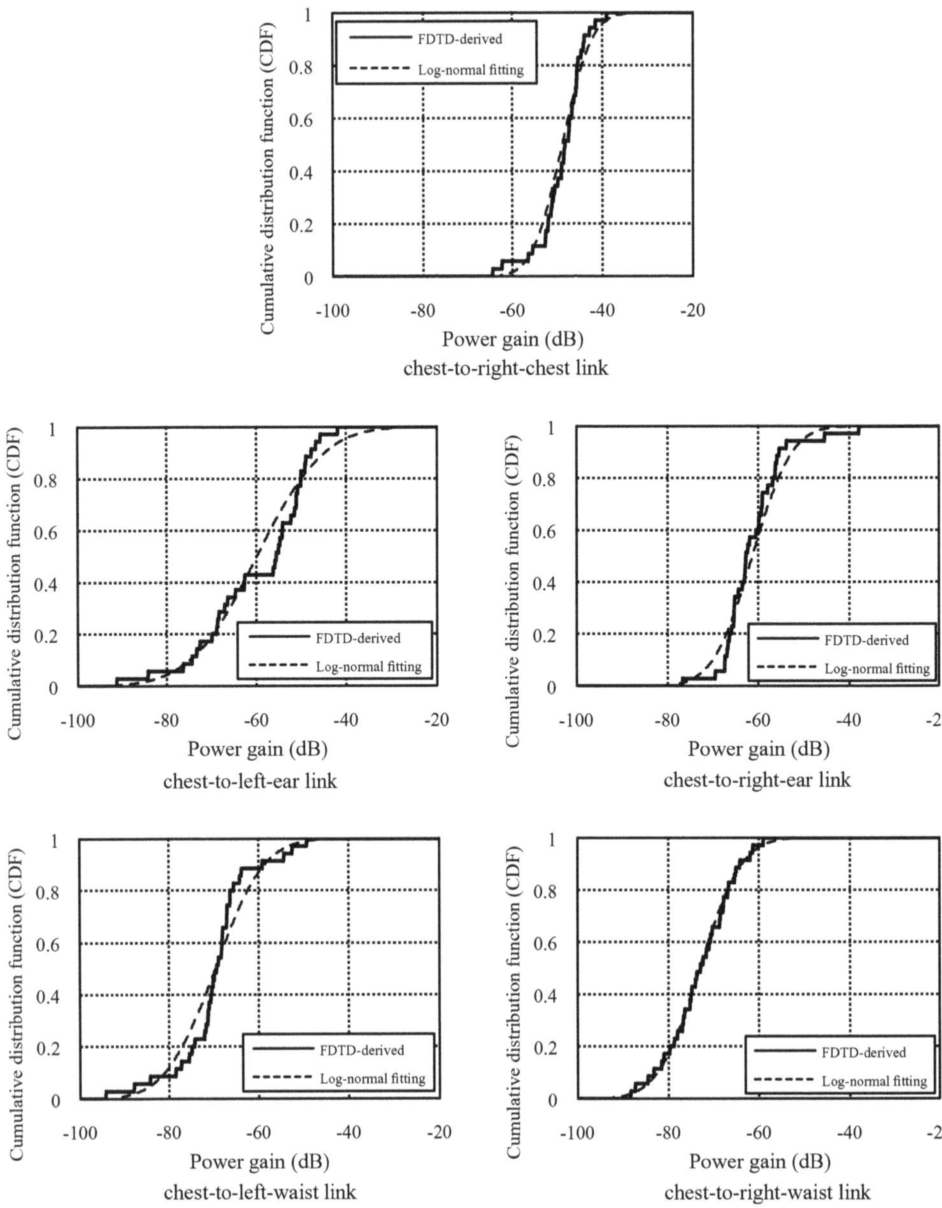

Figure 4.28 Cumulative distribution functions of the first-path power gain

4.3.2.4 Inter-Path Delay Distribution

An impulse response can be represented by a tapped delay line model. In the on-body UWB transmission, adjacent taps may be influenced by a single physical multi-path component, which suggests a correlation. In this case, it is possible to realize an

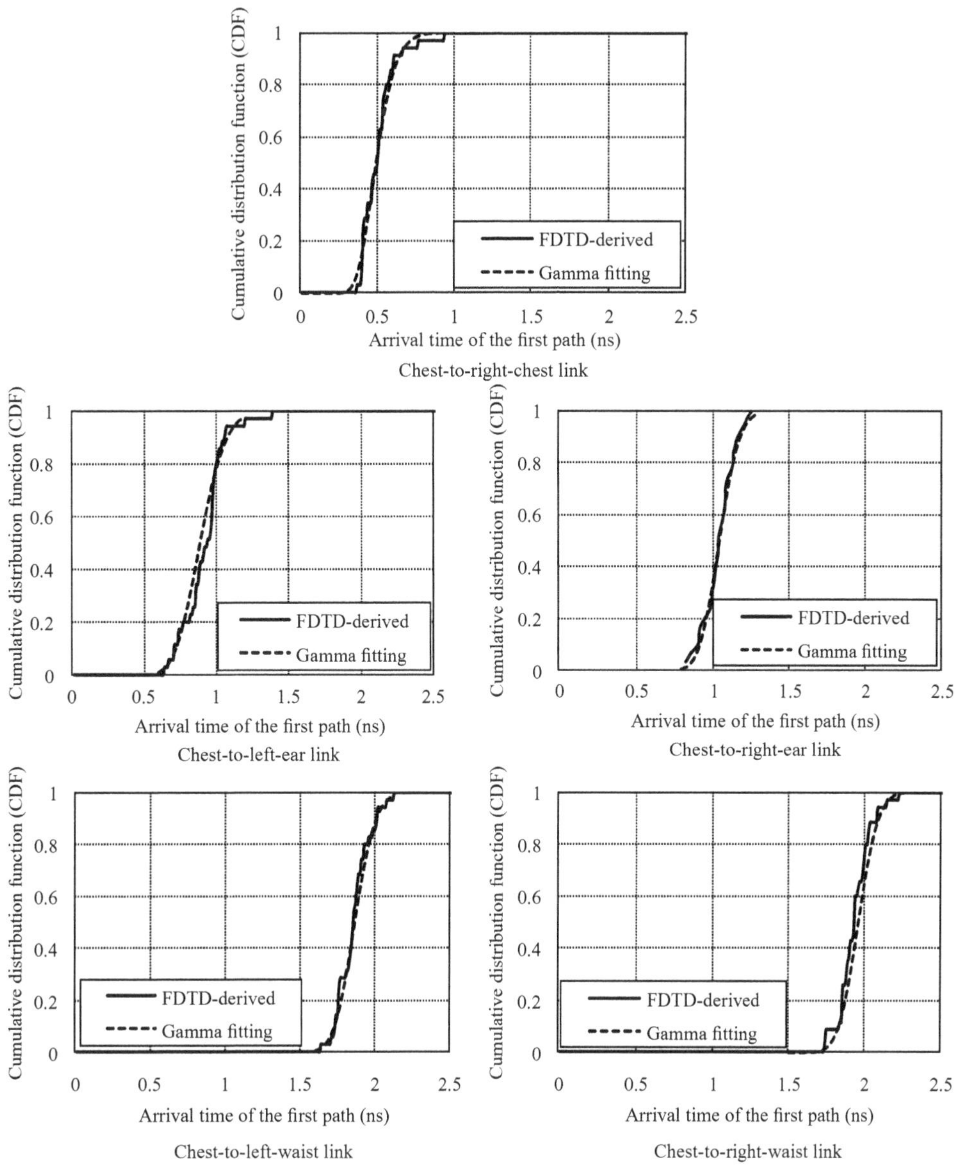

Figure 4.29 Cumulative distribution functions of arrival time of the first path

impulse response based on a uniformly spaced tapped delay line model (Molisch *et al.*, 2006). In the present case, however, the dominant multipath components correspond to parts of the body such as the arms, legs and so on due to their varying positions. Most of the multipaths are distinguishable due to the high time resolution of 0.14 ns. Moreover, the simulation results indicate that the correlation coefficient between the first and the second distinguishable multipaths is as weak as 0.2, which allows us to characterize the inter-path delay without resorting to uniformly spaced taps.

The inter-path delay, which corresponds to the temporal delay between two successive paths, represents the characteristics of the arrival time for all multipaths in the on-body transmission channel. In view of the above observations, we can derive the statistical model for the inter-path delay. At first, we identify the delay time of each path from the corresponding peak in the PDP, and then the difference between the arrival times of two successive multipaths can be calculated in order to obtain the inter-path delay. The inter-path delay data obtained in this way are fitted to some candidate statistical distributions, such as the exponential distribution, Weibull distribution, log-normal distribution and inverse Gaussian distribution.

The inverse Gaussian distribution is a two-parameter family of continuous probability distributions. Its probability density function (PDF) is given by

$$f(x,\mu,\lambda) = \left[\frac{\lambda}{2\pi x^3}\right]^{\frac{1}{2}} \exp\frac{-\lambda(x-\mu)^2}{2\mu^2 x} \tag{4.46}$$

where μ is the mean and λ is the shape parameter. The variance is equal to μ^3/λ. As λ tends to infinity, the inverse Gaussian distribution becomes more like a normal distribution. Figure 4.30 shows the results of the inverse Gaussian fitting to the inter-path delay data for all five links. The inverse Gaussian distribution provides a superior fit. Furthermore, Table 4.9 gives a comparison of the two metrics of the second-order AIC_c for the results of the fitting to the chest-to-right-waist FDTD-calculated data. The second-order AIC_c indicates further that the inverse Gaussian distribution is the best fitting model. For the chest-to-right-waist link, the mean value of the inter-path delay is 0.33 ns and the standard deviation is 0.2 ns. For other representative links, the inverse Gaussian distribution also provides a superior fit in comparison with other distributions, although the mean values and the standard deviations are, of course, different.

Table 4.9 Comparison of fitting models of inter-path delay distribution

Model	Δ	w
Inverse Gaussian	0	1
Lognormal	604	0.0
Weibull	792	0.0
Exponential	852	0.0
Gamma	718	0.0

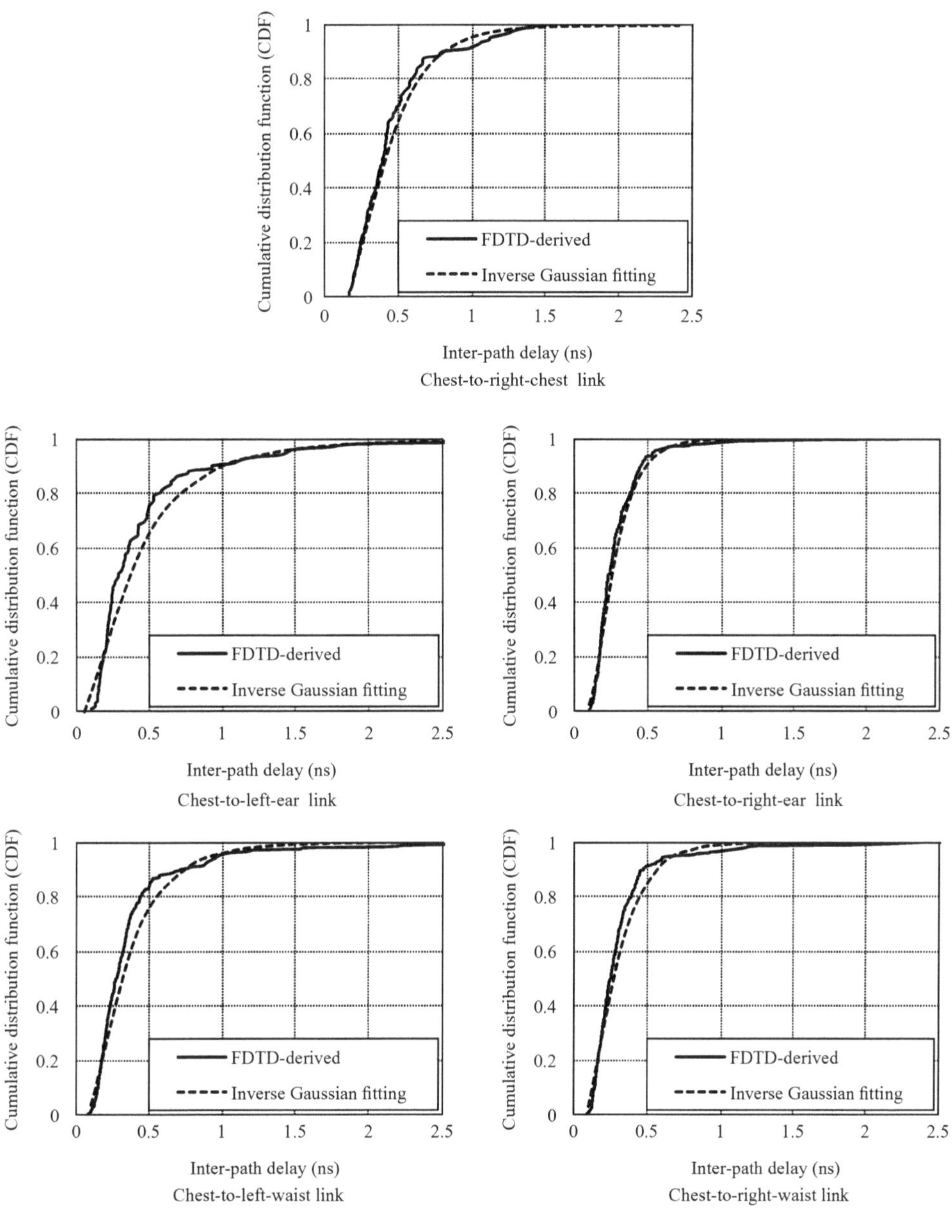

Figure 4.30 Cumulative distribution functions of inter-path delay

4.3.2.5 Summary of the Derived Model Parameters

Based on the above characterizations and parameterizations, we have determined all statistical models and parameters required for the construction of the channel model. These parameters are summarized in Table 4.10.

Table 4.10 Model parameters for five representative transmission links (Wang *et al.*, 2009). Reproduced with permission from Wang Q., Tayamachi T., Kimura I. and Wang J., "An on-body channel model for UWB body area communications for various postures," *IEEE Transactions on Antennas and Propagation*, **57**, 4, 991–998, 2009. © 2009 IEEE

Parameters	Description	Characteristics	Right ear	Left ear	Right chest	Left waist	Right waist
γ (ns)	Time constant for power decay	Exponential law	0.38	0.26	0.21	0.30	0.47
σ (dB)	Standard deviation of power distribution	Log-normal	7.5	12.56	15.6	8.46	7.87
τ_0 (ns)	Average arrival time of first path	Constant	1.05	0.92	0.68	1.89	2.01
$\tau_k - \tau_{k-1}$ (ns)	Distribution of inter-path delay	Inverse Gaussian	$\mu_\tau = 0.30$ $\lambda_\tau = 1.08$	$\mu_\tau = 0.56$ $\lambda_\tau = 0.45$	$\mu_\tau = 0.37$ $\lambda_\tau = 1.43$	$\mu_\tau = 0.38$ $\lambda_\tau = 0.75$	$\mu_\tau = 0.33$ $\lambda_\tau = 0.85$
Ω_0 (dB)	Mean power gain of first path	$\Omega_0 = \mu_\tau G/\gamma$	−62.7	−55.5	−48.3	−69.5	−72.9

In Table 4.10, according to the Saleh–Valenzuela model, the mean power gain Ω_0 of the first path is related to G (the reciprocal of the average path loss in a communication link) as $\Omega_0 = \mu_\tau G/\gamma$, where μ_τ is the mean time interval between two multipaths.

4.3.2.6 Measurement Validation

In order to verify the computational results and the statistically constructed model result, measurement is carried out in a full anechoic chamber. The chest-to-right-waist transmission link is taken as the measurement object. Two small-size and low-profile UWB antennas are mounted on the body surface. The transmitting antenna is fixed on the left side of the chest, and the receiving antenna is fixed on the right side of the waist. The measurement is conducted for 8 people, and with 10 body postures for each person, including standing, walking and sitting.

The measurement method is as follows:

1. The S_{21} parameter for the two antennas on the body is measured by using a network analyzer, where S_{21} is the frequency-domain transfer function.
2. The measured frequency-domain transfer function is converted to the time domain by using an inverse Fourier transform.
3. Based on Equations 4.40 and 4.41, the impulse responses and the APDP are calculated. The APDP is derived from the average over 80 readings (8 people, 10 body postures each).
4. All modeling parameters are extracted from the impulse response and the APDP.

By using the approach described in the previous subsections, a characterization and parameterization of the measurement results for the chest-to-right-waist transmission link is conducted. Table 4.11 presents a comparison of the parameters obtained from the FDTD modeling and the measurements. It is clear that good

Table 4.11 Comparison of parameter values in the FDTD model and measurements (Wang *et al.*, 2009). Reproduced with permission from Wang Q., Tayamachi T., Kimura I. and Wang J., "An on-body channel model for UWB body area communications for various postures," *IEEE Transactions on Antennas and Propagation*, **57**, 4, 991–998, 2009. © 2009 IEEE

Parameter	Model	Measurements
γ (ns)	0.47	0.41
σ (dB)	7.87	8.87
τ_0 (ns)	2.01	—
$\tau_k - \tau_{k-1}$ (ns)	$\mu_\tau = 0.33$	$\mu_\tau = 0.30$
	$\lambda_\tau = 0.85$	$\lambda_\tau = 2.14$

agreement has been achieved, even though it appears that the λ_τ values of the inverse Gaussian distribution of the inter-path delay are somewhat different. In fact, λ_τ refers to the shape of the PDF of the inverse Gaussian distribution, and in spite of the twofold difference between the values, there is only a small difference between the shapes of the respective PDFs. In short, the model parameters are in good agreement with the parameters obtained from the measurement, which validates this modeling method.

4.3.2.7 Channel Model Implementation

Based on the modified Saleh–Valenzuela model and the channel characterization, a discrete time impulse response function applied to these five transmission links is written as

$$h(t) = \sum_{k=0}^{K} \alpha_k \delta(t - \tau_k) \tag{4.47}$$

where α_k is the multipath power gain, and τ_k is the delay of the kth multipath component relative to the arrival time of the first path.

First, the time delays of the multipaths are induced as follows: the first path is generated at a fixed arrival time, and then a temporal delay between two successive paths is generated according to the inverse Gaussian distribution and added to the arrival time of the previous path.

Next, the gain coefficient for each path is defined as follows

$$\alpha_k = p_k \beta_k \tag{4.48}$$

where p_k takes a value of $+1$ or -1 with equal probability, and

$$10 \log_{10}\left(\beta_k^2\right) \propto \text{Normal}\left(\mu_k,\ \sigma^2\right) \tag{4.49}$$

since α_k belongs to a log-normal distribution. From Equation 4.42, we have

$$E\left[\beta_k^2\right] = \Omega_0 e^{-\frac{\tau_k - \tau_0}{\gamma}} \tag{4.50}$$

and μ_k, the mean in Equation 4.49, is written as

$$\mu_k = \frac{10 \ln(\Omega_0) - 10\tau_k/\gamma}{\ln(10)} - \frac{\sigma^2 \ln(10)}{20}. \tag{4.51}$$

All parameters required for the modeling of this simplified impulse response function have already been described in the previous section, and are summarized in Table 4.10.

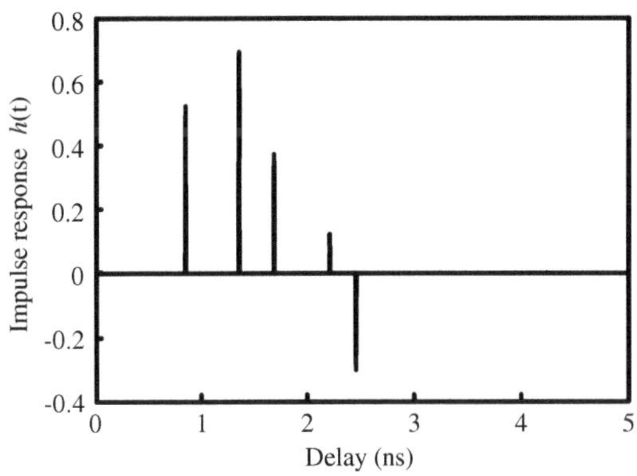

Figure 4.31 Sample impulse response

The propagation models for all five transmission links can be implemented in MATLAB®, and Figure 4.31 shows an implemented impulse response sample for the chest-to-right-waist link with $\Omega_0 = 1$ for the sake of simplicity. A MATLAB® code can be found at www.wiley.com/go/wang/bodyarea.

We can compare the FDTD-calculated channel and the modeled channel using the root mean square (RMS) delay spread σ_τ and the mean excess delay τ_m, which measure the effective duration of the channel impulse response. They are two kinds of representations of the impulse response profile and are frequently used to verify channel models. The two metrics are defined as follows (Hashemi, 1993):

$$\sigma_\tau = \sqrt{\frac{1}{P_R}\int_0^\infty (\tau - \tau_m)^2 p(\tau) d\tau} \tag{4.52}$$

$$\tau_m = \frac{1}{P_R}\int_0^\infty \tau^2 p(\tau) d\tau \tag{4.53}$$

where P_R is the multipath mean power. The above expressions show that σ_τ is defined as the square root of the second central moment of the PDP and τ_m is defined as the first moment of the PDP. σ_τ is a good measure of multipath spread and it gives an indication of the potential inter-symbol interference.

Figures 4.32 and 4.33 show the CDFs of the RMS delay spread and the mean excess delay for two transmission links. As can be seen from the figures, the model matches closely the FDTD-calculated results, and therefore adequately characterizes the transmission links.

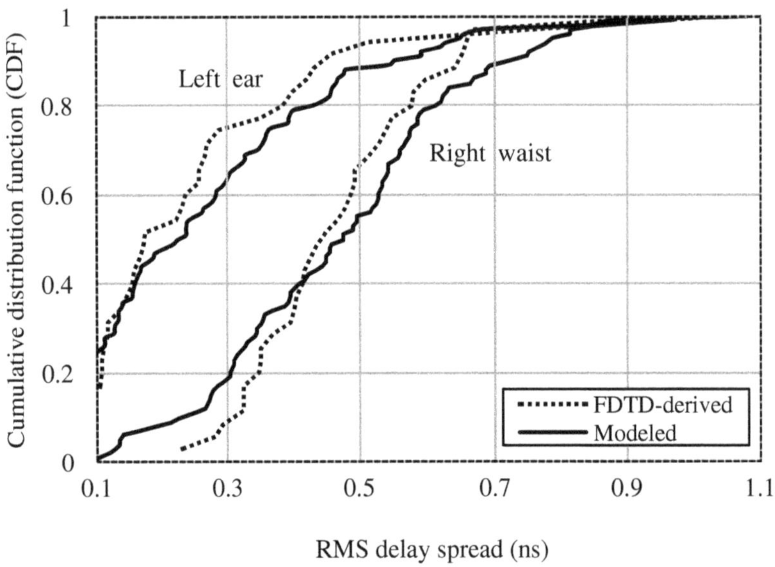

Figure 4.32 Comparison of the FDTD-derived and modeled RMS delay spread distribution (Wang *et al.*, 2009). Reproduced with permission from Wang Q., Tayamachi T., Kimura I. and Wang J., "An on-body channel model for UWB body area communications for various postures," *IEEE Transactions on Antennas and Propagation*, **57**, 4, 991–998, 2009. © 2009 IEEE

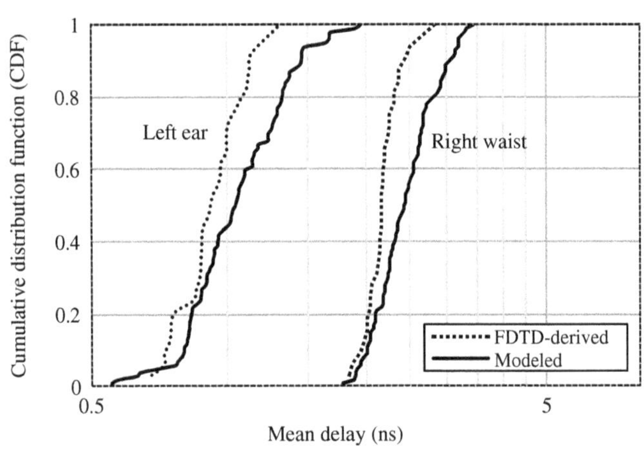

Figure 4.33 Comparison of the FDTD-derived and modeled mean delay distribution

4.3.3 In-Body UWB Channel Model

4.3.3.1 In-Body to On-Body Impulse Response

The same methodology as in on-body UWB channel modeling can also be used for in-body to on-body UWB channel modeling. Based on the simulation set-up for capsule endoscope shown in Figure 4.6, we can get 99 impulse responses at each receiving location for the in-body to on-body channel in total. During the process of inverse Fourier transform to get $h(t)$, the Hamming window with a coefficient of 2 is usually applied in the frequency domain in order to limit the transmitted pulse signal to effective frequency components. Since the UWB low band has a bandwidth of 1.4 GHz, the corresponding time resolution of the impulse response $h(t)$ or the PDP $p(t)$ will be 1.43 ns. In other words, multipath components within such a time width cannot be resolved even if more than one multipath arrives and thus have to be simply added up when deriving the impulse responses.

Due to this limited time resolution, we have to divide the time axis into many bins where each bin has a width of 1.43 ns. So in each impulse response or PDP, the first multipath is identified from the first peak of bins, and the second and third multipath components are identified from the successive bins, respectively. Figure 4.34 shows a typical FDTD-derived PDP for the in-body to on-body multipath channel. In this case, the receiving antenna is located in front of the human abdomen as Rx1, while the transmitting antenna is at one of the 33 locations. The first multipath component can be considered as a direct path between the transmitter and receiver. As a consequence, the successive multipath components can be assumed to correspond to the paths diffracted by or scattering from various tissues or organs of the human body. Moreover, the arrival time of the first multipath component can also be estimated by calculating the division of in-body to on-body local distance and the propagation speed in the human body (about a quarter of light speed in free space). The result of

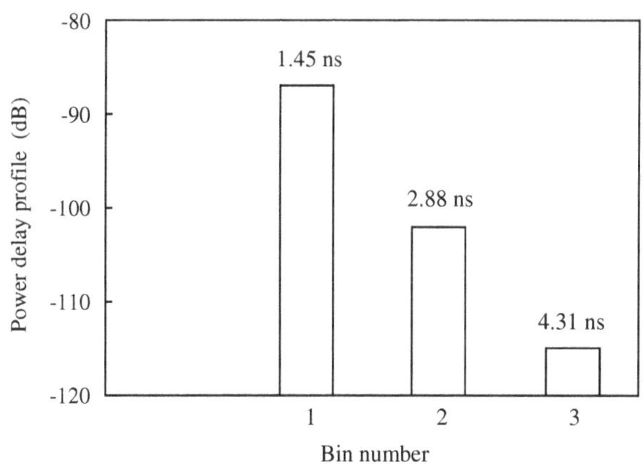

Figure 4.34 Example of FDTD-derived power delay profile

about 1.45 ns provides a good agreement with the first multipath component observed in Figure 4.34.

In addition, also from other results of the derived impulse responses and PDPs for different transmitting and receiving pairs, the direct path always turned out to be the strongest path in comparison with the successive two multipath components. Since the power of the third multipath component is more than 25 dB lower than the first one, which is weak enough to be neglected, we take the first two multipath components with a fixed bin width into account as the dominating multipath components. Therefore, we get an approximated discrete time impulse response channel model with only two multipath components, the first multipath component corresponds to the direct path, and the second one corresponds to a dominating diffracted or scattered path.

Figure 4.35 shows an example of an approximated two-path impulse response model at the receiver location Rx1. The amplitude of this two-path impulse response model is calculated from the root of the integrated power within one bin in the corresponding PDP, and the time interval between the two paths is assumed to be the same as one bin width. It can be concluded that a two-path impulse response model is sufficient to produce an appropriate approximation to the received UWB low band signals. It can also be concluded that the in-body to on-body channel characteristics mainly depend on the large attenuation with less influence by the multipath.

Based on such a well approximated two-path model, the multipath power gain distributions are investigated to provide the parameters in statistical terms. Here the multipath powers for the two paths are fitted to some candidate statistical distributions. As shown in Figure 4.36, the normal distribution provides a superior fit to both of the two-path magnitudes in decibels based on the second-order AIC. That is to say, the log-normal distribution fits the multipath powers well. The average standard deviations of the power variation are 16.8 dB. At the same time, a difference of less than 15 dB between the two multipath components exists when the CDF is 0.5.

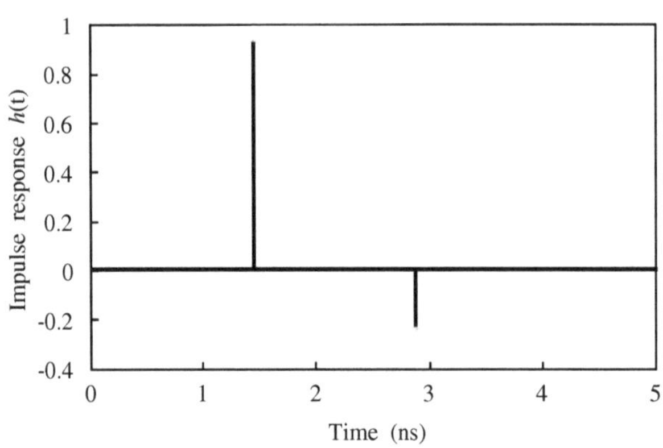

Figure 4.35 Example of approximated two-path impulse response model

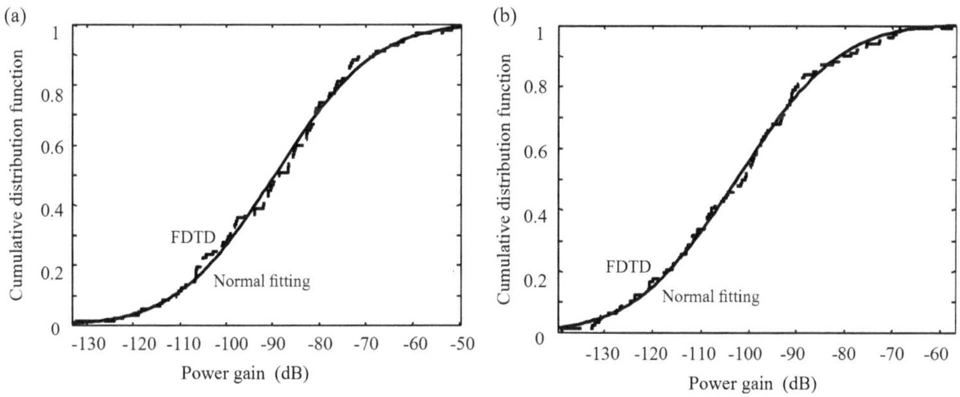

Figure 4.36 Cumulative distribution function of power gain for (a) the first and (b) the second paths

The arrival time of the first multipath component from the in-body to on-body transmission is significantly determined by direct path transmission, but varies at different transmitter locations as well as with different polarizations. From the FDTD-simulated results, 99 different arrival times of direct path are obtained at each receiver location. It is found that the inverse Gaussian distribution fits well the derived arrival times of direct path also based on the second-order AIC. The mean μ and the standard deviation σ of the inverse Gaussian distribution is 1.3 and 0.54 ns, respectively. Moreover, since the inter-path delay is assumed as a fixed bin width, the second path has the same statistical distribution but a longer mean arrival time of 2.73 ns. Based on the above characterization, the parameters for the two-path impulse response model are summarized in Table 4.12 for the UWB in-body to on-body wireless link.

Table 4.12 Main parameters for in-body to on-body UWB impulse response channel model

Description	Characteristics	Parameter
1st path		
Power	Log-normal	$\mu_1 = -89.2$ dB $\sigma_1 = 16.8$ dB
Arrival time	Inverse Gaussian	$\mu_1 = 1.3$ ns $\sigma_1 = 7.5$ ns
2nd path		
Power	Log-normal	$\mu_2 = -102.3$ dB $\sigma_2 = 16.8$ dB
Arrival time	1st path + 1.43 ns	

4.3.3.2 In-Body to Off-Body Impulse Response

With the same approach, for the simulation set-up in Figure 4.9, we can get 180 impulse responses and PDPs for the in-body to off-body channel. Again, due to the limited time resolution, we have to divide the time axis into many bins where each bin has a width of 1.43 ns in each PDP. Figure 4.37 shows a typical FDTD-derived PDP. The first multipath component can be considered as a direct path between the transmitter and receiver, and the others can be assumed to correspond to the paths diffracted by or scattering from various tissues or organs of the human body. The PDP characterizes the mean power of different multipaths. It is found to decay exponentially with the arrival time. With the exponential fitting to the result in Figure 4.37, it is found that the mean power gain is $\Omega_0 = -60.4$ dB, the arrival time of the first path is $\tau_0 = 0.9$ ns and the multipath decay time constant is $\gamma = 0.23$ ns.

In the PDPs, the multipath power is characterized by the bins. The bins that have power which is not 25 dB lower than the maximum value are taken into account in order to extract the channel parameters, while the other components are small enough to be ignored. Then the power gain distribution for each path is investigated, and some possible candidates are considered for the power gain model. Similarly, the second-order AIC is used to rank the fitting results from best to worst. As a result, the log-normal distribution provides a superior fit to the power gain distribution again. Figure 4.38 shows the fitting result for the first and the second paths. Since the amplitudes are expressed in decibels, the normal distribution is applied. The standard deviations of the amplitude variation are $\sigma_1 = 2.32$ dB and $\sigma_2 = 2.04$ dB for the first and second paths, respectively. The small variation is just because of the slight position variation of the transmitter inside the chest.

The arrival time of the first path, determined mainly by the direct transmission distance, varies with the location of the implant transmitter. The inverse Gaussian

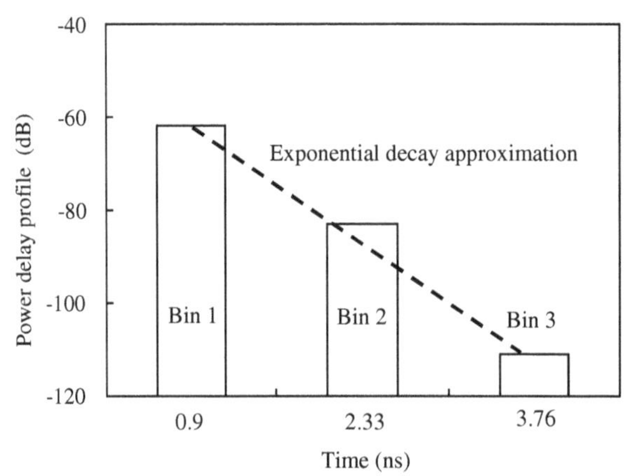

Figure 4.37 Example of FDTD-derived power delay profile

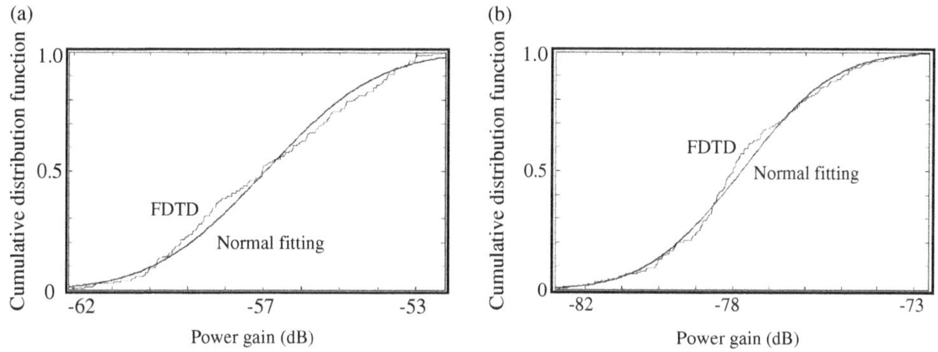

Figure 4.38 Cumulative distribution function of power gain for (a) the first and (b) the second paths (Wang, Masami, and Wang, 2011). Reproduced with permission from Wang J., Masami K. and Wang Q., "Transmission performance of an in-body to off-body UWB communication link," *IEICE Transactions on Communications*, **E94-B**, 1, 150–157, 2011

distribution is fitted to the arrival time of the first path with the mean value $\mu = 0.91$ ns and shape parameter $\lambda = 67.2$ ns, also in accordance with the AIC. The fitting result is shown in Figure 4.39.

Based on the above characterization and parameterization, the two-path impulse response model and corresponding parameters, which are required for implementing the channel model, are summarized in Table 4.13.

By using the established channel model, the discrete time impulse response function can be obtained. First, the time delays of the two paths are generated. For the

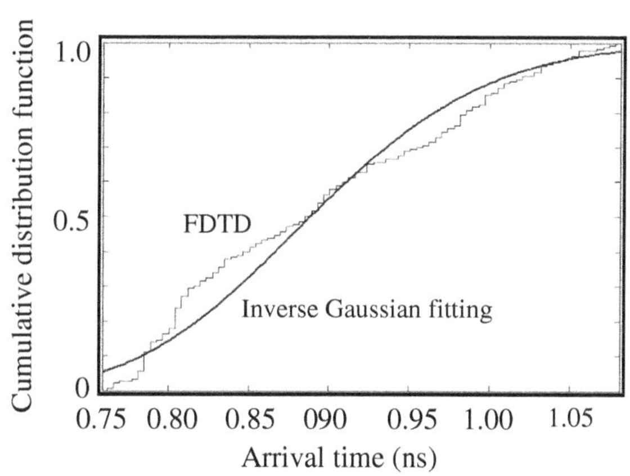

Figure 4.39 Cumulative distribution function of arrival time of the first path (Wang, Masami, and Wang, 2011). Reproduced with permission from Wang J., Masami K. and Wang Q., "Transmission performance of an in-body to off-body UWB communication link," *IEICE Transactions on Communications*, **E94-B**, 1, 150–157, 2011

Table 4.13 Main parameters for in-body to off-body UWB channel model

Description		Characterization	Parameter
Power decay		Exponential law	Ω_0=−60.4 dB γ=0.24 ns
1st path	Power	Log-normal	$\sigma_1 = 2.32$ dB
	Arrival time	Inverse Gaussian	$\mu_1 = 0.91$ ns, $\lambda_1 = 67.2$ ns
2nd path	Power	Log-normal	$\sigma_2 = 2.04$ dB
	Arrive time	1st path + 1.43 ns	

first path, the arrival time is generated according to the inverse Gaussian distribution with the mean $\mu = 0.91$ ns and the shape parameter $\lambda = 67.2$ ns. Afterwards the arrival time of the second path is generated to be the arrival time of the first path plus 1.43 ns. Next, the gain coefficient for each path is defined according to the log-normal distribution. As an alternative method in the in-body to on-body case, the mean is determined from the exponential power decay formula in Equation 4.42, and the standard deviations are taken from Table 4.13.

To examine the validity of the generated impulse response, the mean delay τ_m and the RMS delay spread σ_τ are calculated and compared with the FDTD-calculated result. Figure 4.40 shows the CDFs of the mean delay and delay spread. As can be seen, the model matches closely the FDTD-calculated result, and therefore adequately characterizes the in-body transmission.

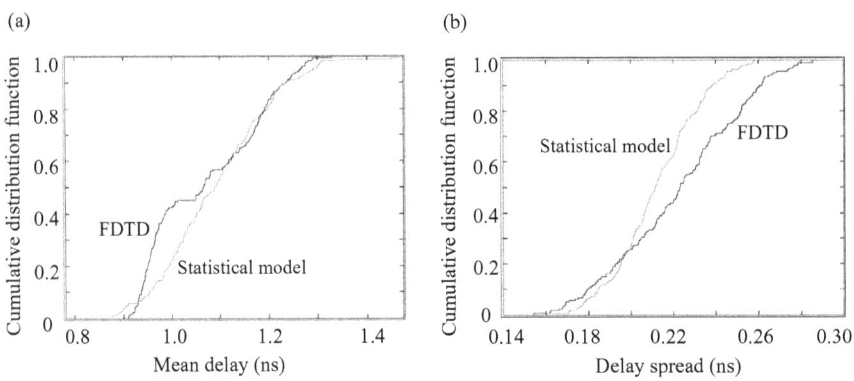

Figure 4.40 Comparison of (a) mean delay and (b) delay spread (Wang, Masami, and Wang, 2011). Reproduced with permission from Wang J., Masami K. and Wang Q., "Transmission performance of an in-body to off-body UWB communication link," *IEICE Transactions on Communications*, **E94-B**, 1, 150–157, 2011

References

Akaike, H. (1973) Information theory as an extension of the maximum likelihood principle. Proceedings of the 2nd International Information Theory Symposium, pp. 267–281.

Aoyagi, T., Takizawa, K., Kobayashi, T. *et al.* (2010) Development of an implantable WBAN path-loss model for capsule endoscopy. *IEICE Transactions on Communications*, **E93-B** (4), 846–849.

Barlow, H.M. and Brown, J. (1962) *Radio Surface Waves*, Oxford University Press, Oxford, pp. 6–28.

Benedetto, D.M.-G. and Giancola, G. (2004) *Understanding Ultra Wide Band Radio Fundamentals*, Prentice Hall, New Jersey.

Burnham, K.P. and Anderson, D.R. (2002) *Model Selection and Multimodel Inference: A Practical Information-Theoretic Approach*, 2nd edn, Springer-Verlag, New York.

Chavez-Santiago, R., Khaleghi, A., Balasingham, I., and Ramstad, T.A. (2009) Architecture of an ultra wideband wireless body area network for medical applications. Proceedings of the 2nd International Symposium on Applied Sciences in Biomedical and Communication Technologies, Bratislava, Slovakia.

Cho, N., Yoo, J., Song, S.-J. *et al.* (2007) The human body characteristics as a signal transmission medium for intrabody communication. *IEEE Transactions on Microwave Theory and Techniques*, **55** (5), 1080–1086.

Conroy, J.T., LoCicero, J.L., and Ucci, D.R. (1999) Communication techniques using monopulse waveforms. Proceedings of IEEE Military Communications Conference, vol. 2, pp. 1181–1185.

Fort, A., Desset, C., De Doncker, P. *et al.* (2006) An ultra-wideband body area propagation channel model - From statistics to implementation. *IEEE Transactions on Microwave Theory and Techniques*, **54** (4), 1820–1826.

Gabriel, C. (1996) Compilation of the dielectric properties of body tissues at RF and microwave frequencies. Brooks Air Force Technical Report AL/OE-TR-1996-0037.

Ghavami, M., Michael, L.B., Haruyama, S., and Kohno, R. (2002) A novel UWB pulse shape modulation system. *Wireless Personal Communications*, **23** (1), 105–120.

Hashemi, H. (1993) The indoor radio propagation channel. *Proceedings of the IEEE*, **81** (7), 943–968.

Hämäläinen, M., Hovinen, V., Iinatti, J., and Latva-aho, M. (2001) In-band interference power caused by different kinds of UWB signals at UMTS/WCDMA frequency bands. Proceedings of the IEEE Radio and Wireless Conference, pp. 97–100.

IEEE P802.15 Working Group for Wireless Personal Area Networks (2009) Channel model for body area network (BAN), IEEE P802.15-08-0780-09-0006.

Khaleghi, A. and Balasingham, I. (2009) Improving in-body ultra wideband communication using near-field coupling of the implanted antenna. *Microwave and Optical Technology Letters*, **51** (3), 585–589.

Molisch, A.F., Cassiolo, D., Chong, C.-C. *et al.* (2006) A comprehensive standardized model for ultra wideband propagation channels. *IEEE Transactions on Antennas and Propagation*, **54** (11), 3151–3166.

Qureshi, W.A. (2004) Current and future applications of the capsule camera. *Nature Reviews Drug Discover*, **3**, 447–450.

Rice, L.P. (1959) Radio transmission into buildings at 35 and 150 MHz. *Bell System Technical Journal*, **38** (1), 197–210.

Saleh, A.A.M. and Valenzuela, R.A. (1987) A statistical model for indoor multipath propagation. *IEEE Journal on Selected Areas in Communications*, **5** (2), 128–137.

Shi, J. and Wang, J. (2010) Channel characterization and diversity feasibility for in-body to on-body communication using low-band UWB signals. Proceedings of the 3rd International Symposium on Applied Science in Biomedical and Communication Techniques, Rome, Italy.

Tang, P.K., Chew, Y.H., Ong, L.C. *et al.* (2006) Small-scale transmission statistics of UWB signals for body area communications. Proceedings of the 64th IEEE Vehicular Technology Conference, pp. 1–5.

Taparugssanagorn, A., Rabbachin, A., Hamalainen, M. *et al.* (2008) A review of channel modeling for wireless body area network in wireless medical communications. Proceedings of the 11th International Symposium on Wireless Personal Multimedia Communications.

Wang, J., Nishikawa, Y., and Shibata, T. (2009) Analysis of on-body transmission mechanism and characteristic based on an electromagnetic field approach. *IEEE Transactions on Microwave Theory and Techniques*, **57** (10), 2464–2470.

Wang, J., Masami, K., and Wang, Q. (2011) Transmission performance of an in-body to off-body UWB communication link. *IEICE Transactions on Communications*, **E94-B** (1), 150–157.

Wang, Q., Tayamachi, T., Kimura, I., and Wang, J. (2009) An on-body channel model for UWB body area communications for various postures. *IEEE Transactions on Antennas and Propagation*, **57** (4), 991–998.

Win, M.Z. and Scholtz, R.A. (2000) Ultra-wide bandwidth time-hopping spread-spectrum impulse radio for wireless multiple-access communications. *IEEE Transactions on Communications*, **48** (4), 679–691.

Zasowski, T., Meyer, G., Althaus, F., and Wittneben, A. (2006) UWB signal propagation at the human head. *IEEE Transactions on Microwave Theory and Techniques*, **54** (4), 1836–1845.

Zhao, Y., Hao, Y., Alomainy, A., and Parini, C. (2006) UWB on-body channel modeling using Ray theory and subband FDTD method. *IEEE Transactions on Microwave Theory and Techniques*, **54** (4), 1827–1835.

5

Modulation/Demodulation

5.1 Introduction

In order to transmit a digital signal over a body area communication band, the digital signal is often impressed onto a carrier signal. The corresponding process is called modulation in which the digital information signal varies one or more properties of the carrier signal, including amplitude, phase and its frequency. The inverse process of modulation is called demodulation where the original digital information is extracted from the modulated carrier signal. In this chapter, we will describe summarily the feasible modulation and demodulation schemes in body area communications. For various available frequency bands described in Chapter 1, different schemes will be covered. Moreover, based on the body area communication channel characteristics, improvements and simplifications on the receiver structure will also be introduced.

In digital modulation and demodulation, the carrier signal is typically a high frequency sinusoid waveform, while a pulse train may also be utilized in wideband communications. The digital signal may be either binary with two levels or M-ary with multiple levels where each level will represent a discrete pattern of information bits. The fundamental digital modulation methods include amplitude-shift keying (ASK), frequency-shift keying (FSK), and phase-shift keying (PSK). There are also various variations based on these three fundamental modulation schemes. Note that these modulation schemes are essentially narrow-band schemes in which the bandwidth of the modulated signal depends on the bandwidth of the baseband signal. These fundamental narrow-band schemes are well qualified for MICS band, ISM band or HBC band communication. In addition to the fundamental modulation methods, wideband schemes required in certain body area communication scenarios using UWB band will also be covered in this chapter. In fact, two typical UWB

Body Area Communications: Channel Modeling, Communication Systems, and EMC, First Edition.
Jianqing Wang and Qiong Wang.

schemes, impulse radio UWB (IR-UWB) and multi-band orthogonal frequency-division multiplexing (MB-OFDM), will be presented in this chapter. For UWB communication, the commonly used receivers are based on the pulse energy detection or correlation demodulation with the template waveform. This still applies in body area UWB communications. However, as described in Chapters 2 and 4, the complicated transmission mechanisms in the body area will give rise to a multipath affected propagation channel so that a received signal may end up being the superimposition of several attenuated, delayed, time-varying and eventually distorted replicas of a transmitted signal. Given the body area multipath channel model characteristics derived in Chapter 4, RAKE reception comes in handy to counter the effects of multipath propagation, which could very well result in better communication performance in a multipath environment than in a "clean" environment. In addition, diversity reception can be applied at the receiver to improve the reception performance via multiple receivers/antennas as well as different polarizations.

In this chapter, the modulation schemes for digital shift keying as well as IR-UWB and MB-OFDM will be described first, followed by the corresponding demodulation schemes and error probability analysis. RAKE reception will be separately presented and the error probability with RAKE receiver in body area multipath channel will be given in detail. Finally, we will present the diversity reception application in the in-body communications where multiple receivers with different optimizations will be applied and the diversity effect will be presented in Chapter 6.

5.2 Modulation Schemes

5.2.1 ASK, FSK and PSK

In digital modulation, a continuous carrier signal is modulated by a discrete digital signal. There are three fundamental modulation schemes in digital modulation: ASK, FSK and PSK. They make use of the amplitude, frequency and phase of the carrier signal, respectively, to transmit the digital baseband signal.

In ASK, the amplitude of the carrier signal varies in accordance with the digital modulating signal. The time-domain ASK signal can be expressed in the form:

$$S_{ASK}(t) = \left[\sum\nolimits_n a_n g(t - nT_s)\right] \cdot A \cos \omega_c t \tag{5.1}$$

where $g(t)$ is the baseband signal waveform, T_s is the signal interval, ω_c is the carrier angular frequency, a_n is the amplitude of the digits and A is the amplitude of the carrier signal. The simplest form is the binary ASK, in which the carrier signal acts as an On or Off switch via the binary logics "1" and "0." In the modulated signal, logic "1" is represented by the presence of a carrier and logic "0" is represented by

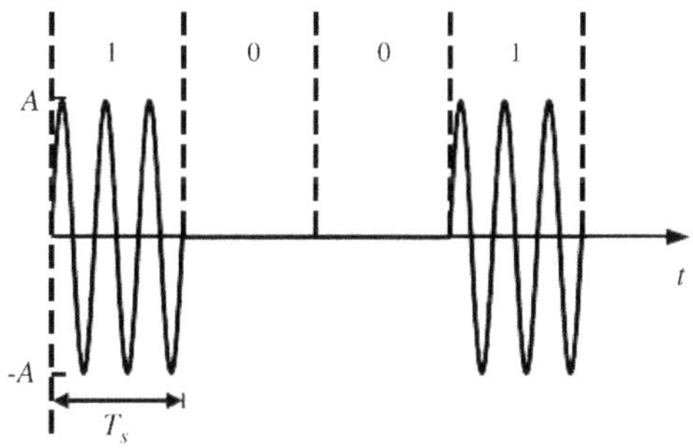

Figure 5.1 Typical waveform of binary ASK or OOK

the absence of a carrier, thus giving on/off keying (OOK) operation. The OOK signal is represented by

$$S_{OOK}(t) = a_n \cdot A \cos \omega_c t \tag{5.2}$$

$$\text{with } a_n = \begin{cases} 1, & \text{with probability } P \\ 0, & \text{with probability } 1 - P. \end{cases}$$

A typical OOK waveform is shown in Figure 5.1.

For the M-ary ASK (MASK) signal, the carrier amplitude has M possible values rather than two digits in OOK, which means $a_n \in \{A_i\}$, where $i = 0, 1, 2, \cdots\cdots, M - 1$. Each amplitude a_n is transmitted in one symbol interval T_s. The modulation of MASK is similar to the binary ASK. Generally, the binary series is divided into sets of n bits, $n = \log_2 M$ and then transformed to M baseband signal waveforms. The M baseband signal waveforms act as the input modulating signals. The amplitude of the modulated signal is proportional to the amplitude of the baseband signal waveform.

In FSK, the frequency of the carrier signal varies in accordance with the digital modulating signal. Similarly, the time-domain binary FSK signal can be expressed in the form

$$S_{FSK}(t) = \left[\sum\nolimits_n a_n g(t - nT_s)\right] \cdot A \cos \omega_1 t + \left[\sum\nolimits_n \bar{a}_n g(t - nT_s)\right] \cdot A \cos \omega_2 t \tag{5.3}$$

where the carrier frequencies ω_1 and ω_2 change with the modulating signal a_n and its anti-code $\bar{a}_n$. If the binary "1" corresponds to carrier angular frequency ω_1 and "0" corresponds to carrier angular frequency ω_2, with the simplest case that $g(t)$ is the single rectangular pulse, a typical BFSK waveform is then as shown in Figure 5.2.

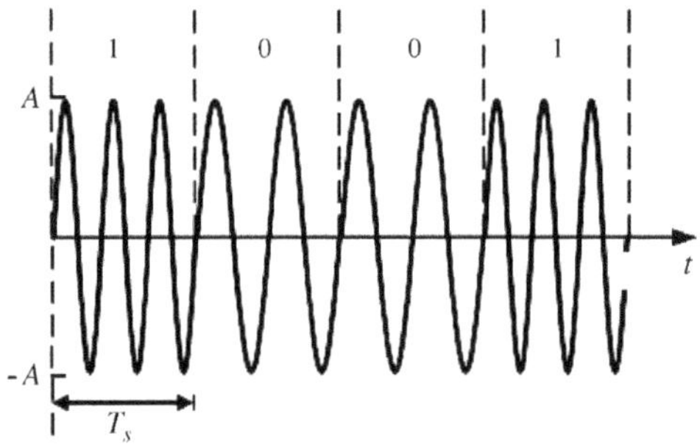

Figure 5.2 Typical waveform of FSK

In M-ary FSK (MFSK), the M symbols can be expressed as

$$S_i(t) = \sqrt{\frac{2E_s}{T_s}} \cos \omega_i t, \quad 0 \leq t \leq T_s, \; i = 0, 1, 2, \cdots\cdots, M-1 \tag{5.4}$$

where E_s is the signal energy in one symbol interval T_s and ω_i is the carrier angular frequency with M possible values. Generally, we have the difference of the carrier frequency $N/2T_s$ (where N is a positive integer), which means M signals are orthogonal to each other.

Moreover, a special type of BFSK with modulation index of 0.5 is known as the minimum shift keying (MSK). The modulation index of 0.5 corresponds to a frequency difference $|f_1 - f_2| = 1/2T_s$, which is the smallest possible difference if the signals of the two frequencies are to be orthogonal over one bit interval. The advantage of MSK is that there are no phase discontinuities in the modulated signal waveforms because the frequency variations occur at the carrier zero crossing points. Thus it is a continuous phase scheme. Prior to the MSK modulation, a Gaussian filter is often used to further reduce the bandwidth of a baseband pulse train. This is known as Gaussian filtered MSK or GMSK. GMSK gives a smoothed phase trajectory of MSK signals and thus limits the instantaneous frequency variations. This feature results in a narrower bandwidth than for MSK.

In PSK, the phase of the carrier signal varies in accordance with the digital modulating signal. In binary PSK (BPSK), the binary "0" and "1" are generally represented by phase 0° or 180°, respectively. The time-domain BPSK signal can be expressed in the form

$$S_{BPSK}(t) = \left[\sum\nolimits_n a_n g(t - nT_s)\right] \cdot A \cos \omega_c t \tag{5.5}$$

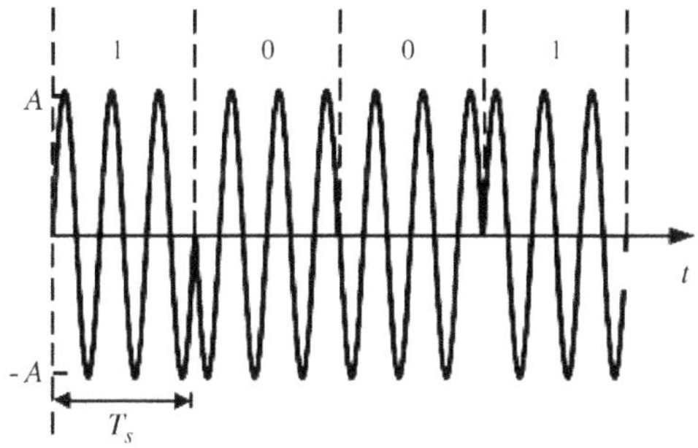

Figure 5.3 Typical waveform of BPSK

where a_n is different from that in binary ASK and binary FSK. Here

$$a_n = \begin{cases} +1, & \text{with probability } P \\ -1, & \text{with probability } 1 - P. \end{cases}$$

Therefore, in one signal interval T_s, given the $g(t)$ is a rectangular pulse with width T_s, we have

$$S_{BPSK}(t) = \pm A \cos \omega_c t = A \cos(\omega_c t + \varnothing_i), \ \ \varnothing_i = 0 \text{ or } \pi. \tag{5.6}$$

Figure 5.3 shows a typical waveform for BPSK.

For M-ary PSK (MPSK) signal, the carrier phase has M possible values and the M symbols can be expressed as

$$S_i(t) = \sqrt{\frac{2E_s}{T_s}} \cos(\omega_c t + \varphi_i), \quad 0 \le t \le T_s, \quad i = 0, 1, 2, \cdots\cdots, M - 1 \tag{5.7}$$

where E_s is the signal energy in one symbol interval T_s and φ_i is the carrier phase with M possible values.

5.2.2 *IR-UWB*

The most common and traditional UWB signal uses extremely short pulses with duration on the order of nanoseconds to transmit information, rather than continuous waveforms, with typically no RF modulation. This transmission technique has been extensively used and goes under the name of impulse radio (IR). IR-UWB is characterized by a low duty cycle of pulses so that the transmitted power can be very small. Moreover, carrier modulation is not required and therefore there is no up and down conversion. A RF power amplifier is also not needed. As a result, the IR-UWB

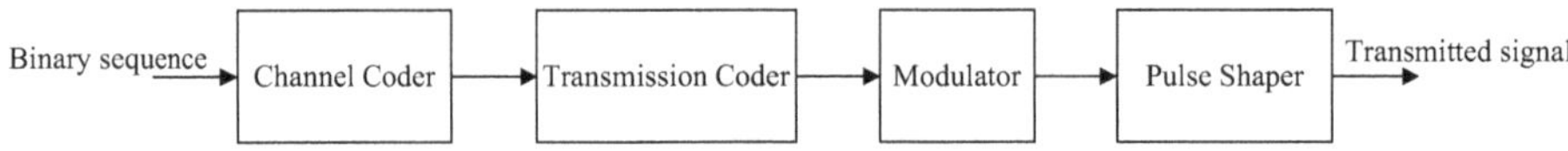

Figure 5.4 IR-UWB transmission flow chart

transceiver has a simple architecture with low cost. For IR-UWB, the pulses are the information carrier and the way in which the information data modulate the pulses may vary. OOK modulation, pulse position modulation (PPM) and pulse amplitude modulation (PAM) are the commonly adopted modulation schemes in IR-UWB. In addition to modulation and in order to shape the spectrum of the generated signal, the data bits are often encoded using pseudorandom or pseudo-noise (PN) codes. The period and duty cycle of the pulse will vary in a pseudorandom manner under the control of an encoded sequence. The encoded data symbols will introduce a time delay on generated pulses leading to the so-called time-hopping UWB (TH-UWB). Time hopping is one of the spread spectrum techniques. Combining it with PPM is known as the PPM-TH-UWB. Another common spread spectrum technique is the amplitude modulation of basic pulses by the encoded data symbols, that is, the so-called PAM direct-sequence UWB (PAM-DS-UWB). The generation of the common OOK-UWB, PPM-TH-UWB and PAM-DS-UWB signals will be discussed in the next section.

The system model for an IR-UWB transmitter is shown in Figure 5.4 (Benedetto and Giancola, 2004). The binary information data are first encoded by a channel coder, followed by a transmission coder, and then modulated by an OOK, PPM or PAM, and finally transmitted via the antenna as a pulse train. The channel coder adopts the repetition coder which is easy to implement although it is a relatively simple method of encoding data across a channel. The transmission coder provides a pseudorandom code to encode the position or the amplitude of the pulses.

Figure 5.5 shows the transmission scheme for IR-UWB with OOK modulation. When the information bit is “1” the UWB pulses are transmitted, while nothing is transmitted when the bit is “0”. This scheme is especially effective for low-power design but it suffers from low SNR within the bit “0” duration because no signal energy is transmitted within that duration. A useful method to cope with this

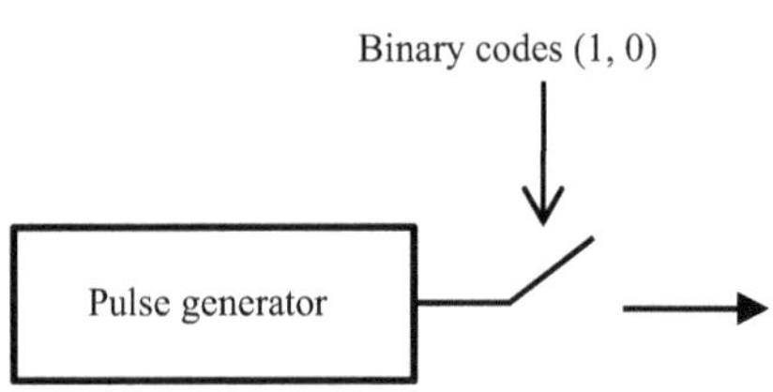

Figure 5.5 Transmitter for IR-UWB with OOK modulation

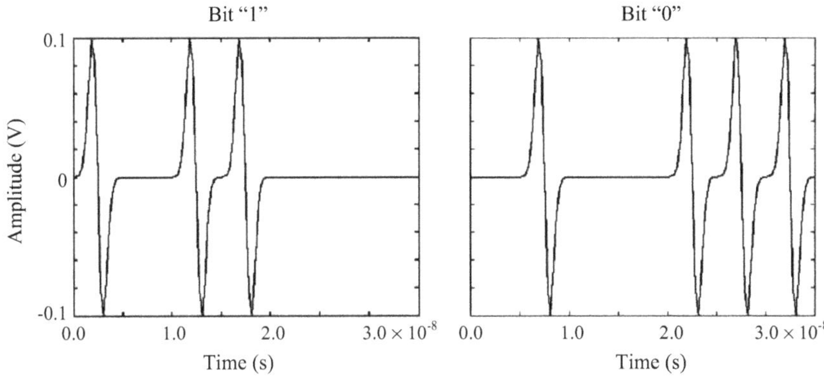

Figure 5.6 An example of an IR-UWB signal with encoded OOK modulation

problem is to encode the bits "1" and "0" with a code containing both "1" and "0." For example, if we encode the bits "1" and "0" into {1011000} and {0100111}, and then perform the OOK modulation according to the encoded codes, we can always have some pulses transmitted both within the bit "1" and bit "0" periods. Figure 5.6 shows such an encoded OOK modulated signal waveform.

Figure 5.7 shows the transmission scheme for a PPM-TH-UWB signal. The input binary sequence at a data rate of $1/T_b$ bits/s (bps) is first repeated N_s times and results in a binary sequence at a data rate of $1/T_s = N_s/T_b$ bps. The pseudorandom TH code, acting as the transmission code, applies an integer-valued code c to the binary sequence. The generic element of the TH code c_i is an integer value with uniform distribution on the interval $[0, N_h - 1]$ where N_h represents the cardinality of the TH code. The bit interval T_s is divided into N_h chips with each chip time $T_c = T_s/N_h$. The periodicity of the TH code is represented by N_p and the periodicity of the TH code commonly coincides with the length of the repetition code, that is, $N_p = N_s$. The coded real-valued sequence then enters the PPM modulator. The modulator generates a sequence of Dirac pulses at a rate of $1/T_s$ pulses/s. These pulses locate at times $iT_s + c_iT_c$ where iT_s is the nominal pulse position while c_iT_c is the time shift introduced by the TH code. After the PPM modulation, the pulses occur at times $iT_s + c_iT_c + a_i\varepsilon$ where ε is the time shift introduced by PPM modulation and a_i is the binary sequence following the code repetition coder. In general the time shift introduced by the PPM modulator is usually much smaller than the shift

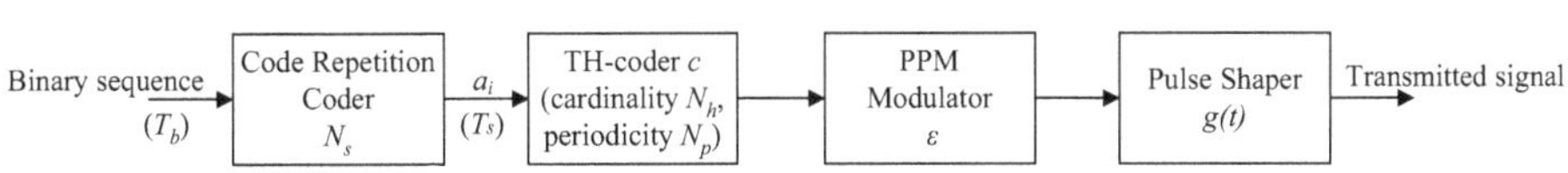

Figure 5.7 Transmission scheme for a PPM-TH-UWB signal

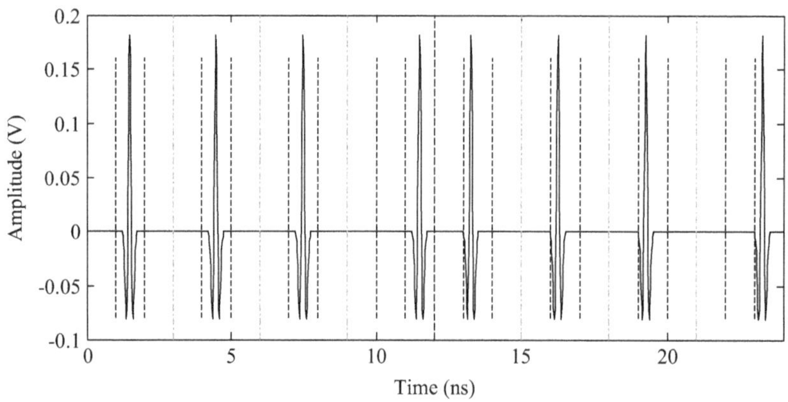

Figure 5.8 Example of a PPM-TH-UWB signal with binary bits [1, 0] and TH code [1 1 1 2]

introduced by the TH code, that is, $\varepsilon < T_c$. The last block is the pulse shaper with impulse response $g(t)$.

Based on the above system, the PPM-TH-UWB signal can be expressed as:

$$S(t) = \sum_{i=-\infty}^{+\infty} g(t - iT_s - c_i T_c - a_i \varepsilon). \tag{5.8}$$

In the PPM-TH-UWB signal, we can note that pulses occur at times $(iT_s + c_i T_c + a_i \varepsilon)$. Figure 5.8 shows an example of a PPM-TH-UWB signal with binary bits [1, 0] and TH code [1 1 1 2]. Note that $T_s = 3$ ns, $N_s = 4$, $T_c = 1$ ns, $N_p = 4$, $N_h = 3$, and $\varepsilon = 0.25$ ns.

As noted, the binary information is encoded in the pulse position, and this will induce an inherent disadvantage to PPM as it is sensitive to the multipath affected channel. This is because in multipath affected channels the multiple paths of each transmitted pulse will interfere with the accurate determination of the pulse position. That is, the signal energy associated with each symbol is spread out in time and therefore the transmitted symbol will occur in the next symbol period and this will result in the inter-symbol interference in the receiver.

Similarly to PPM-TH-UWB, the transmission scheme for a PAM-DS-UWB signal is shown in Figure 5.9. After the code repetition coding, the binary sequence is first converted into a positive- and negative-valued sequence. The transmission coder applies a DS code c_i to the antipodal sequence a_i. The DS code c_i has periodicity N_p and N_p is commonly assumed to be equal to N_s. Like the PPM modulator, the PAM modulator generates a sequence of Dirac pulses at a rate of $1/T_s$ pulses/s, while these pulses locate at times iT_s. Finally the output of the modulator enters the pulse shaper with impulse response $g(t)$.

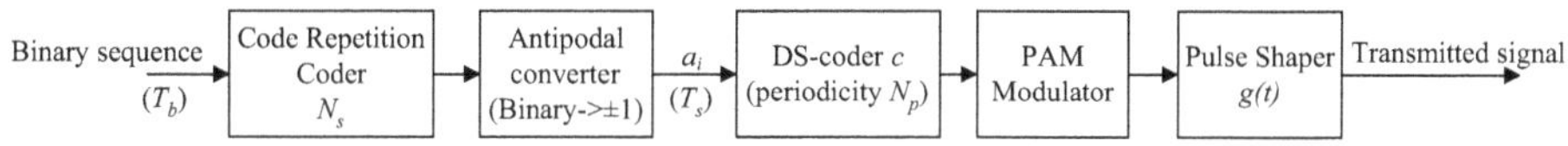

Figure 5.9 Transmission scheme for a PAM-DS-UWB signal

The PAM-DS-UWB signal can be expressed as

$$S(t) = \sum_{i=-\infty}^{+\infty} a_i c_i g(t - iT_s). \tag{5.9}$$

Note that the definitions of a_i and c_i are the same as in the above PPM-TH-UWB signal except that the a_i sequence here is a positive- and negative-valued sequence. In the PAM-DS-UWB signal, all pulses are located at time iT_s. Figure 5.10 shows an example of a PAM-DS-UWB signal with binary bits [1, 0] and DS code [−1 −1 1 1 1]. Note that $T_s = 2$ ns, $N_s = 5$, $T_c = 2$ ns, and $N_p = 5$.

5.2.3 MB-OFDM

Based on the UWB definition by the FCC (FCC, 2002), if the bandwidth is larger than 500 MHz, it is also a UWB signal. The multi-band (MB) solution divides the whole UWB band from 3.1 to 10.6 GHz into smaller frequency sub-bands of at least 500 MHz. In each sub-band, different modulation schemes can be adopted for data modulation. The modulation can be any form of modulation used with digital data, but the most common modulations are BPSK, quadrature PSK (QPSK), and quadrature amplitude modulation (QAM). The sub-band bandwidth can be occupied by an appropriate data rate. The most common approach is OFDM. The basic idea of OFDM is to use a large number of parallel closely spaced orthogonal subcarriers

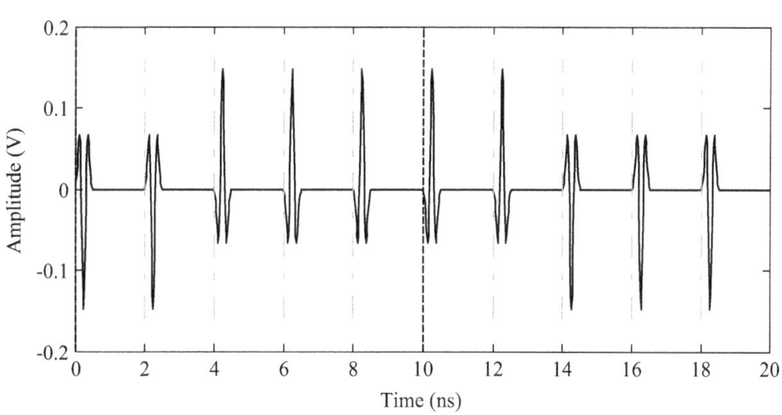

Figure 5.10 Example of a PAM-DS-UWB signal with binary bits [1, 0] and DS code [−1 −1 1 1 1]

instead of a single wide-band carrier to transmit information. Each subcarrier corresponds to a quadrature modulation of a signal. These subcarriers are equally spaced by Δf in the frequency domain. If T_0 is the time to transmit each symbol, then $\Delta f = 1/T_0$ can keep the orthogonality between different subcarriers. This is because the resulting sinc frequency response function of the signals is such that the first nulls occur at the subcarrier frequencies on the adjacent channels. Orthogonal subcarriers all have an integer number of cycles within the symbol period.

In an OFDM modulator, the input serial binary sequence is first divided into multiple data streams of K bits to generate blocks of N symbols. The N symbols have L possible values, and $K = N \log_2 L$ hence holds. Each symbol will then modulate a different orthogonal subcarrier. The outputs of all the modulators are linearly summed, and the result is the signal to be transmitted. It could be up-converted and amplified if needed. A guard interval T_G is introduced at the beginning of each symbol to eliminate the inter-symbol interference in a multipath propagation channel. The total OFDM symbol duration is thus $T = T_0 + T_G$. The guard interval is usually a copy of the end section of the OFDM symbol, the so-called cyclic prefix (CP). Appending the CP does not cause any discontinuities and the original frequency of the orthogonal subcarriers still hold. The CP acts as a buffer region where delayed information from the previous symbols can get stored. The receiver has to exclude samples from the CP which is corrupted by the previous symbol when choosing the samples for an OFDM symbol. The reason that the guard interval consists of a copy of the final section of the OFDM symbol is to maintain the carrier synchronization at the receiver. Of course, the flipside of adding the CP is the loss in data rate as we are conveying redundant information. Given that transmission of the CP reduces the data rate, the CP duration should be minimized as much as possible. Typically, the CP duration is determined by the expected duration of the multipath channel in the operating environment. For example, for the indoor wireless multipath channel, the typically expected multipath channel is of around 0.8 μs duration, hence the CP is chosen to be 0.8 μs in the IEEE 802.11a specification. However, for the on-body area UWB multipath channel, the delay spread is around 0.4 ns.

The complex envelope of the OFDM signal corresponding to a block of N symbols can be expressed as (Benedetto and Giancola, 2004):

$$s(t) = g_T(t) \sum_{m=0}^{N-1} c_m e^{j2\pi f_m t} \tag{5.10}$$

where $g_T(t)$ is the baseband rectangular waveform with finite duration T

$$g_T(t) = \begin{cases} \sqrt{1/T} & \text{for } -T_G = T_0 - T \le t \le T_0 \\ 0 & \text{elsewhere} \end{cases} \tag{5.11}$$

and $c_m = a_m + jb_m$ indicates the point in the constellation and f_m is the corresponding subcarrier frequency.

In order to cover a wide bandwidth, hundreds or even thousands of parallel subcarriers might be used. To implement that with hardware is a challenge even with modern semiconductor technology. In fact, the whole process in the OFDM modulator can be accomplished in computer hardware by using the inverse fast Fourier transform (IFFT). The IFFT converts an input signal from the time domain by mapping its baseband frequencies onto their representative phases and amplitudes in preparation for modulation to the passband. All the individual carriers with modulation are in digital form and then subjected to an IFFT mathematical process, creating a single composite signal that can be transmitted. The fast Fourier transform (FFT) at the receiver sorts all the signal components into the individual sine-wave elements of specific frequencies and amplitudes to recreate the original data stream.

The digital representation of the complex envelope is

$$s[n] = s(nt_c) = g_T(nt_c) \sum_{m=0}^{N-1} c_m e^{j\frac{2\pi f_m n T_0}{N}} \tag{5.12}$$

where $t_c = T_0/N$ is the sampling period. It represents the samples at multiple t_c. In fact, the summation item in Equation 5.12 corresponds to the nth element of vector $\boldsymbol{C}$ representing the IFFT of the vector $\{c_0, \ldots, c_n, \ldots, c_{N-1}\}$, which can be further expressed as:

$$s[n] = g_T(nt_c)(-1)^n c_n. \tag{5.13}$$

Figure 5.11 shows the block diagram of a MB-OFDM transmitter. A fast serial data stream $s[n]$ is first divided by means of a serial-to-parallel conversion into N parallel slower data streams. Each of the N parallel data streams is then mapped to a symbol stream through the digital constellation mapping scheme (QPSK, BPSK, QAM, etc.). And then each symbol set modulates different subcarriers via the IFFT block. The resulting complex time-domain signal is divided into real and imaginary analog components by means of digital-to-analog converters (DACs) and then used to modulate a main RF carrier.

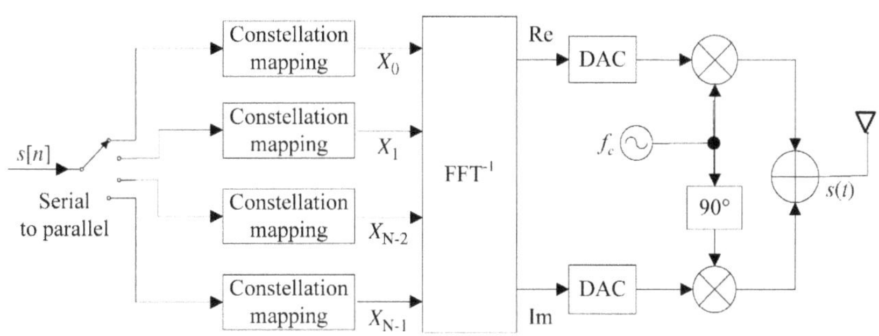

Figure 5.11 Block diagram of an OFDM transmitter

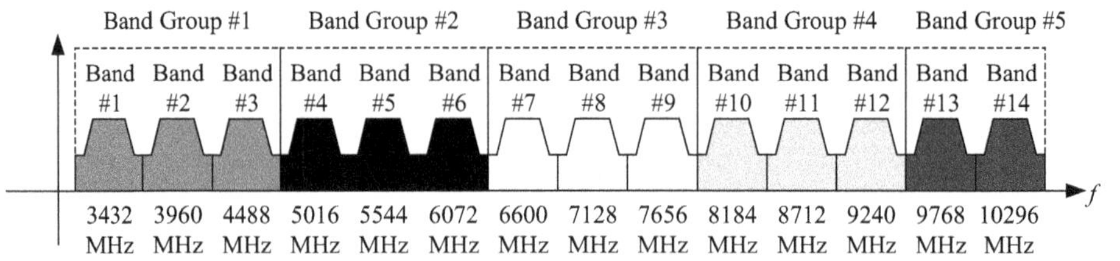

Figure 5.12 Diagram of the band-group allocation over the 3.1–10.6 GHz band

The IEEE 802.15.3a working group provides a higher speed UWB PHY standard ECMA-368 (ECMA, 2005) based on the MB-OFDM solution (IEEE P802.15, 2004). ECMA-368 is a widely adopted international specification for MB-OFDM UWB. The MB-OFDM UWB employs the OFDM technology and interleaves OFDM symbols to hop in frequency domain, the 7.5 GHz bandwidth is segmented into five band groups, each band group having 2–3 sub-bands which occupy 528 MHz. Figure 5.12 shows the officially issued band-group allocation over the 3.1–10.6 GHz band. Information bits are interleaved across all bands to better exploit frequency diversity over the whole frequency spectrum. Although a wide band of frequencies could be used, certain practical considerations limit the frequencies that are normally used for MB-OFDM UWB. Based on existing semiconductor technology, use of the spectrum from 3.1 to 4.8 GHz is considered to be optimal for initial deployments. Limiting the upper bound simplifies the design of the radio and analog front end circuitry as well as reducing interference with other services. Additionally the frequency band from 3.1 to 4.8 GHz, that is, band group 1, is sufficient for three sub-bands of 500 MHz when using MB-OFDM UWB. Band group 1 is divided into three sub-bands and frequency hopping is employed.

Based on the specifications of the ECMA-368 standard, the main parameters to produce a MB-OFDM UWB signal are listed in Table 5.1. As can be seen, each OFDM symbol period of 312.5 ns consists of a guard time of 70.1 ns and an information period of 242.4 ns. The OFDM period is used for transmitting the 100 information symbols and 28 pilot symbols which are used for time synchronization to avoid inter-symbol interference and frequency synchronization to avoid inter-carrier interference. Each symbol of the OFDM block (both data and pilot symbols) modulates the corresponding subcarrier for a period of 242.4 ns, resulting in a frequency spacing of 4.125 MHz. The OFDM bandwidth is thus 528.5 MHz. Figure 5.13 shows two OFDM symbols with the simulation parameters as shown in Table 5.1.

Compared with IR-UWB, the main advantage of the OFDM approach is that the symbol rate in each sub-band can be greatly lowered which can reduce the inter-symbol interference and enhance the ability to resist multipath transmission. Meanwhile, a large number of subcarriers convey multiple low rate symbols simultaneously, which does not reduce the whole data rate.

Table 5.1 OFDM simulation parameters from ECMA-368

Parameter	Value
Number of subcarriers (FFT size)	128
Number of data subcarriers	100
FFT sampling frequency (GHz)	50
Subcarrier frequency spacing (MHz)	4.125
Symbol interval (ns)	312
Cyclic prefix duration, T_p (ns)	60.6
Guard time, T_G (ns)	70.1
Symbol duration, T_0 (ns)	242.4
OFDM bandwidth (MHz)	528.5
Modulation	QPSK
Band group	1

The straightforward disadvantage of the MB-OFDM is that it is complex, making it more expensive to implement. In addition, MB-OFDM is also sensitive to carrier frequency variations. To overcome this problem, OFDM systems transmit pilot carriers along with the subcarriers for synchronization at the receiver. Another disadvantage is that an MB-OFDM signal has a high peak to average power ratio. As a result, the complex MB-OFDM signal requires linear amplification. This means greater inefficiency in the RF power amplifiers and more power consumption.

5.3 Demodulation and Error Probability

5.3.1 Optimum Demodulation for ASK, FSK and PSK

In a practical digital transmission system, optimum demodulation for a digital modulated signal generally requires maximal SNR and minimal error probability. Under

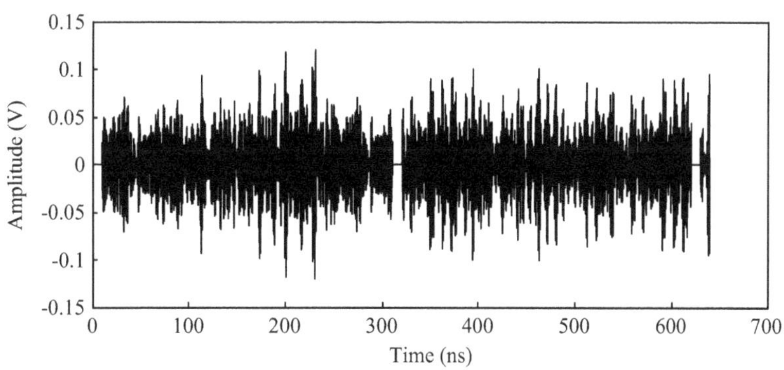

Figure 5.13 Two OFDM symbols with parameters shown in Table 5.1

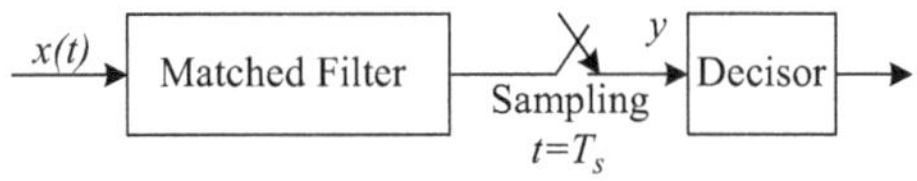

Figure 5.14 Matched filter demodulation

the maximal SNR criterion, the matched filter (MF), as the optimal linear filter, is employed to maximize the SNR in the presence of additive white Gaussian noise (AWGN). The impulse response of the matched filter is the mirror and translation of the input signal and therefore it can be expressed as

$$h(t) = KS(T_s - t) \tag{5.14}$$

where K is a constant. Its transmission function is proportional to the complex conjugate of the signal spectrum $S(f)$ as in Equation 5.15, hence the name "matched filter".

$$H(f) = KS^*(f)e^{-j2\pi f T_s}. \tag{5.15}$$

Figure 5.14 gives the schematic diagram of the matched filter demodulation. At the sampling moment T_s, the output signal reaches the maximum value and the maximum SNR is therefore achieved.

Under the minimal error probability criterion, the correlator is employed which consists of a multiplier and an integrator as shown in Figure 5.15.

In fact, the transmitted signal generally occurs in $(0, T_s)$ and therefore in the MF demodulation the output signal in Figure 5.14 can be expressed as:

$$y_s(t) = x(t) * h(t) = K\int_0^{T_s} x(t-\tau)S(T_s-\tau)d\tau. \tag{5.16}$$

When $t = T_s$, we have

$$y_s(T_s) = K\int_0^{T_s} x(t)S(t)dt. \tag{5.17}$$

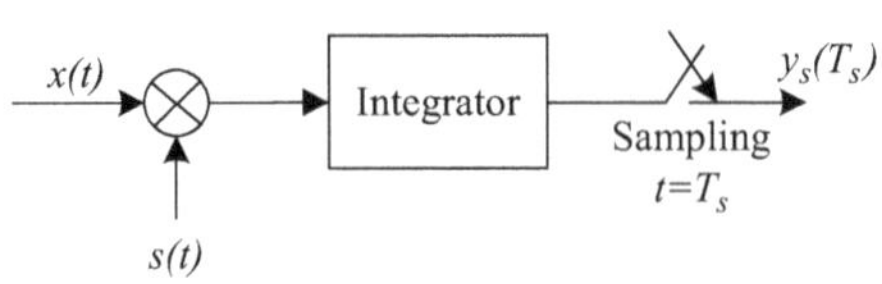

Figure 5.15 Correlator demodulation

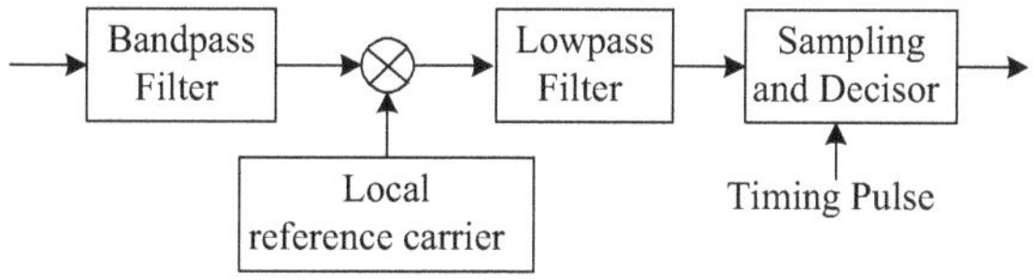

Figure 5.16 Optimum demodulation for an ASK signal

This expression is equivalent to the correlator demodulation as shown in Figure 5.15. In fact, the MF demodulation and the correlator demodulation are consistent. Therefore, the correlator demodulation is usually taken as the optimum demodulation. Actually, only if the filter (either passband or baseband) is designed on the basis of the MF requirements, the correlator demodulation is the real optimum demodulation.

Figure 5.16 shows a diagram of the optimum demodulation for the ASK signal. It can be seen that a reference carrier has to be generated locally at the receiver.

The optimum demodulation for the FSK signal is similar to that for ASK, but with multiple circuits corresponding to different carrier frequency.

The demodulation of the BPSK signal must employ a local carrier with the same carrier frequency and same phase. The carrier recovery can be achieved based on nonlinear transformation from the modulated signal. Figure 5.17 shows the optimum correlator demodulation for the BPSK signal. There are phase ambiguity problems of the recovered carrier, so differentially encoded BPSK (DBPSK) is often used in practice. The DBPSK signal conveys data based on the relative phase variation instead of the absolute phase of the carrier. This kind of encoding may be demodulated in the same way as for nondifferential PSK but the phase ambiguities can be ignored.

In binary digital demodulation, assuming that the two signals are expressed as $S_1(t)$ and $S_2(t)$ and the corresponding signal energies during the symbol interval $0 \leq t \leq T_s$ are

$$E_{S1} = \int_0^{T_s} S_1^2(t)dt \tag{5.18}$$

$$E_{S2} = \int_0^{T_s} S_2^2(t)dt, \tag{5.19}$$

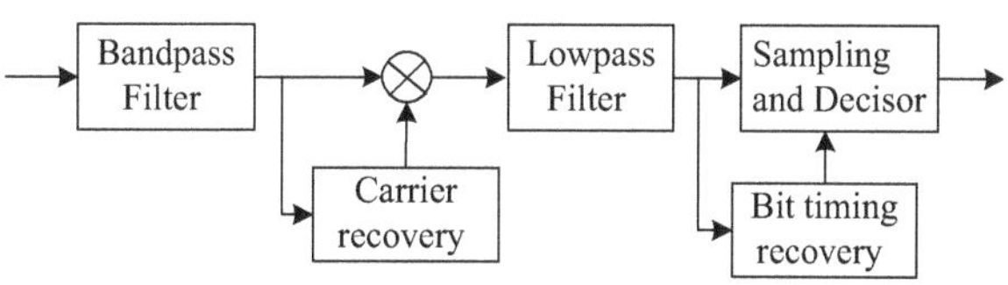

Figure 5.17 Optimum demodulation for a PSK signal

the correlation coefficient between $S_1(t)$ and $S_2(t)$ is therefore expressed as:

$$\rho = \frac{\int_0^{T_s} S_2(t)S_1(t)dt}{\sqrt{E_{S1}E_{S2}}} \tag{5.20}$$

which describes the similarity between the two signals and the ρ ranges in (−1,1).

In optimum demodulation, the bit error probability is (Haykin, 1988)

$$P_b = Q\left(\sqrt{\frac{E_{S1} + E_{S2} - 2\rho\sqrt{E_{S1}E_{S2}}}{2N_0}}\right) \tag{5.21}$$

where

$$Q(x) = \frac{1}{\sqrt{2\pi}}\int_x^{\infty} e^{-t^2/2}dt = \frac{1}{2}\text{erfc}\left(\frac{x}{\sqrt{2}}\right) \tag{5.22}$$

is usually named as the Q function and erfc is the complementary error function. N_0 is the noise unilateral power spectral density.

If the two signals have the same energy, that is, $E_{S1} = E_{S2} = E_b$ where E_b represents the signal energy in one bit, then we have

$$P_b = Q\left(\sqrt{\frac{E_b}{N_0}(1-\rho)}\right). \tag{5.23}$$

This is the minimal bit error probability for binary modulation in optimum demodulation, either in MF demodulation or correlator demodulation.

For the binary ASK or OOK signal, $E_{S1} = 0$, $E_b = E_{S2}/2$, therefore we have

$$P_{b,ASK} = Q\left(\sqrt{\frac{E_b}{N_0}}\right). \tag{5.24}$$

For the BPSK signal, $S_1(t) = -S_2(t)$ and $\rho = -1$ can be derived, based on Equation 5.23, we have

$$P_{b,BPSK} = Q\left(\sqrt{\frac{2E_b}{N_0}}\right). \tag{5.25}$$

For the binary FSK signal, if the difference between the two carrier frequencies f_1 and f_2 is $N/2T_s$ (where N is a positive integer), then the two signals keep orthogonal and $\rho = 0$ can be derived, then the bit error probability for the binary FSK signal with $\rho = 0$ is

$$P_{b,FSK} = Q\left(\sqrt{\frac{E_b}{N_0}}\right). \tag{5.26}$$

In another case, if the difference between the two carrier frequencies gets close to infinity, the correlation coefficient ρ is close to zero too.

For the M-ary shift keying signal (M>2), the principle of the optimum demodulation is the same as the binary shift key signal. However, the error probability of M-ary shift keying signal is usually expressed in terms of symbol error probability rather than bit error probability.

The optimum symbol error probability for the MASK signal in correlator demodulation is approximately

$$P_{S,MASK} = \frac{2(M-1)}{M} Q\left(\sqrt{\frac{6\log_2 M}{M^2-1} \cdot \frac{E_b}{N_0}}\right) \tag{5.27}$$

where the signal amplitude A_m takes the discrete levels, that is,

$$A_m = 2m - 1 - M \quad m = 1, 2, \ldots, M. \tag{5.28}$$

Similarly, the optimum symbol error probability of the MPSK signal is approximately

$$P_{S,MPSK} = 2Q\left(\sqrt{2\log_2 M \sin^2\left(\frac{\pi}{M}\right) \cdot \frac{E_b}{N_0}}\right) \tag{5.29}$$

when $E_b/N_0 \gg 1$.

5.3.2 Noncoherent Detection for ASK, FSK and PSK

The correlator demodulation requires generation of a synchronous reference carrier signal at the receiver. In this sense it is considered to be a coherent detection. Compared with coherent detection, the significant advantage of noncoherent detection is there is no need to generate the reference carrier signal at the receiver, which largely simplifies the receiver structure. Noncoherent detection is a suboptimal demodulation scheme.

For the ASK signal, envelope detection is used to extract the amplitude information of the modulated signal. A half-wave rectifier or full-wave rectifier acts as the envelope detection device. Figure 5.18 shows a diagram of the noncoherent

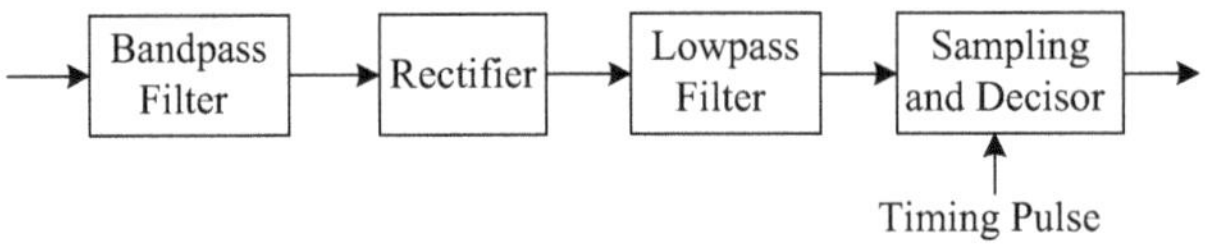

Figure 5.18 Noncoherent detection for an ASK signal

detection scheme for the ASK signal. Actually, noncoherent detection for the ASK signal is more common than the correlator demodulation.

The bit error probability for the binary ASK signal in noncoherent detection can be expressed as

$$P_{b,NCASK} = \frac{1}{2} e^{-E_b/2N_0}. \tag{5.30}$$

For the binary FSK signal with $\rho = 0$, the modulated signals are orthogonal to each other. The two transmitted signals have the same energy if equal amplitude modulation is assumed. The diagram of the noncoherent detection scheme for the FSK signal is shown in Figure 5.19. The bandpass MF corresponds to signals with different frequencies.

The bit error probability for the binary FSK signal in noncoherent detection can be expressed as

$$P_{b,NCFSK} = \frac{1}{2} e^{-E_b/2N_0}. \tag{5.31}$$

It has the same bit error probability as the binary ASK. Actually, Equation 5.30 is the bit error probability expression for any binary orthogonal signal in optimum noncoherent detection.

When the orthogonal MFSK signal is demodulated using the noncoherent detection scheme, the symbol error probability is (Lindsey and Simon, 1973):

$$P_{S,NCMFSK} = \sum_{k=1}^{M-1} \frac{(-1)^{k+1}}{k+1} C_{M-1}^{k} e^{-\mathrm{k} \log_2 M E_b/(k+1)N_0}. \tag{5.32}$$

For the BPSK signal, as described in Section 5.3.1, demodulation must use the correlator demodulation while the carrier recovery at the receiver has a

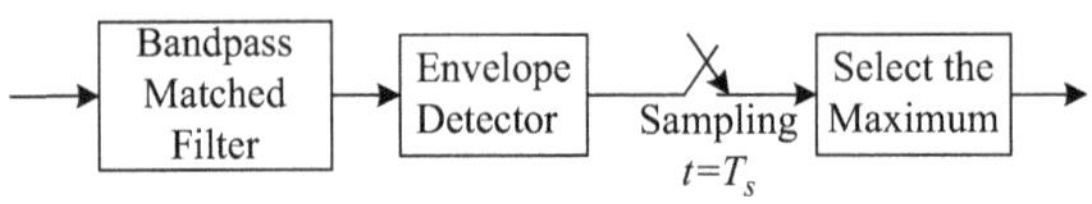

Figure 5.19 Noncoherent detection for a FSK signal

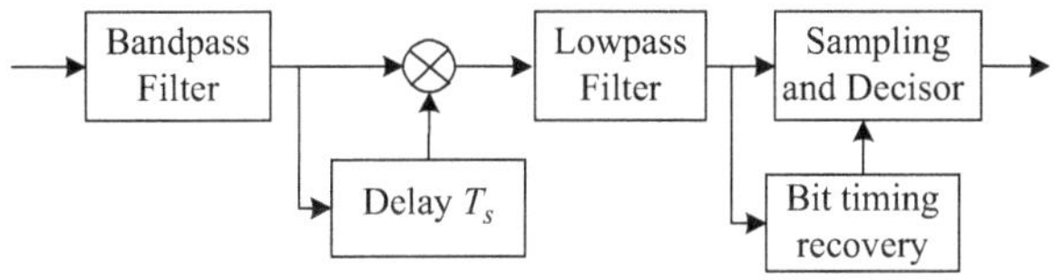

Figure 5.20 Noncoherent detection for a DBPSK signal

phase ambiguity problem. DBPSK is therefore often used in practice (Smith, 1985). Noncoherent detection for the DBPSK signal is shown in Figure 5.20. It takes the delayed signal as the reference signal, in which the delayed period is the symbol interval.

The bit error probability for the noncoherent detected DBPSK signal can be expressed as

$$P_{b,NCDPSK} = \frac{1}{2}e^{-E_b/N_0}. \tag{5.33}$$

Compared with Equations 5.30 and 5.31, it can be seen that the bit error probability for noncoherent detected DBPSK exceeds that for noncoherent detected binary ASK and FSK.

For the M-ary noncoherent detected DBPSK signal, the symbol error probability is (Arthurs and Dym, 1962)

$$P_{S,NCMPSK} = 2Q\left(\sqrt{2\log_2 M \sin^2\left(\frac{\pi}{\sqrt{2}M}\right)\cdot\frac{E_b}{N_0}}\right). \tag{5.34}$$

Figure 5.21 compares the error probability performance of coherent and noncoherent detection of binary ASK, FSK and PSK, respectively. Note that the probability of error for coherent BPSK outperforms that for binary ASK and FSK. At the same bit error probability, the required E_b/N_0 is 3 dB lower than for binary ASK and FSK which implies the transmitted signal energy can be cut in half. The error probability for noncoherent FSK is the same as that for noncoherent ASK. The performance of the noncoherent detection is poor compared with coherent detection. At a bit error probability of 10^{-5}, the difference is around 1 dB. The error probability performance of noncoherent DBPSK is also shown in Figure 5.21 and it can be noted that the noncoherent DBPSK is better than noncoherent binary ASK and FSK.

5.3.3 Optimum Demodulation for IR-UWB

The optimum demodulation for the PPM-TH-UWB and PAM-DS-UWB signal is essentially similar to correlator demodulation for the ASK signal. As shown in Figure 5.22, the optimum demodulation for PPM-TH-UWB consists of a multiplier,

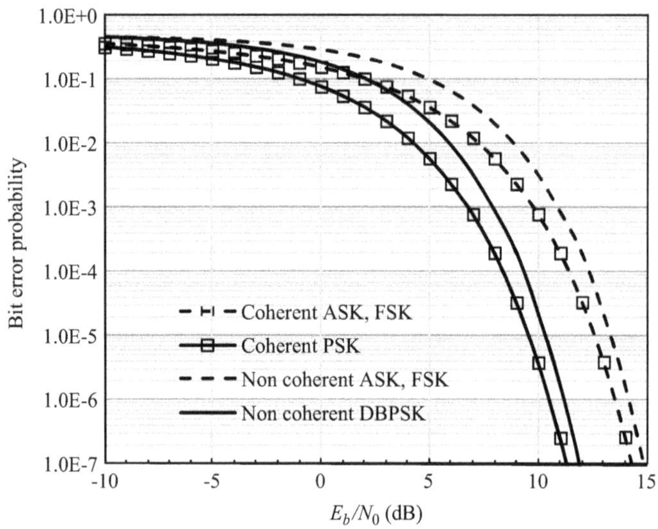

Figure 5.21 Bit error probability of binary modulation

an integrator and sampling decisor. The local correlation mask is the TH pulse template $g(t - iT_s - c_iT_c)$. Through the correlator, the difference induced by the PPM time shift $a_i\varepsilon$ can be detected and the transmitted bit can be estimated at the output. If the PPM time shift ε is larger than the pulse duration T_M, the two transmitted signals corresponding to bit 1 and bit 0 are orthogonal to each other. For the binary orthogonal PPM signal, the minimal bit error probability can be expressed as follows

$$P_{b,PPM} = Q\left(\sqrt{\frac{E_b}{N_0}}\right) = Q\left(\sqrt{\frac{N_S E_{RX}}{N_0}}\right) \tag{5.35}$$

where E_b is the received energy per bit, E_{RX} is the received energy per pulse, and N_S is the code repetition redundancy. Note that at the cost of increasing the number of pulses per bit, the received energy is increased by a factor N_S and the bit error

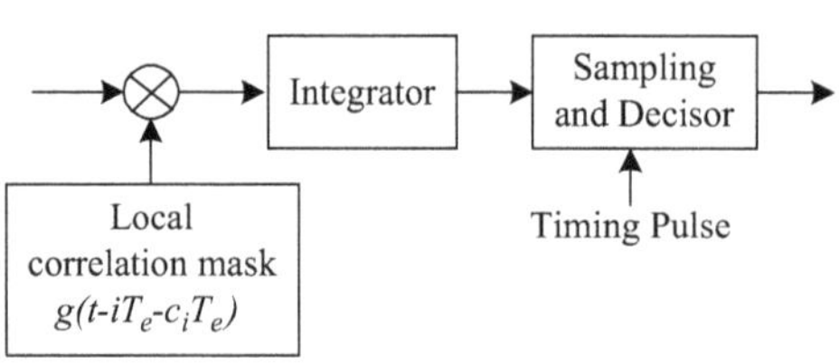

Figure 5.22 Optimum demodulation for a PPM-TH-UWB signal

probability will consequently reduce. This is the benefit due to the presence of N_S pulses per bit from the application of a code repetition coder before modulation.

Equation 5.35 is the minimal bit error probability which corresponds to an optimum decision detection. Actually, in the presence of multiple pulses per symbol, two possible decision strategies can be adopted at the receiver. One is the soft decision detection, which considers N_S pulses per bit as a single multi-pulse signal; the other is the hard decision detection, which implements N_S independent decisions over the N_S pulses per bit. For the code repetition coding scheme, the soft decision is better than the hard decision in the AWGN channel. In fact, the soft decision in general outperforms hard decision for a variety of common code families. Equation 5.35 is the bit error probability in soft decision detection.

Figure 5.23 compares the bit error probability versus E_{RX}/N_0 for a binary PPM signal with different N_S values. We can observe that the bit error probability significantly decreases when increasing the redundancy of the code repetition coder.

The average symbol error probability for M-ary PPM can be written as follows

$$P_{S,MPPM} = \frac{1}{\sqrt{2\pi}} \int_{-\infty}^{\infty} e^{-\frac{1}{2}\left(x-\sqrt{\frac{2E_{RX}}{N_0}}\right)^2} \left[1 - \left(\frac{1}{\sqrt{2\pi}} \int_{-\infty}^{x} e^{-\frac{x^2}{2}} dx\right)^{M-1}\right] dx. \quad (5.36)$$

The optimum demodulation for PAM-DS-UWB is similar to that for PPM-TH-UWB. As shown in Figure 5.24, the difference lies in the local correlation mask which is here the DS pulse template $c_i g(t - iT_s)$.

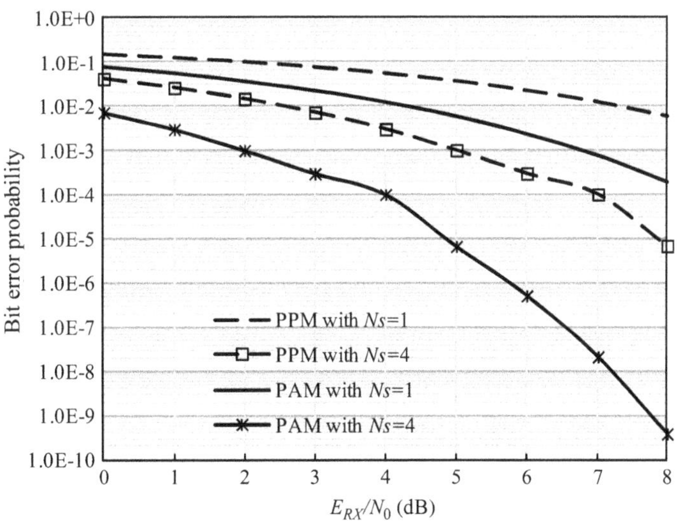

Figure 5.23 Bit error probability versus E_{RX}/N_0 for PPM-TH-UWB and PAM-DS-UWB signals with one pulse per bit ($N_s = 1$) and four pulses per bit ($N_s = 4$)

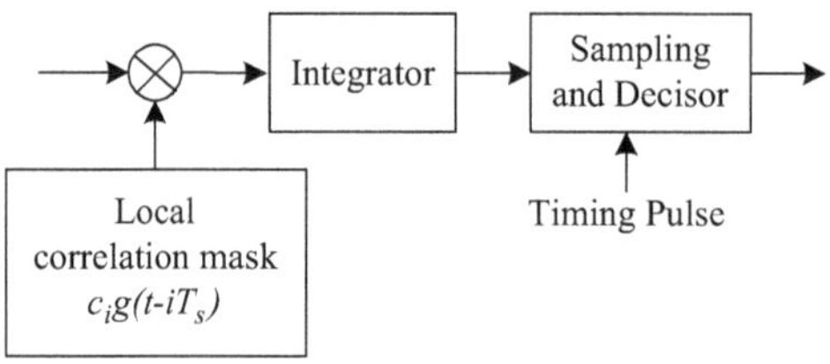

Figure 5.24 Optimum demodulation for a PAM-DS-UWB signal

For the binary orthogonal PAM signal, the minimal bit error probability can be expressed as follows

$$P_{b,PAM} = Q\left(\sqrt{\frac{2E_b}{N_0}}\right) = Q\left(\sqrt{\frac{2N_S E_{RX}}{N_0}}\right). \tag{5.37}$$

Comparing Equations 5.37 and 5.35, it can be seen that binary orthogonal PPM signals require a factor of two in energy to achieve the same error probability as binary antipodal PAM signals. Consequently, the bit error probability of binary orthogonal signals is 3 dB worse than for binary antipodal signals. It can be observed from the error probability in Figure 5.23 that at any given error probability, the SNR required for PPM is 3 dB more than that for PAM.

The average symbol error probability for M-ary PAM can be written as follows

$$P_{S,MPAM} = \frac{2(M-1)}{M} Q\left(\sqrt{\frac{6}{M^2-1}\cdot\frac{E_s}{N_0}}\right) = \frac{2(M-1)}{M} Q\left(\sqrt{\frac{6\log_2 M}{M^2-1}\cdot\frac{E_b}{N_0}}\right) \tag{5.38}$$

where E_s is the energy per symbol.

5.3.4 Noncoherent Detection for IR-UWB

The optimum demodulator may be quite complex in order to induce an appropriate template signal in the receiver especially in a multipath channel. The hardware itself may also consume a lot of power. Noncoherent energy detection is a suboptimal demodulation scheme. It is possible to capture the multipath energy easily, since the multipath combining does not require any knowledge with respect to the phases or polarities. Although this advantage can only be obtained at the expense of a higher SNR, the envelope detection is a useful alternative especially for low power, low complexity and low data rate systems.

The noncoherent energy detection can be used for IR-UWB with OOK or PPM. Figure 5.25 shows the block diagram of the energy detector for OOK and PPM. The detector consists of a squarer circuit, an integrator with integration time T and a decisor. The received signal is first squared and then integrated over time T. Here we assume one pulse for one bit and the integration time T is large enough to collect the

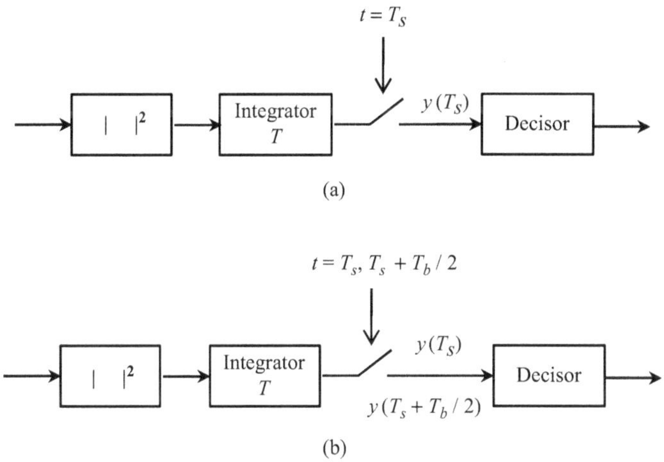

Figure 5.25 Block diagram of energy detector for IR-UWB with (a) OOK and (b) PPM

transmitted energy over one bit for ease of explanation. In the OOK scheme, since the energy detector output $y(T_s)$ for bit "1" should be larger than that for bit "0," we can assume a threshold to decide the transmitted bit to be "1" or "0" in the decisor. On the other hand, in the PPM scheme, the energy detector outputs the sampling values at both the first and last half of one bit duration. If the value $y(T_s)$ in the first half is larger than $y(T_s + T_b/2)$ in the last half, the pulse signal should be transmitted at the first position, otherwise the pulse signal is transmitted $T_b/2$ s later where T_b is the bit duration. Based on this information the decision circuit can also decide the transmitted bit to be "1" or "0."

To derive the error probability of energy detection for IR-UWB, we have to know the noise PDF at first. However, due to the square behavior of the energy detector, the noise PDF is obviously not Gaussian at the detector outputs. Assuming a wide communication bandwidth B, we may consider the integrator output as the sum of $2M = 2BT + 1$ independent random variables. Hence, the resulting decision variable y is a sum of $2M$ independent variables with chi-square distribution. According to the central limit theorem, the PDF of y tends to be a Gaussian function when $2M$ tends to infinity. In general, when $2M$ becomes higher than 40, the density of probability of y becomes Gaussian with a confidence better than 5%. In other words, as long as the product of bandwidth B and integration time T is larger than 20, the Gaussian approximation is valid. This condition is basically satisfied in body area UWB communications, because the data rate is usually below 10 Mbps and the bandwidth is at least 500 MHz.

Now let us assume that the information bits are equally probable and the mean energy of the transmitted signals is E_b over one bit. Then for OOK, the signal for bit "1" will have energy of $2E_b$, whereas for bit "0" no energy is transmitted. Under the

Gaussian approximation for the noise at the output of the energy detector, the bit error probability of OOK can be simply expressed as (Dubouloz *et al.*, 2005)

$$P_{b,EDOOK} = Q\left(\frac{2E_b/N_0}{\sqrt{BT} + \sqrt{BT + 4E_b/N_0}}\right). \tag{5.39}$$

It should be noted that the optimal threshold in the OOK decisor depends on the SNR level, and the above bit error probability is given at the optimal threshold. For a fixed threshold in actual implementation, 0.6 times of $2E_b$ may be a reasonable choice, and the corresponding bit error probability will degrade compared with that in optimal threshold.

On the other hand, for PPM, no threshold setting is required because we only need to compare the magnitude of the integration value in the first half and the last half of a bit duration. The transmitted energy per bit is always E_b for both "1" and "0." Based on the same approximation, the bit error probability can be expressed as

$$P_{b,EDPPM} = Q\left(\frac{E_b/N_0}{\sqrt{2BT + 2E_b/N_0}}\right). \tag{5.40}$$

Figures 5.26 and 5.27 show the bit error probability versus E_b/N_0 for OOK and PPM with noncoherent energy detection, respectively. The basic bit error probability is at a similar level for the two modulation schemes for the same E_b/N_0. Due to the adoption of the optimum threshold, OOK seems to outperform PPM slightly. Moreover, an obvious degradation of error probability is observed, for both OOK and PPM, with the increase of the integration time T for a fixed bandwidth. This is

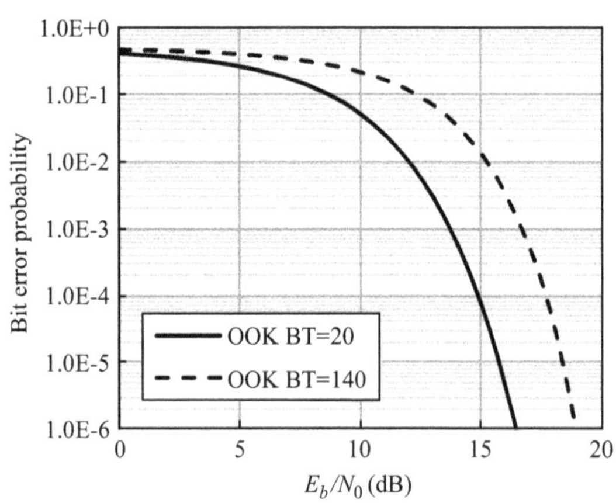

Figure 5.26 Bit error probability versus E_b/N_0 for OOK energy detection

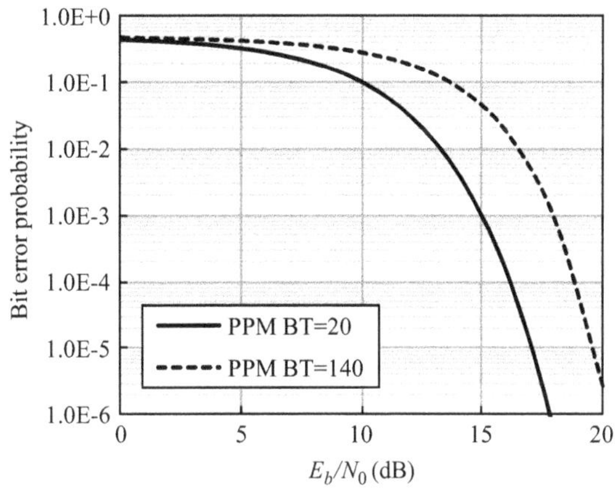

Figure 5.27 Bit error probability versus E_b/N_0 for PPM energy detection

attributed to the increased noise energy. In order to improve the bit error probability, it is therefore effective to collect all the signal energy at an integration time T as short as possible.

5.3.5 MB-OFDM Demodulation

As described in Section 5.2.3, OFDM modulation includes CP insertion and IFFT operation. Thus the OFDM demodulation mostly revolves around CP removal and FFT operation.

Figure 5.28 shows the corresponding OFDM receiver for the OFDM transmitter in Figure 5.11. The received signal is then quadrature-mixed down to baseband using cosine and sine waves at the carrier frequency. Since this process also creates signals at two times carrier, low-pass filters are used to reject the high frequency components. The baseband signals are then sampled using ADC, and a forward FFT is used to convert them back to the frequency domain. The FFT process returns

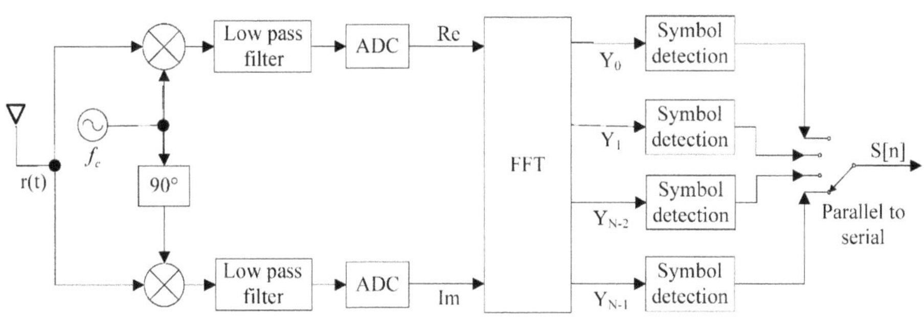

Figure 5.28 Block diagram of an OFDM receiver

N parallel streams. These streams are respectively converted to a binary stream using an appropriate symbol detector and are then recombined into a serial stream, that is, the demodulated binary data.

For the bit error probability of the MB-OFDM UWB signal, we can express the probability of error for each subcarrier for the case of OFDM with QAM modulation as follows

$$P_{s,OFDM\ with\ QAM} \cong 2\left(1-\frac{1}{\sqrt{M}}\right)erfc\left[\frac{1}{\sigma_n\left(\sqrt{M}-1\right)}\right] \tag{5.41}$$

$$\text{with} \quad \sigma_n^2 = \frac{\lambda^2}{2}\left(1+\frac{T_G}{T_0}\right)\frac{2N_0}{E_M}$$

where $\lambda^2 = 2\left(M^{1/2}+1\right)/3\left(M^{1/2}-1\right)$ is the constellation power, and $E_M = (\lambda^2 T)/(2T_0)$ is the energy per symbol. When $M = 2$ and $M = 4$, Equation 5.41 is in fact the probability of error for OFDM with BPSK and QPSK modulation, respectively.

5.4 RAKE Reception

As shown in Chapter 4, various body postures and movements will result in a multipath propagation channel in body area. The multipath propagation channel makes electromagnetic waves traveling along the body reflected, diffracted, and scattered, so that a received signal ends up being the superimposition of several attenuated, delayed, and eventually distorted replicas of a transmitted waveform. This can result in either constructive or destructive interference, amplifying or attenuating the signal power at the receiver. This effect is called fading, and is referred to as multipath fading. Strong destructive interference may result in temporary failure of communication due to a severe drop in the channel SNR. In order to counter the effects of multipath fading, a RAKE receiver has been proposed (Price and Green, 1958). The RAKE receiver consists of multiple correlators, in which the received signal is multiplied by time-shifted versions of a locally generated code sequence. The intention is to separate signals such that each finger only sees signals coming in over a single path. The spreading code is chosen to have a very small autocorrelation value for any nonzero time offset. This avoids crosstalk between fingers.

The RAKE receiver is designed to optimally detect a spread spectrum signal transmitted over a multipath channel. It is essentially an extension of the concept of the MF in Section 5.3.1. In the MF receiver, the signal is correlated with a locally generated copy of the signal waveform. If, however, the received signal is distorted by the channel, the receiver should correlate the incoming received signal by a copy of the expected received signal, rather than by a copy of the transmitted waveform.

Thus the receiver should estimate the delay profile of the channel, and adapt its locally generated copy according to this estimate. In Chapter 4, the body area UWB multipath channel model has been determined statistically based on the classical Saleh–Valenzuela model, which has a tapped-delay-line format. This enables the receiver to have the knowledge of the channel delay profile characteristics, including multipath delays, phases of the multipath components, amplitudes of the multipath components and number of multipath components.

There are two primary methods to combine the RAKE-receiver finger outputs. One method weights each output equally and is therefore called equal gain combining (EGC). The second method uses the data to estimate weights which maximize the SNR of the combined output. This technique is known as maximal ratio combining (MRC). As can be seen, with the EGC method, the different contributions are first aligned in time and then added without any particular weighting. In MRC, the different contributions are weighted before the combination and the weights are determined to maximize the SNR before the decision process.

Figure 5.29 shows the structure of the RAKE receiver, which consists of a parallel bank of N_R correlators, followed by a combiner that determines the variable to be used for the decision on the transmitted signal. Each correlator is locked on one of the different replicas of the transmitted signal. $\{m_1(t), m_2(t), \ldots m_R(t)\}$ are the correlator masks which are the delayed replicas of the transmitted signal. $\{\omega_1, \omega_2, \ldots \omega_R\}$ are the statistically independent time-invariant weighting factors which depend on the combining method implemented at the receiver. In the case of EGC, all factors are equal to 1, that is, the combiner simply adds the outputs of the correlators without applying any weighting. In the MRC case, the output of each branch is multiplied by a weighting factor, which is proportional to the signal amplitude on that branch.

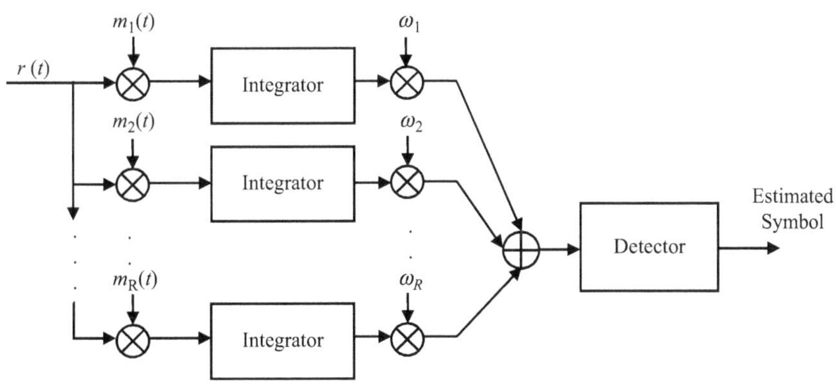

Figure 5.29 RAKE receiver with N_R parallel correlators

The RAKE receiver can take advantage of multipath to improve the SNR at the receiver, yet the adoption of a RAKE considerably increases the complexity of the receiver. This complexity increases with the number of multipath components analyzed and combined before decision, and can be reduced by decreasing the number of components processed by the receiver. However, a reduction of the number of paths leads to a decrease of energy collected by the receiver. A quasi-analytical investigation of the existing trade-off between receiver complexity and percentage of captured energy in a RAKE receiver for IR-UWB systems can be found in Win and Scholtz (1998).

As described in Chapter 4, the body area UWB multipath channel can be modeled by a tapped delay line with statistically independent time-invariant tap weights. Hence, a RAKE receiver can process the received signal in an optimum manner which will achieve the performance of an equivalent L-order diversity communication system.

In the multipath fading propagation channel, the instantaneous SNR is a random variable at the receiver output that includes the effect of fading. The variation of the SNR is a random process related to the fading characteristics. If the received SNR is no longer fixed, the bit error performance in optimum demodulation no longer applies. Therefore, we must average the bit error probability in the AWGN channel over the PDF of the SNR (Proakis, 2007). That is, we must evaluate the integral

$$P_b(\bar{\gamma}_b) = \int_0^\infty P_0(\gamma_b)p(\gamma_b)d\gamma_b \tag{5.42}$$

where $\bar{\gamma}$ represents the average SNR, $P_b(\bar{\gamma}_b)$ is the average bit error probability in a multipath fading channel, $P_0(\gamma_b)$ is conditional error probability in AWGN channel and $p(\gamma_b)$ denotes the PDF of γ_b when the SNR is random. As we can see here, the term "average" in the error probability refers to statistical averaging over the probability distribution of the SNR. Therefore, we need to derive the probability distribution of the SNR under the RAKE reception in the body area multipath fading channel.

Assuming the signal power gain of the kth resolvable multipath

$$\Omega_k = \alpha_k^2 \tag{5.43}$$

then we have the instantaneous SNR on the kth multipath channel γ_k

$$\gamma_k = \frac{E_b}{N_0}\alpha_k^2 \tag{5.44}$$

where E_b/N_0 corresponds to the SNR per bit.

As described above, the RAKE receiver can process separately multiple received multipaths and achieve the SNR of an equivalent L-order diversity communication system. The received SNR per bit can then be represented as

$$\gamma_b = \sum\nolimits_{k=1}^{L} \gamma_k = \frac{E_b}{N_0} \sum\nolimits_{k=1}^{L} \alpha_k^2 \tag{5.45}$$

in which L is the number of RAKE fingers and also the number of resolvable multipaths. Now we must determine the PDF of γ_b, $\mathrm{p}(\gamma_\mathrm{b})$. The whole derivation process can be generalized into two steps.

1. Derive the statistical characteristics of $\sum_{k=1}^{L} \alpha_k^2$
   In the indoor/outdoor wireless multipath propagation environment, α_k is Rayleigh distributed and α_k^2 has a chi-square probability distribution with two degrees of freedom. The $p(\gamma_b)$ can be solved based on the characteristic function of the chi-square probability distribution. While in the body area multipath propagation environment, α_k^2 is log-normal distributed. It is not so easy to get a solution for the probability distribution of a sum of log-normal-distributed variables via the characteristic function because the simplest representation for the characteristic function of log-normal distribution is not a closed-form solution. In fact, the characteristic function of log-normal distribution is expressed by Taylor series expansion and contains infinite items.
   A summary for the distribution of sums of log-normally distributed random variables is given in Dufresne (2009). In wireless systems the sum of log-normals can be approximated by a single log-normal. Now we start to derive the distribution of the sum of log-normals analytically via the moment generating function (MGF) method in Wu, Mehta, and Zhang (2005).
   The MGF of a random variable is an alternative definition of its probability distribution. In fact, the MGF can be interpreted as a weighted integral of the PDF. Thus, it provides the basis of an alternative route to analytical results compared with working directly with PDFs. Assuming a log-normal random variable X, then $X = 10^{0.1Y}$, $Y \sim N(\mu, \sigma^2)$ where μ and σ are the mean and standard deviation of the Gaussian variable Y. While no general closed-form expression for the log-normal MGF is available, it can be readily expressed by a series expansion based on Gauss–Hermite integration

$$\hat{\Psi}_X(s; \mu, \sigma) = \sum\nolimits_{n=1}^{N} \frac{w_n}{\sqrt{\pi}} \exp\left[-s \exp\left(\frac{\sqrt{2}\sigma a_n + \mu}{\xi}\right)\right] \tag{5.46}$$

   where N is the Hermite integration order, $\xi = 10 / \ln 10$ is a scaling constant, w_i and a_i are the weight and abscissa, respectively, and tabulated in Abramowitz and Stegun (1972) for $N \leq 20$. The accuracy of the Gaussian–Hermite representation of the MGF is subject to the value of N. It can be found that the log-normal MGF can be accurately approximated by its Gauss–Hermite expansion with $N = 6$.

Assume $X_k = \alpha_k^2$, and then $X_k = 10^{0.1Y_k}$, where $Y_k \sim N(\mu_{Y_k}, \sigma_{Y_k}^2)$. The log-normal sum $X = \sum_{k=1}^{L} \alpha_k^2 = \sum_{k=1}^{L} X_k$ can now be similarly approximated by $X = 10^{0.1Y}$ and $Y \sim N(\mu_Y, \sigma_Y^2)$. The MGF of the log-normal sum equals the product of the MGF of each log-normal variable. Therefore, we can set up a system of two independent equations to calculate μ_Y and σ_Y^2 by matching the MGF of X with the MGF of $\sum_{k=1}^{L} X_k$ via two different values s_1 and s_2, as follows

$$\sum_{n=1}^{N} \frac{w_n}{\sqrt{\pi}} \exp\left[-s_m \exp\left(\frac{\sqrt{2}\sigma_Y a_n + \mu_Y}{\xi}\right)\right] = \prod_{k=1}^{L} \hat{\Psi}_{X_k}(s_m; \mu_{Y_k}, \sigma_{Y_k})\ m = 1, 2. \tag{5.47}$$

The left items of the above two equations are the MGFs of X and the right item is the product of the MGFs of X_k. The μ_Y and σ_Y^2 on the left-hand side can thus be derived based on the known μ_{Y_k} and $\sigma_{Y_k}^2$ of each multipath. Note that the above equations are nonlinear equations and can be solved numerically using the *fsolve* function in MATLAB®. The approximation accuracy also depends on the values of s_1 and s_2. Increasing s_1 and s_2 will result in more accurate approximation of the tail portion of the sum PDF. On the contrary, decreasing s_1 and s_2 will approximate more accurately the head portion of the sum PDF. In the body area propagation environment, the reasonable multipath is in a limited number and therefore the log-normal sum keeps a limited value. The head portion of the PDF needs to be computed accurately. More detailed guidelines for choosing s_1 and s_2 can be found in Wu, Mehta, and Zhang (2005).

Based on the above derivation, it can be summarized that $X = \sum_{k=1}^{L} \alpha_k^2$ is log-normally distributed with parameters (μ_Y, σ_Y^2) which can be numerically determined from Equation 5.47.

2. Derive the statistical characteristics of γ_b

From

$$X = \sum_{k=1}^{L} \alpha_k^2$$

and

$$X = 10^{0.1Y}, \quad Y \sim N(\mu_Y, \sigma_Y^2) \tag{5.48}$$

we have

$$\ln X = \frac{\ln 10}{10} Y \tag{5.49}$$

that is, $\ln X$ is a normal distribution

$$\ln X \sim N(\mu_X, \sigma_X^2) = N\left(\frac{\ln 10}{10}\mu_Y, \left(\frac{\ln 10}{10}\sigma_Y\right)^2\right) \tag{5.50}$$

with

$$\mu_X = \frac{\ln 10}{10}\mu_Y \tag{5.51}$$

$$\sigma_X = \frac{\ln 10}{10}\sigma_Y. \tag{5.52}$$

So the mean and variance of the log-normal distribution X is

$$E(X) = e^{\mu_x + \frac{\sigma_x^2}{2}} \tag{5.53}$$

$$D(X) = (e^{\sigma_x^2} - 1)e^{2\mu_x + \sigma_x^2}. \tag{5.54}$$

Since

$$\gamma_b = \frac{E_b}{N_0}\sum_{k=1}^{L} \alpha_k^2 = \frac{E_b}{N_0}X \tag{5.55}$$

we have

$$E(\gamma_b) = \frac{E_b}{N_0}E(X) \tag{5.56}$$

$$D(\gamma_b) = \left(\frac{E_b}{N_0}\right)^2 D(X) \tag{5.57}$$

where $E(\gamma_b)$ and $D(\gamma_b)$ are the mean and variance of the log-normal distributed γ_b.

Assuming $\gamma_b = e^Q$ then we have $Q \sim N(\mu, \sigma^2)$. The parameters μ and σ can be obtained based on the known mean and variance of γ_b. The relationship is as follows

$$\mu = \ln E(\gamma_b) - \frac{1}{2}\ln\left(\frac{D(\gamma_b)}{E(\gamma_b)^2} + 1\right) \tag{5.58}$$

$$\sigma^2 = \ln\left(\frac{D(\gamma_b)}{E(\gamma_b)^2} + 1\right). \tag{5.59}$$

The parameters μ and σ are just the parameters required in the PDF expression of γ_b

$$p(\gamma_b) = \frac{1}{\sqrt{2\pi}\sigma\gamma_b} e^{-\frac{(\ln \gamma_b - \mu)^2}{2\sigma^2}}. \tag{5.60}$$

Now we have derived the PDF of the SNR under the RAKE reception in the body area multipath fading channel. The average bit error probability of a RAKE receiver can be therefore calculated via the numerical integral of Equation 5.42. It can be seen that the bit error performance with a RAKE receiver depends on the delay spread characteristics of the multipath channel as well as the bit error probability in AWGN.

5.5 Diversity Reception

Diversity is a technique for improving the communication performance by using two or more communication channels (or branches) with different characteristics. It is based on the consideration that multiple versions of the same signal may be received and combined in the receiver because the same information may reach the receiver on statistically independent channels. The simplest case is a receiver with two antennas. By setting the receiver to always choose the antenna which has larger instantaneous power, the SNR statistic at the detector input is changed so that it has a smaller probability of being low. Such a combining scheme is known as selection combining (SC) of diversity.

Diversity can exploit multipath propagation to result in a diversity gain for combating the multipath fading. It is most effective when the different transmission channels carry independent fading copies of the same signal. This means that the correlation between the fading of the channels is low. The correlation coefficient characterizes the correlation between received signals on different branches. Assuming x and y as the signal envelopes on two different branches, the correlation coefficient is given as

$$\rho = \frac{\sum_{n=1}^{N} (x_n - \bar{x})(y_n - \bar{y})}{\sqrt{\sum_{n=1}^{N} (x_n - \bar{x})^2} \sqrt{\sum_{n=1}^{N} (y_n - \bar{y})^2}}. \tag{5.61}$$

For two statistically independent receiver signals, the correlation coefficient is zero, which gives the best diversity effectiveness. In general, when the correlation coefficient is below a certain threshold, typically 0.5 or 07, the receiver signals are considered as decorrelated. This requirement is met in most body area communication scenarios, and a diversity effect is thus expected to provide an effective improvement on the bit error probability performance.

The promising diversity schemes in body area communications are spatial diversity and polarization diversity. Both spatial diversity and polarization diversity generally employ multiple receiving antennas. In spatial diversity, the antennas are separately placed on the human body to receive signals over different propagation paths. In polarization diversity, the antennas are arranged with different polarization such as vertical polarization and horizontal polarization to receive different signals. Since both schemes provide an acceptable correlation coefficient below 0.5 (Shi and Wang, 2010), a diversity combining technique is applicable on the receiver to obtain a diversity gain.

The diversity combining techniques can be distinguished into three kinds.

The first kind is SC diversity. Of the *M* received signals, the signal with the maximum instantaneous SNR is selected. The corresponding SNR is

$$\gamma_{SC} = \mathrm{Max}(\gamma_1, \gamma_2, \cdots, \gamma_M) \tag{5.62}$$

where γ_m is the SNR at the *m*th branch. In reality, the "instantaneous SNR" is an average SNR over a time period which is longer than the reciprocal of the signal bandwidth. Figure 5.30 shows the block diagram of SC diversity. The envelope levels are actually used to measure the sum of signal and noise for selecting the branch.

The second kind is EGC diversity in which all the received signals are summed coherently. Figure 5.31 shows the block diagram of EGC diversity with two diversity branches. Each branch corresponds to a single fading channel. Defining the received signal at the *m*th diversity branch as $|v_m|e^{j\phi_m}$ $(m = 1, 2, \ldots, M)$, the summed signal is then given by

$$v = \sum_{m=1}^{M} |v_m| e^{j\phi_m}. \tag{5.63}$$

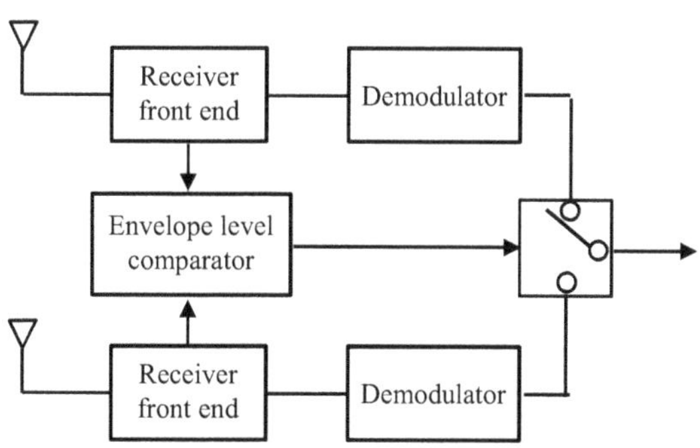

Figure 5.30 Block diagram of SC diversity

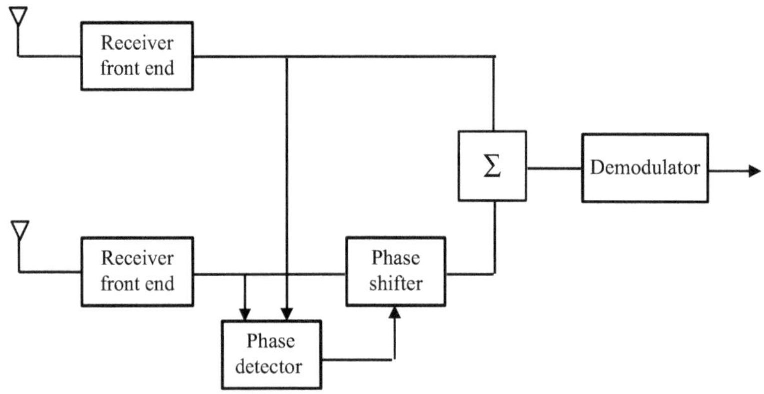

Figure 5.31 Block diagram of EGC diversity

Under the decorrelated condition, the received fading signal $|v_m|e^{j\phi_m}$ is approximately considered as independent. The EGC diversity equally weighs each branch and sums them to produce the decision statistic. That is

$$v = \sum_{m=1}^{M} w_m |v_m| e^{j\phi_m} \tag{5.64}$$

where w_m denotes the diversity weight of the mth branch, which is set to $w_m = e^{-j\phi_m}$, that is, the amplitudes are set to one and the phases are adjusted to the same for all branches. Then we have the combined signal as

$$v = \sum_{m=1}^{M} |v_m|. \tag{5.65}$$

On the other hand, the signal will be perturbed by AWGN at each branch. Assuming the AWGN is statistically independent and has the same noise power N, the mean SNR at the output of the EGC diversity receiver will be given by

$$\gamma_{EGC} = \frac{\left(\sum_{m=1}^{M} |v_m|\right)^2}{\sum_{m=1}^{M} N} = \frac{\left(\sum_{m=1}^{M} \sqrt{\gamma_m}\right)^2}{M} \tag{5.66}$$

because $\gamma_m = |v_m|^2/N$.

Since EGC diversity does not require the amplitude information of the received signals, in other words, it only needs the phase information to make the outputs of the different diversity branches be co-phased before being summed, it is

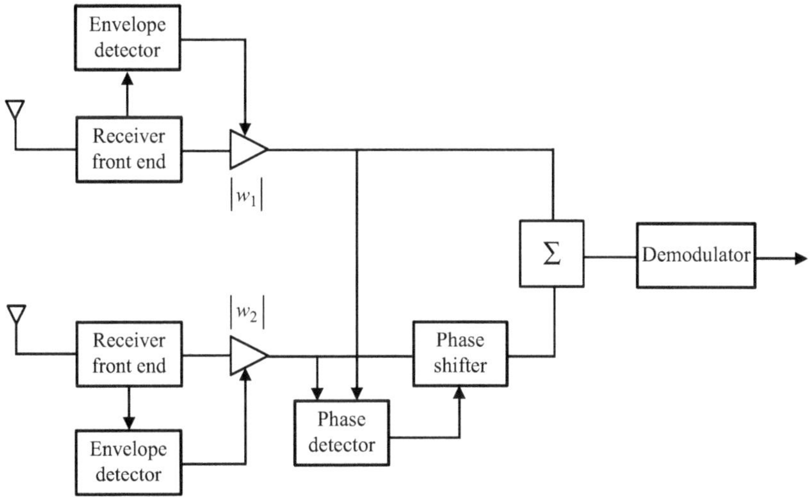

Figure 5.32 Block diagram of MRC diversity

advantageous in terms of the simple implementation and better performance compared with the SC diversity.

The third kind of combining technique of diversity is MRC diversity. Figure 5.32 shows the block diagram for MRC diversity. The received signals are not only phase-corrected but also weighted by the amplitude. The weighting factor for the amplitude is proportional to

$$\frac{|v_m|}{\sqrt{\sum\limits_{m=1}^{M} |v_m|^2}}$$

in the mth branch so that

$$w_m = \frac{|v_m|}{\sqrt{\sum\limits_{m=1}^{M} |v_m|^2}} e^{-j\phi_m}. \tag{5.67}$$

Then from Equation 5.64 the combined receive signal in MRC diversity is

$$v = \sum_{m=1}^{M} w_m |v_m| e^{j\phi_m} = \sum_{m=1}^{M} \frac{|v_m|^2}{\sqrt{\sum\limits_{m=1}^{M} |v_m|^2}} = \sqrt{\sum_{m=1}^{M} |v_m|^2} \tag{5.68}$$

and the combined noise power is

$$N_{MRC} = \sum_{m=1}^{M} |w_m|^2 N = N. \tag{5.69}$$

So the mean SNR for the MRC diversity is given by

$$\gamma_{MRC} = \frac{\sum_{m=1}^{M} |v_m|^2}{N} = \sum_{m=1}^{M} \gamma_m. \tag{5.70}$$

The MRC diversity obviously performs better than EGC. However, it suffers from the necessity of detecting the signal envelopes in all branches, which yields a significant increase in receiver complexity.

The CDF is a useful index to demonstrate the diversity effect. Since the received signals vary according to the log-normal distribution in body area communications, the combined signal also approximately follows the log-normal distribution as described in the previous section. We plot the CDF for single branch, two-branch EGC and two-branch MRC as a function of normalized received signal level in Figure 5.33 as an example in the log-normal fading environment. We assume the received signal levels to be v_1 and v_2 in two log-normal channels; the equivalent received signal level for EGC diversity and MRC diversity are equal to $(v_1 + v_2)/2^{1/2}$ and $\left(v_1^2 + v_2^2\right)^{1/2}$, respectively, based on Equations 5.66 and 5.70.

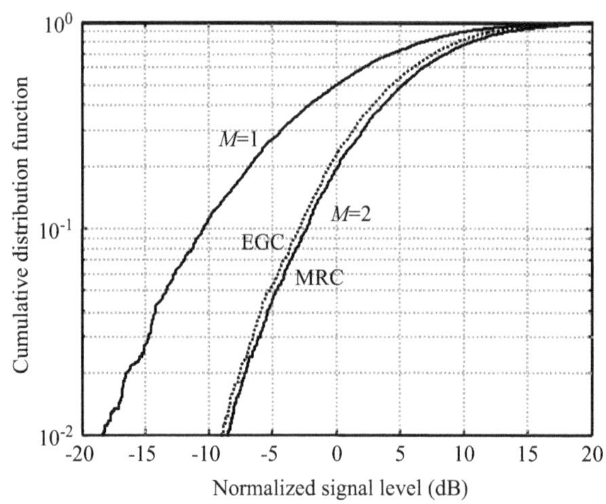

Figure 5.33 CDF of diversity in log-normal environment

At 1% CDF value, for example, both the EGC and MRC improve the signal level by nearly 10 dB. However, the noise power in EGC diversity is twice that in MRC diversity. This means MRC has superior performance to EGC.

Assuming the average SNR at each branch is the same, the average bit error probability for EGC diversity and MRC diversity are given, respectively, by

$$P_{b,EGC}(\bar{\gamma}_{EGC}) = \int_0^{\infty} P_0(\gamma_{EGC}) p(\gamma_{EGC}) d\gamma_{EGC} \tag{5.71}$$

and

$$P_{b,MRC}(\bar{\gamma}_{MRC}) = \int_0^{\infty} P_0(\gamma_{MRC}) p(\gamma_{MRC}) d\gamma_{MRC} \tag{5.72}$$

where $\bar{\gamma}_{EGC}$ or $\bar{\gamma}_{MRC}$ is the average SNR at each branch, $P_0(\gamma_{EGC})$ or $P_0(\gamma_{MRC})$ is the bit error probability in AWGN channel, and $p(\gamma_{EGC})$ or $p(\gamma_{MRC})$ is the PDF on γ_{EGC} or γ_{MRC} when $\bar{\gamma}_{EGC}$ or $\bar{\gamma}_{MRC}$ is given. The PDFs can be derived by using the MGF method as described in the previous section.

A spatial diversity reception is especially useful in in-body communication for improving the shadow fading, as shown in the next chapter.

References

Abramowitz, M. and Stegun, I. (1972) *Handbook of Mathematical Functions with Formulas, Graphs, and Mathematical Tables*, 9th edn, Dover, Washington, DC.

Arthurs, E. and Dym, H. (1962) On the optimum detection of digital signals in the presence of while Gaussian noise – A geometric interpretation and a study of three basic data transmission systems. *IRE Transactions on Communications Systems*, **10** (4), 336–372.

Benedetto, D.M.-G. and Giancola, G. (2004) *Understanding Ultra Wide Band Radio Fundamentals*, Prentice Hall, New Jersey.

Dubouloz, S., Denis, B., de Rivaz, S., and Ouvry, L. (2005) Performance analysis of LDR UWB non-coherent receivers in multipath environments. Proceedings of IEEE International Conference on Ultra-Wideband, pp. 491–496.

Dufresne, D. (2009) *Sums of Lognormals, Centre for Actuarial Studies*, University of Melbourne, Melbourne.

ECMA (2005) Standard ECMA-368: High data rate ultra wideband PHY and MAC standard, December 2005 [Online]. http://www.ecma-international.org/publications/files/ECMA-ST/ECMA-368.pdf.

FCC (2002) Revision of Part 15 of the Commission's rules regarding ultra-wideband transmission system: first report and order, Technical Report FCC 02-48 (Adopted February 14, 2002; Released April 22, 2002).

Haykin, S. (1988) *Digital Communications*, John Wiley & Sons, Ltd, New York.

IEEE P802.15 (2004) Multi-band OFDM physical layer proposal for IEEE 802.15 Task Group 3a (Doc. Number P802.15-03/268r3).

Lindsey, W.C. and Simon, M.K. (1973) *Telecommunications Systems Engineering*, Prentice Hall, New Jersey.

Price, R. and Green, P.E. Jr (1958) A communication technique for multipath channels. *Proceedings of IRE*, **46**, 555–570.

Proakis, J.G. (2007) *Digital Communications*, 4th edn, McGraw-Hill, Boston.

Shi, J. and Wang, J. (2010) Channel characterization and diversity feasibility for in-body to on-body communication using low-band UWB signals. Proceedings of the 3rd International Symposium on Applied Science in Biomedical and Communication Techniques, Rome, Italy.

Smith, D.R. (1985) *Digital Transmission Systems*, Van Nostrand Reinhold, New York.

Win, M.Z. and Scholtz, R.A. (1998) On the energy capture of ultrawide bandwidth signals in dense multipath environments. *IEEE Communications Letters*, **2** (9), 245–247.

Wu, J., Mehta, N., and Zhang, J. (2005) A flexible lognormal sum approximation. Proceedings of IEEE Global Telecommunications Conference, pp. 3413–3417.

6

Body Area Communication Performance

6.1 Introduction

The communication performance of a body area communication system is directly influenced by the channel it operates in. Body area channels differ from traditional wireless indoor/outdoor channels on account of the fact that great changes take place in signal propagation on or in or off the body. As a special transmission medium for wireless communication, human body tissue brings new challenges to channel modeling as well as to communication performance evaluation. Chapter 4 has generalized body area channels in the form of a static shadow fading channel and a dynamic multipath fading channel. Body area channels are characterized by different communication frequencies and communication channel position. The channel models have been further classified into different kinds according to the frequency band as well as the channel position in the body area. In this chapter, we will evaluate the communication performance in terms of frequency and communication channel position, that is, according to the following four major types: on-body UWB communication; in-body UWB communication; in-body MICS band communication; and human body communication, respectively. For in-body UWB communication, two application scenarios will be covered: in-body to on-body transmission for capsule endoscope application and in-body to off-body transmission for cardiac pacemaker application. Finally, a dual mode scheme will be described which can combine both the human body communication function and the in-body to on-body UWB/MICS communication function.

The methodology for evaluating the average bit error rate (BER) performance will be first derived in both the static channel and dynamic multipath channel. For each communication type, the average BER performance in both the static shadow fading

Body Area Communications: Channel Modeling, Communication Systems, and EMC, First Edition.
Jianqing Wang and Qiong Wang.

channel and dynamic multipath fading channel will be presented in detail. Link budget analysis will then be described in terms of maximum system safety margins based on the derived average BER performance. Last but not the least, we will discuss several communication parameters including the communication distance, data rate as well as the required transmit power for all communication types in each respective section.

6.2 On-Body UWB Communication

The on-body UWB path loss model derived in Section 4.2.2 and the on-body UWB dynamic multipath fading channel model derived in Section 4.3.2 enable the analysis and evaluation of on-body UWB communication performance. In this section, we will first introduce the methodologies for evaluating the average BER performance in a static shadow fading channel and in a dynamic multipath fading channel. Then a link budget analysis will be addressed based on the derived BER performance in the on-body UWB channel. In addition, the BER performance versus communication distance will also be presented.

6.2.1 Bit Error Rate

6.2.1.1 Average BER in a Static Shadow Fading Channel

The on-body UWB path loss model of Equation 4.12 accounts for the shadowing effect on the human body. The shadowing is mainly induced by the diffraction in the shadowed regions of the body. It can directly result in variation of the received signal at the receiver front end. The instantaneous SNR is a random variable at the receiver which includes the effect of shadowing. In fact, the variation of the SNR is the direct indicator of the shadowing effect.

Equation 5.42 describes the average BER in a dynamic multipath fading channel, in which the instantaneous SNR is also a random variable at the receiver. This equation averages the bit error probability in AWGN channel over the PDF of the SNR. In fact, the conditional error probability in Equation 5.42 applies as long as the received SNR is a statistical random variable with a certain PDF. Now let us derive the average BER for arbitrary on-body receiver locations. In this case the received SNR varies in a specific statistical distribution due to the shadow fading. First we write the conditional error probability expression again here as

$$P_b(\bar{\gamma}) = \int_0^\infty P_0(\gamma)p(\gamma)d\gamma \tag{6.1}$$

where $\bar{\gamma}$ represents the average SNR, $P_b(\bar{\gamma})$ is the average BER in a static shadow fading channel, $P_0(\gamma)$ is the BER in a AWGN channel, and $p(\gamma)$ denotes the PDF of γ. As noted here, the term "average" in the error probability refers to statistical averaging over the probability distribution of the SNR. Therefore, we need to know the probability distribution of the SNR for deriving the average BER in a static shadow fading channel.

The derivation process is as follows:

1. Derive the statistical characteristics of path loss
   According to the log-distance path loss model in Equation 4.12, we can rewrite the path loss as follows
   $$PL_{dB} = PL_{ave,dB} + S_{dB} \tag{6.2}$$
   in which
   $$PL_{ave,dB} = PL_{0,dB} + 10n\log_{10}\frac{d}{d_0} \tag{6.3}$$
   and S_{dB} is normal distributed with standard deviation σ_{dB}.
   Then PL_{dB} is also normal distributed with mean value $PL_{ave,dB}$ and standard deviation σ_{dB}. Since
   $$PL_{dB} = 10\log_{10} PL = \frac{10}{\ln 10}\ln PL \tag{6.4}$$
   we have
   $$\ln PL = \frac{\ln 10}{10} PL_{dB} = \alpha PL_{dB} \quad \alpha = \frac{\ln 10}{10}. \tag{6.5}$$
   According to the relationship between $\ln PL$ and PL_{dB}, it can be deduced that $\ln PL$ is also normal distributed with mean value $\alpha PL_{ave,dB}$ and standard deviation $\alpha\sigma_{dB}$, where α is a constant shown in Equation 6.5, $PL_{ave,dB}$ and σ_{dB} are known in the derived path loss model.
2. Derive the statistical characteristics of E_b/N_0
   The received power P_r is related to the transmitted power P_t via the path loss PL
   $$P_r = P_t/PL \tag{6.6}$$
   Then the received energy per bit E_b is $E_b = P_r/f_b$ in which f_b is the data rate. Therefore the energy per bit to noise power spectral density ratio can be written as follows
   $$\frac{E_b}{N_0} = \frac{P_r/f_b}{N_0} = \frac{P_t}{N_0 f_b}\cdot\frac{1}{PL} \tag{6.7}$$
   where N_0 is the thermal noise unilateral power spectral density. E_b/N_0 is an important figure of merit and is also often utilized in SNR calculation.

Equation 6.7 can be rewritten as

$$\ln\left(\frac{E_b}{N_0}\right) = \ln\left(\frac{P_t}{N_0 f_b}\right) - \ln PL = A - \ln PL \tag{6.8}$$

where A is a constant. The ln (E_b/N_0) is thus normal distributed with mean value

$$\mu_s = A - \alpha PL_{ave,dB} \tag{6.9}$$

and standard deviation

$$\sigma_s = \alpha\sigma_{dB}. \tag{6.10}$$

Consequently, E_b/N_0 is log-normal distributed with mean value

$$\mu_\gamma = e^{\mu_s + \frac{\sigma_s^2}{2}} \tag{6.11}$$

and standard deviation

$$\sigma_\gamma = \mu_\gamma\sqrt{e^{\sigma_s^2} - 1}. \tag{6.12}$$

3. Parameterization for average BER
   According to the statistical characteristics of E_b/N_0, γ is log-normal distributed with

$$\mu_\gamma = \bar{\gamma} \tag{6.13}$$

$$\sigma_\gamma = \bar{\gamma}\sqrt{e^{\alpha^2\sigma_{dB}^2} - 1}. \tag{6.14}$$

It is easily shown that

$$p(\gamma) = \frac{1}{\sqrt{2\pi}\sigma\gamma} e^{-\frac{(\ln\gamma - \mu)^2}{2\sigma^2}} \tag{6.15}$$

where

$$\mu = \ln\mu_\gamma - \frac{\sigma^2}{2} = \ln\bar{\gamma} - \frac{1}{2}\alpha^2\sigma_{dB}^2 \tag{6.16}$$

$$\sigma = \sqrt{\ln\left[1 + \left(\frac{\sigma_\gamma}{\mu_\gamma}\right)^2\right]} = \alpha\sigma_{dB}. \tag{6.17}$$

4. Solve the average BER $P_b(\bar{\gamma}) = \int_0^\infty P_0(\gamma)p_\gamma(\gamma)d\gamma$ using numerical integration

   First, assume the average E_b/N_0 in decibels to be $\bar{\gamma}_{dB,m}$. For example,

$$m = 1 : \bar{\gamma}_{dB,m} = 0\,\text{dB}$$

$$m = 2 : \bar{\gamma}_{dB,m} = 2\,\text{dB}$$

$$m = 3 : \bar{\gamma}_{dB,m} = 4\,\text{dB}$$

$$\ldots\ldots$$

   Then calculate the statistical parameters for the log-normal distributed E_b/N_0.

$$\mu_{\gamma m} = \bar{\gamma}_m = 10^{\bar{\gamma}_{dB,m}/10} \tag{6.18}$$

$$\sigma_{\gamma m} = \bar{\gamma}_m \sqrt{e^{\alpha^2 \sigma_{dB}{}^2} - 1} \tag{6.19}$$

   Finally, solve the numerical integration

$$P_b(\bar{\gamma}_m) = \sum_{n=1}^{N} P_0(\gamma_n) p_m(\gamma_n)(\gamma_n - \gamma_{n-1}) \tag{6.20}$$

   where

$$p_m(\gamma_n) = \frac{1}{\sqrt{2\pi}\sigma_m \gamma_n} e^{-\frac{(\ln\gamma_n - \mu_m)^2}{2\sigma_m{}^2}} \tag{6.21}$$

$$\mu_m = \ln \bar{\gamma}_m - \frac{1}{2}\alpha^2 \sigma_{dB}{}^2 \tag{6.22}$$

$$\sigma_m = \alpha \sigma_{dB}. \tag{6.23}$$

The above four steps describe the detailed derivation process for the average BER in a static shadow fading channel. As can be seen, the statistical characteristic of E_b/N_0 is dependent on the shadowing characteristic of the path loss model. The BER

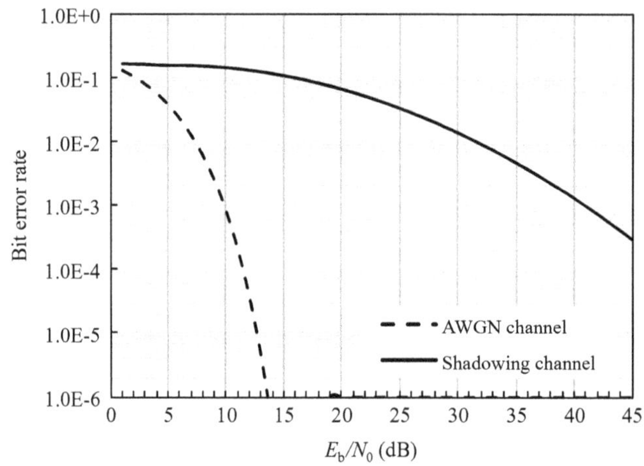

Figure 6.1 BER comparison in AWGN channel and on-body UWB static shadow fading channel ($\sigma_{dB} = 10$dB) for IR-UWB with PPM scheme

performance degradation is inevitable in a static shadow fading channel compared with an AWGN channel. The standard deviation value of the shadowing will primarily determine the communication performance degradation with reference to the AWGN channel.

Using the derived shadowing parameters in Section 4.2.2 and the derived bit error probabilities for IR-UWB in an AWGN channel in Section 5.3.3, we can obtain the average BER performance as described above. Figures 6.1 and 6.2 show the

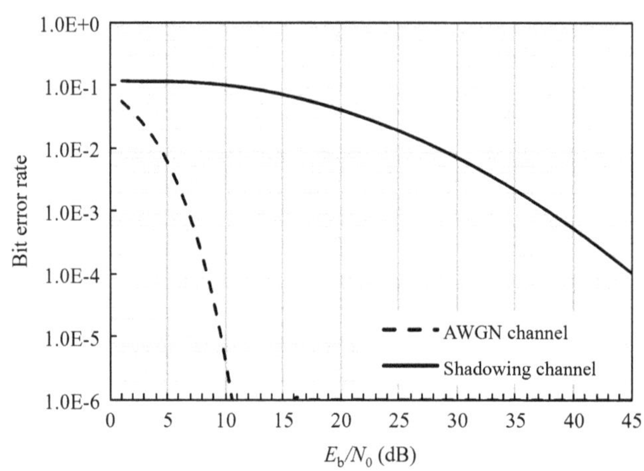

Figure 6.2 BER comparison in AWGN channel and on-body UWB static shadow fading channel ($\sigma_{dB} = 10$ dB) for IR-UWB with PAM scheme

comparisons between an ideal BER in an AWGN channel and a deteriorated average BER for binary PPM and PAM. As noted, the average BER performance of PPM in on-body UWB static shadow fading channel is 3 dB poorer than PAM. Moreover, the above derivation applies for the traditional optimum demodulation scheme. From the results, it can be noted that the average BER performance in the on-body UWB static shadow fading channel has deteriorated significantly.

6.2.1.2 Average BER in a Dynamic Multipath Fading Channel

On-body UWB communication performances are mainly decided by the on-body UWB multipath propagation channel. As shown in Section 4.3.2, the multipaths in on-body UWB communication are essentially induced by the reflection due to movement of one or more parts of the body. The multipath channel makes electromagnetic waves traveling along the body reflected, diffracted, and scattered, so that a received signal ends up being the superimposition of several attenuated, delayed, and eventually distorted replicas of a transmitted waveform. System performance is therefore bound to be degraded by the distortion of pulses due to propagation over such a real channel. However, as described in Chapter 5, the performance degradation due to multipath propagation can be mitigated via multiple correlators in RAKE reception or CP in MB-OFDM demodulation if a detailed characterization of the multipath channel can be achieved at the receiver. Section 4.3.2 provides detailed parameterization for typical on-body UWB multipath channels on the upper body, which potentiates the implementation of the anti-multipath technique as well as the analysis of communication performances.

As described in Section 4.3, the RMS delay spread is a good measure of multipath spread and it gives an indication of the potential for inter-symbol interference. In fact, it is the multipath spread that restricts the practical communication performance, including data rate, communication distance, and so on. The most direct behavior is that the delay spread will put a constraint on the potential data rate. In a case where a high data rate is expected for a known on-body UWB channel, the error bits will increase and the BER performance will deteriorate.

Based on the findings in Section 4.3, the characteristics of the on-body dynamic multipath fading channels differ considerably for different propagation links due to the variability of the link geometry. Now let us focus on one of the five representative transmission links shown in Figure 4.25, that is, the chest-to-right-waist channel. The RMS delay spread for the chest-to-right-waist channel is summarized in Table 6.1. As can be seen, the RMS delay spread is around 0.47 ns. The BER performance is shown in Figure 6.3 with increasing data rate using the conventional correlation receiver described in Section 5.3. Table 6.2 shows the simulation specifications. More than 10 discrete impulse response functions are used for the BER simulation to have a statistically average performance since the BER result depends on the particular realization of the channel impulse response.

Table 6.1 FDTD-derived and modeled RMS delay spread

	FDTD-derived	Modeled
Mean	0.469	0.470
Standard deviation	0.145	0.177

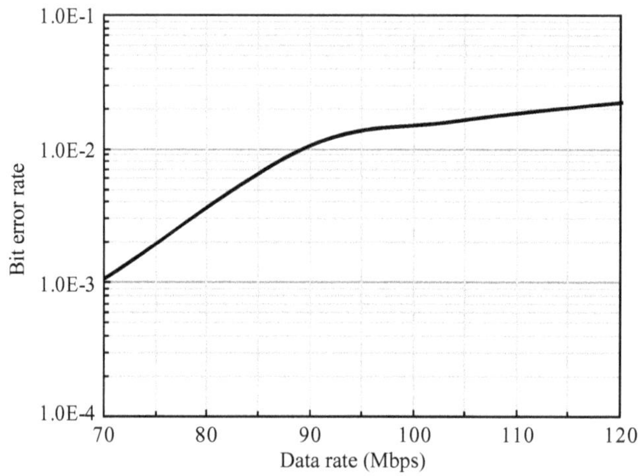

Figure 6.3 BER versus data rate performance with a correlation receiver (Wang and Wang, 2010). Reproduced with permissions from Wang Q. and Wang J., "Performance of ultra wideband on-body communication based on statistical channel model," *IEICE Transactions on Communications*, **E93-B**, 4, 833–841, 2010

The BER performance is obtained using the communication simulation tool Advanced Design System (ADS, Agilent) (Wang and Wang, 2010). The BER performance in Figure 6.3 is under the FCC permissible maximum UWB transmitting power. As expected, when the data rate goes up to 90 Mbps, the BER becomes even worse than 0.01.

In error correction theory, the probability of error 0.01 or 0.001 is usually taken as a criterion threshold in which the error correction code can work effectively. For

Table 6.2 Data rate simulation specifications

UWB frequency band (GHz)	3.1–10.6
Pulse	Second derivative Gaussian pulse
Chips per bit	16
Modulation scheme	PPM
Receiver	Correlation detection

example, in Bic, Duponteil, and Imbeaux (1991) it is shown that the intersection point of BER with coding and BER without coding, that is, the point of coding gain equal to zero, is between BER = 0.01 and BER = 0.001 for forward error correction. It is therefore reasonable to use 0.01 and 0.001 as indices. That is to say, as long as the BER is not worse than 0.01 or 0.001, the error bits can be detected and then corrected using specific error correction codes. In turn, if the error probability is very high and surpasses the threshold, the error correction coding may not be practical.

Based on the error correction theory, we can conclude that the maximum desirable data rates for the chest-to-right-waist channel with correlation receiver are 70 Mbps and 90 Mbps, which correspond to a BER threshold of 0.001 and 0.01, respectively.

At the two maximum data rates, the BER performance with a correlation receiver is given in Figure 6.4 to compare with the ideal performance under AWGN channel. Meanwhile, the BER performance with correlation receiver at 10 Mbps is also shown in Figure 6.4 as a representative data rate. It can be noted that the performance of the correlation receiver is not so satisfactory in the multipath-affected on-body channel since the BER performance deteriorated badly. Actually, the conventional correlation receiver is not appropriate for the multipath-affected on-body channel since its structure foresees the presence of a correlator that is matched to one single waveform, while in the multipath-affected on-body channel the superposition of multiple signals is not evitable. An advanced receiver structure is therefore needed to improve the communication performance. Yet, the simple structure of a correlation receiver still has an advantage for on-body communications.

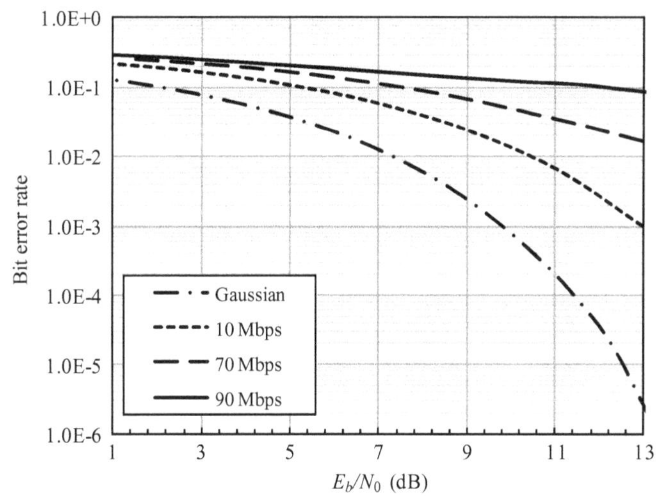

Figure 6.4 BER performance at different data rates with a correlation receiver

As described in Chapter 5, the performance degradation can be mitigated using a RAKE receiver if a detailed characterization of the multipath-affected channel is known. The RAKE receiver can improve the BER performance since it takes advantage of multipaths to enhance the whole SNR. On the other hand, the adoption of a RAKE considerably increases the complexity of the receiver structure. This complexity increases with the number of multipath components to be analyzed and the corresponding finger number, and can be reduced by decreasing the number of components processed by the receiver. However, a reduction of the number of multipaths to be analyzed leads to a decrease in energy collected by the receiver.

For a known on-body UWB dynamic multipath fading channel, it is possible to derive a reasonable finger number in order to have a rational use of the RAKE structure. Based on the previous FDTD-calculated results in Section 4.3.2, the probability of multipaths as well as the captured energy of multipaths can be calculated for the chest-to-right-waist channel. Figure 6.5 shows the probability versus the number of multipaths. Four effective multipaths have the highest probability in all of the possible multipaths. Figure 6.6 shows the energy percentage captured by multipaths. As noted, about 80% of the received energy is captured by the first two multipaths, and 92% of the received energy is captured by the first four multipaths. Therefore, 2-finger and 4-finger RAKE receivers, which correspond to the first two and first four multipath components, can be reasonably employed.

Assume that the optimum MRC scheme is used in the simulation set-up. Figures 6.7–6.9 show the BER performances with 2- and 4-finger RAKE receivers at the data rates 10, 70 and 90 Mbps, respectively. Here 10 Mbps is also evaluated as a representative data rate for on-body communication. As expected, both the 2-finger and 4-finger RAKE receiver significantly improves the BER performance compared with the conventional correlation receiver. At a data rate of 10 Mbps, the degradation in performance of the 2-finger and 4-finger RAKE receivers is less than 2 dB at a BER of 0.01 with respect to the ideal AWGN channel, while at a data rate of 70 and

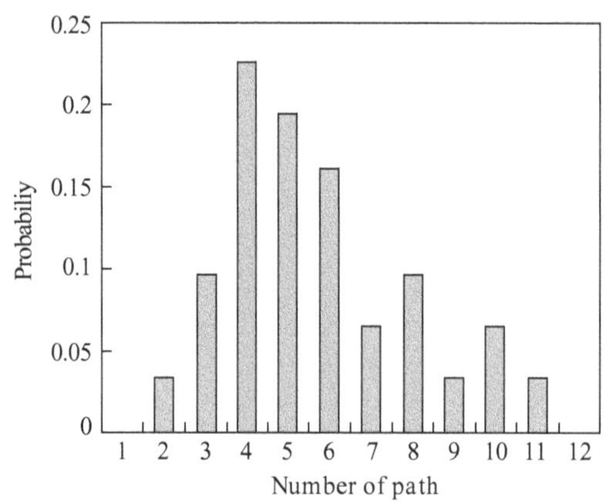

Figure 6.5 Probability versus number of multipaths

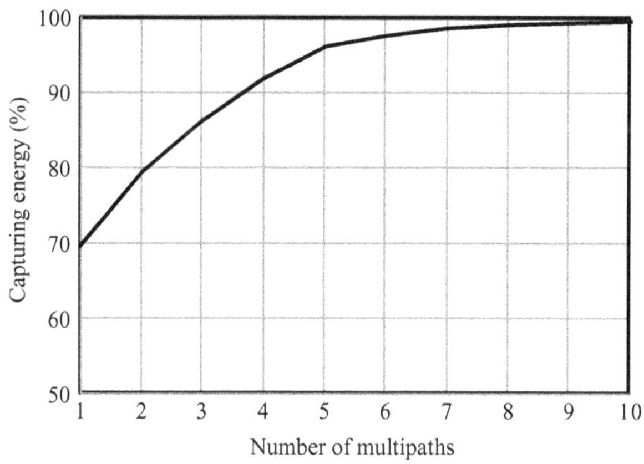

Figure 6.6 Energy percentage captured by multipaths

90 Mbps, the performance degradations are less than 3 and 4 dB, respectively. The BER of the 4-finger RAKE receiver is superior to that of the 2-finger RAKE receiver. At a BER of 0.01, the performance degradation of the 4-finger RAKE receiver for 10 Mbps is about 1 dB and for 70 and 90 Mbps is about 2 dB compared with the ideal Gaussian channel. These results demonstrate that both the 2-finger and 4-finger RAKE receivers are effective in the on-body multipath channel and they can considerably improve the communication performance.

However, as described above, simplicity in the device structure is also an important issue which cannot be ignored. The complicated structure for the 4-finger RAKE receiver may hinder its application in wearable body area communication. In this sense, the 2-finger RAKE receiver can be an option as a compromise between the communication performance and structure complication, since it can get satisfactory

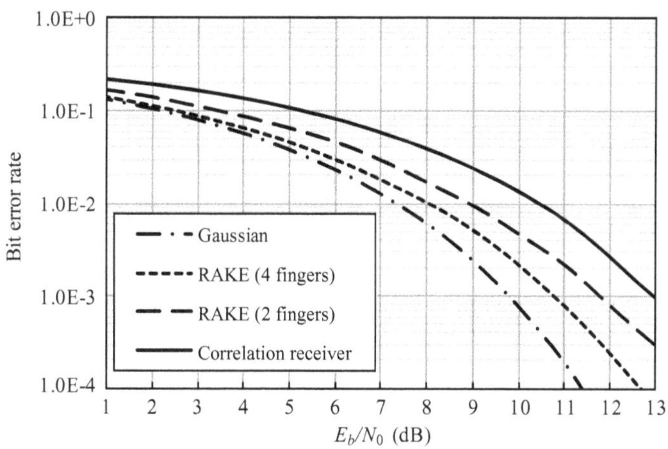

Figure 6.7 BER at a data rate of 10 Mbps

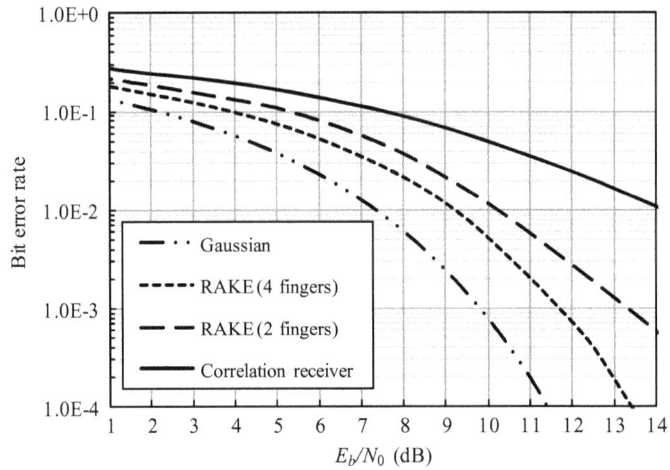

Figure 6.8 BER at a data rate of 70 Mbps (Wang and Wang, 2010). Reproduced with permissions from Wang Q. and Wang J., "Performance of ultra wideband on-body communication based on statistical channel model," *IEICE Transactions on Communications*, **E93-B**, 4, 833–841, 2010

performance whilst having a relatively simple structure compared with the 4-finger RAKE receiver.

Based on the previous analysis taking the chest-to-right-waist channel as the example, now let us take a brief look at the BER performance of the other four

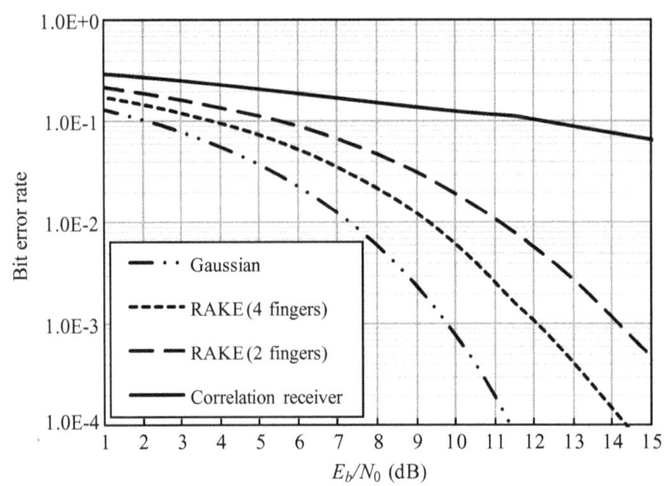

Figure 6.9 BER at a data rate of 90 Mbps (Wang and Wang, 2010). Reproduced with permissions from Wang Q. and Wang J., "Performance of ultra wideband on-body communication based on statistical channel model," *IEICE Transactions on Communications*, **E93-B**, 4, 833–841, 2010

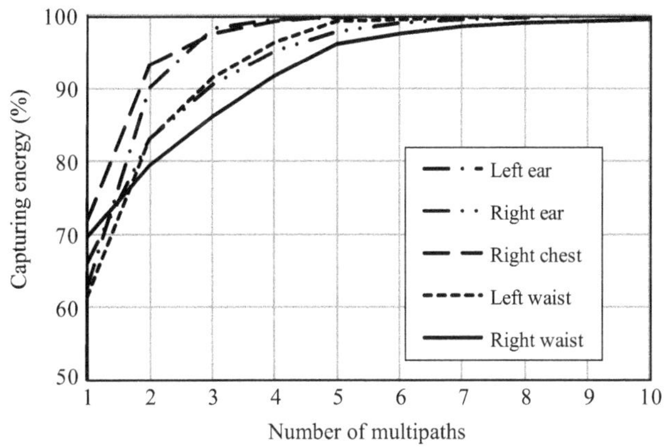

Figure 6.10 Energy percentage captured by multipaths for the five transmission links

channels corresponding to receiver locations at the left ear, right ear, right chest and left waist, respectively. Figure 6.10 shows the energy percentage captured by multipaths. As noted, more than 95% of the received energy can be captured by the first four multipaths for the other four channels. Figure 6.11 compares the BER performance of all five channels using a correlation receiver and 4-finger RAKE receiver, respectively, and the corresponding data rate is 70 Mbps. The BER performance for the five channels is very similar using a 4-finger RAKE receiver since over 95% energy has been captured even though the channel model parameters are different as

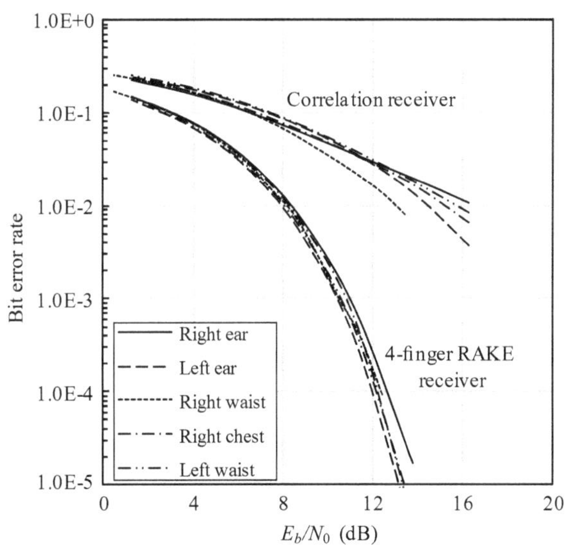

Figure 6.11 BER at a data rate of 70 Mbps for the five communication links

shown in Table 4.10. The BER performance using a correlation receiver differs slightly since it mainly depends on the captured energy by the first multipath.

6.2.2 Link Budget

A link budget is the accounting of all of the gains and losses from the transmitter, through the transmission medium to the receiver in a communication system. It accounts for the attenuation of the transmitted signal due to propagation, as well as the antenna gains, feedline and miscellaneous losses. In order to analyze the link budget for an on-body UWB channel, apart from the derived path loss, the possible signal power as well as the noise characteristics also have to be calculated.

The UWB emission mask issued by the FCC (2002) is designed to make UWB radio signals coexist with other radio signals. This emission mask imposes a limit on the maximum allowed transmitting power at any given frequency. For the whole UWB frequency band, that is, from 3.1 to 10.6 GHz, the maximum emission power density is not allowed to exceed −41.3 dBm/MHz. Note that the FCC emission mask refers to a unilateral PSD $P_M(f) = -41.3$ dBm/MHz; therefore, the maximum allowed transmitting power $P_{M\,max}$ for a signal occupying the whole UWB frequency band is (Benedetto and Giancola, 2004)

$$\begin{aligned} P_{M\,max}(\text{dBm}) &= 10\log_{10} \int_{3.1\times 10^3}^{10.6\times 10^3} P_M(f)df \\ &= -41.3 + 10\log_{10}(10.6\times 10^3 - 3.1\times 10^3) \\ &\cong -2.55\ \text{dBm} \end{aligned} \tag{6.24}$$

that is

$$P_{M\,max} \cong 0.55\ \text{mW}. \tag{6.25}$$

Similarly, if the signal bandwidth is 500 MHz locating within the frequency range 3.1–10.6 GHz, which corresponds to the definition for the MB-UWB solution, the maximum allowed transmitting power $P_{M\,max}$ is

$$P_{M\,max}(\text{dBm}) = -41.3 + 10\log_{10}500 = -14.3\ \text{dBm}. \tag{6.26}$$

The FCC regulation illustrates the fact that the UWB radio signal must meet the emission mask and the maximum allowed transmitting power requirements. Signal power up to the maximum allowed transmitting power can be considered as the best case in performance analysis for the UWB radio link. Given the allowed maximum power, we can now evaluate the system margin when a predetermined probability of error must be guaranteed at the receiver.

First, let us determine the noise characteristics at the receiver. Let us assume that the only noise source at the receiver is AWGN. This noise is typically thermal, introduced by the receiving antenna and the front-end circuit of the receiver. The thermal noise PSD expressed in joules, that is, W/Hz is given by:

$$N_0 = k[T_a + (N_F - 1)T_0] \tag{6.27}$$

where T_a is the receiving antenna temperature, $T_0 = 300\,\mathrm{K}$ is the environment temperature, $k = 1.38 \times 10^{-23}\mathrm{J/K}$ is the Boltzmann constant, and finally, N_F is the noise figure of the receiving device. Note that N_0 here is the unilateral PSD and the corresponding bilateral PSD is half.

Since the receiving device is worn on the human body, a reasonable hypothesis is introduced that $T_a = T_0 = 300$ K. Equation 6.27 can therefore be rewritten as

$$N_0 = kT_0N_F \tag{6.28}$$

and in decibels

$$N_{0,dB} = 10\log_{10}(kT_0) + N_{F,dB}. \tag{6.29}$$

Secondly, the link E_b/N_0 can be determined at a given data rate f_b for the communication system.

Based on Equation 4.2, the received power at the receiver front end can be more accurately rewritten as follows

$$P_{r,dBW} = P_{t,dBW} + G_{TX,dB} - L_{TX,dB} - PL_{dB} - L_{M,dB} + G_{RX,dB} - L_{RX,dB} \tag{6.30}$$

which includes all the gains and losses from the transmitter and receiver as well as the miscellaneous losses and the explanation for each physical quantity can be found in Table 6.3. Then the link SNR or E_b/N_0 in decibels is

$$E_b/N_{0,dB} = P_{r,dBW} - 10\log_{10}f_b - N_{0,dB}. \tag{6.31}$$

Finally, we define a system margin M_s by

$$M_s = \frac{E_b/N_0}{[E_b/N_0]_{spec}} \tag{6.32}$$

where $[E_b/N_0]_{spec}$ denotes the required E_b/N_0 for obtaining a specific probability of error.

In Equation 6.32, if the link E_b/N_0 exceeds the required $[E_b/N_0]_{spec}$, which means system margin $M_s \geq 0$ dB, the wireless communication is feasible. The larger the system margin, the more reliable and robust the communication transmission.

Table 6.3 Parameters for on-body UWB link budget analysis

Transmitter and receiver	
Frequency (GHz)	3.1–10.6
Transmitter output power P_t ([dBm)	−2.55
Transmitter antenna gain G_{TX} ([dBi)	0
Transmitter losses L_{TX} (dB)	0
Miscellaneous losses L_M (dB)	3
Receiver antenna gain G_{RX} (dBi)	0
Receiver losses L_{RX} (dB)	0
Standard temperature T (K)	300
Receiver noise figure N_F (dB)	6
Boltzmann constant k (J/K)	1.38E-23
Signal quality	
Bite error rate	10^{-3}
$[E_b/N_0]_{spec}$ (dB) (static channel with shadowing)	38 (PAM, correlation receiver) 41 (PPM, correlation receiver)
$[E_b/N_0]_{spec}$ [dB] (dynamic multipath fading channel)	11.8 (PPM-TH, 2-finger RAKE 10 Mbps) 10.8 (PPM-TH, 4-finger RAKE 10 Mbps) 13.2 (PPM-TH, 2-finger RAKE 70 Mbps) 11.7 (PPM-TH, 4-finger RAKE 70 Mbps) 14.2 (PPM-TH, 2-finger RAKE 90 Mbps) 12.1 (PPM-TH, 4-finger RAKE 90 Mbps)
Coding gain (dB)	0

Similarly, according to the error correction theory, here we use BER = 0.001 as the predetermined probability of error threshold. For a correlation receiver, the average BER in the on-body UWB static shadow fading channel is shown in Figures 6.1 and 6.2. As noted, the required E_b/N_0 for binary PAM and PPM are 38 and 41 dB, respectively. Based on the system parameters in Table 6.3, Figures 6.12 and 6.13 show the dependence of the system margin on the distance with data rates of 0.1, 1 and 10 Mbps. It can be found that the system has a margin larger than 0 dB for PPM and PAM signals at a communication distance of about 0.3 m and 0.4 m, respectively, for data rate less than 1 Mbps, while for data rate of 10 Mbps, the communication distance is greatly reduced to 0.2 m or less.

On the other hand, in a dynamic multipath fading environment, an advanced receiver structure like the RAKE receiver can be expected to provide a better system

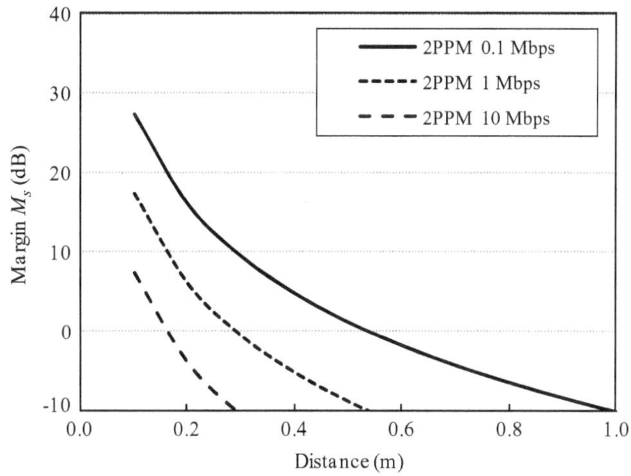

Figure 6.12 Maximum system safety margins as a function of the distance in on-body static shadow fading channel for PPM signals at different data rates

margin. Based on the results in Figures 6.7–6.9, we can summarize the required E_b/N_0 for 2-finger and 4-finger RAKE receivers with different data rates as shown in Table 6.3. The system margin is shown in Figure 6.14. As expected, the communication distance has been largely improved with the 2-finger or 4-finger RAKE receivers. At a communication distance of nearly 0.6 m, the maximum system margin can be larger than 0 dB even at a data rate as high as 90 Mbps. Within a communication distance of 1.3 m, the system margin is always larger than 0 dB at a data rate of 10 Mbps. In fact, a distance of 1.3 m can nearly cover all communication ranges on

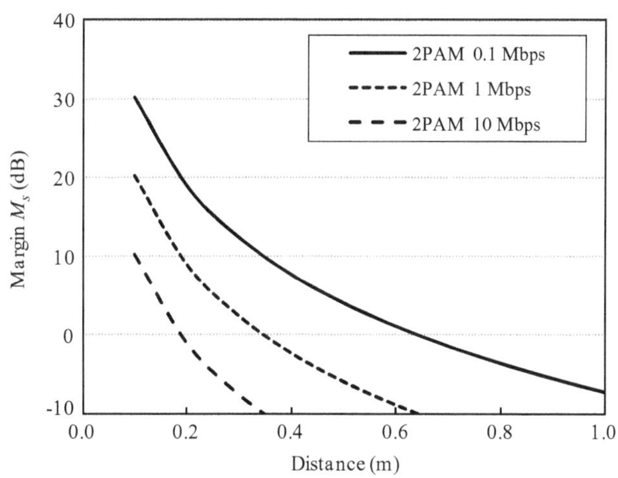

Figure 6.13 Maximum system safety margins as a function of the distance in on-body static shadow fading channel for PAM signals at different data rates

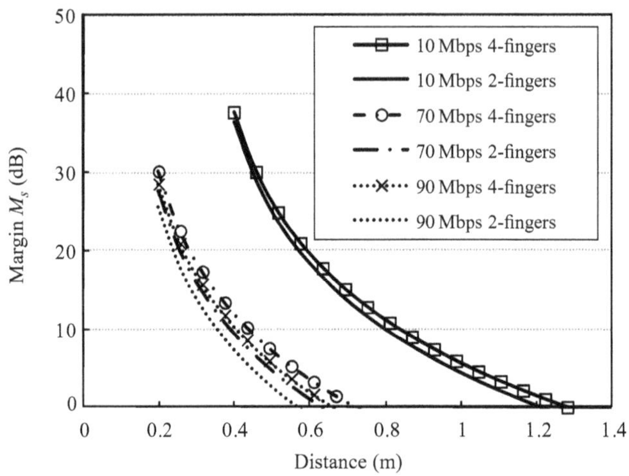

Figure 6.14 Maximum system safety margins as a function of the distance in on-body dynamic multipath fading channel for PPM signals with RAKE receiver

the human body. In the link budget analysis, keep in mind that the available maximum power (−2.55 dBm) is fully exploited. In addition, reducing the data rate can effectively increase the maximum communication distance.

6.2.3 Maximum Communication Distance

Figures 6.7–6.9 describe the relationship between the average E_b/N_0 and the average BER in multipath fading channels. Based on the path loss Equation 4.10 and the maximum allowed transmitting power Equations 6.24 and 6.25, the maximum receiving power can be obtained based on Equation 6.30 and therefore the E_b/N_0 term can be derived as a function of propagation distance d as follows

$$\begin{aligned} E_b/N_{0,dB} &= P_{r,dBW} - 10\log_{10}f_b - N_{0,dB} \\ &= P_{t,dBW} - PL_{dB} - 10\log_{10}f_b - N_{0,dB} \\ &= P_{t,dBW} - PL_{0,dB} - 10n\log_{10}\left[\frac{d}{d_0}\right] - 10\log_{10}f_b - N_{0,dB} \end{aligned} \tag{6.33}$$

where the gains and losses from the transmitter and receiver as well as the miscellaneous losses in Equation 6.30 are assumed to be zero. In this case, based on Equation 6.33, we can get the average BER as a function of propagation distance at a given data rate f_b. These analysis and derivation procedures hold, given that there will not be a big difference in the channel statistical characteristics when the communication distance spreads to the whole body, which is basically reasonable.

Figures 6.15–6.17 show the BER versus communication distance at data rates of 10, 70 and 90 Mbps, respectively, for PPM correlation detection. For ensuring a BER

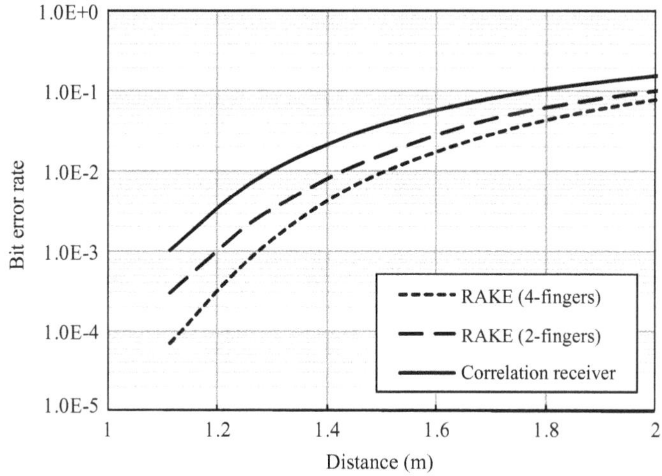

Figure 6.15 BER versus distance at a data rate of 10 Mbps

of 0.001, at a data rate of 10 Mbps, the maximum communication distance with correlation receiver is about 1.1 m while with RAKE receiver (4-finger) is about 1.3 m. At a data rate of 70 Mbps, the maximum communication distance with the correlation receiver deteriorates heavily while with a RAKE receiver it is about 0.7 m. At a data rate of 90 Mbps, similar behavior holds and the distance with correlation receiver deteriorates heavily and the distance with a RAKE receiver is reduced to 0.65 m.

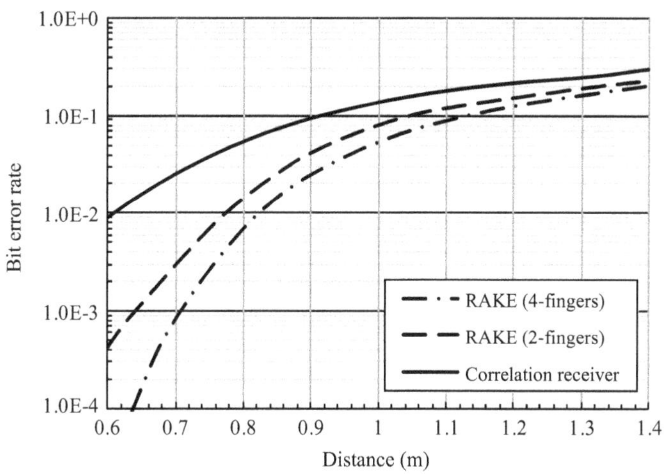

Figure 6.16 BER versus distance at a data rate of 70 Mbps (Wang and Wang, 2010). Reproduced with permissions from Wang Q. and Wang J., "Performance of ultra wideband on-body communication based on statistical channel model," *IEICE Transactions on Communications*, **E93-B**, 4, 833–841, 2010

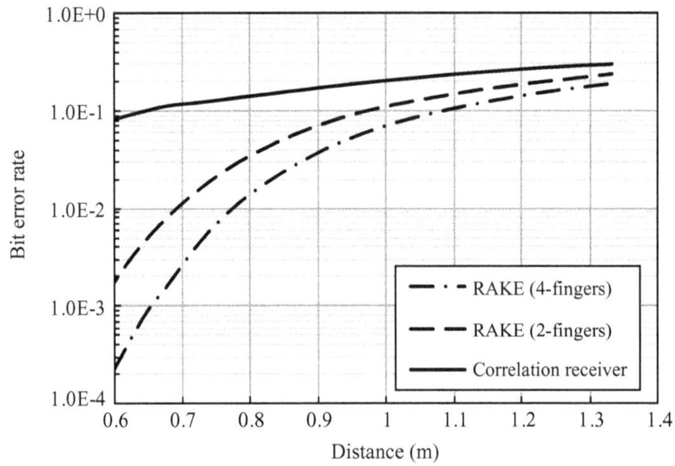

Figure 6.17 BER versus distance at a data rate of 90 Mbps (Wang and Wang, 2010). Reproduced with permissions from Wang Q. and Wang J., "Performance of ultra wideband on-body communication based on statistical channel model," *IEICE Transactions on Communications*, **E93-B**, 4, 833–841, 2010

Table 6.4 gives the quantitative description of the communication distances at two BER thresholds. At the given BER thresholds and the given maximum data rate, the communication distance is around 0.6 m. This means that under the FCC emission spectral mask, the effective communication distance for UWB-PPM-TH with 2-finger and 4-finger RAKE receivers is at least 0.6 m on the human body at the maximum data rate. If a lower data rate such as 10 Mbps is acceptable, the communication distance on the human body could be longer than 1.3 m.

The BER versus distance performance with different data rates in Figures 6.15–6.17 are based on the path loss characteristic mainly on the front side of the human body. If the communication link is over a longer communication distance, the path loss will increase. Under the condition that the maximum transmitting power is maintained, in order to keep the same BER at 0.01 or 0.001, the data rate has to be decreased to compensate for the power loss due to the longer communication distance.

Table 6.4 Maximum communication distances at different data rates for PPM

	10 Mbps		70 Mbps		90 Mbps	
	2-finger	4-finger	2-finger	4-finger	2-finger	4-finger
BER = 0.01	1.42	1.50	0.77	0.83	0.70	0.78
BER = 0.001	1.20	1.28	0.64	0.71	0.57	0.65

6.3 In-Body UWB Communication

The in-body UWB channel models enable the analysis and evaluation of communication performances for capsule endoscope and cardiac pacemaker applications. In this section, with the same methodology as described in the previous section, we will derive the average BER performances for the in-body UWB channels. We will also conduct a link budget analysis based on the derived BER performances, and discuss the relationship of the data rate versus required transmitting power.

6.3.1 Bit Error Rate

6.3.1.1 In-Body to On-Body Transmission for Capsule Endoscope Application

The in-body to on-body channel model can be used for the scenario of a capsule endoscope application. In this application, the SNR variation is mainly induced by the transmitter locations inside the body. The transmitted signals propagate through different tissues and organs due to the different transmitting locations of the swallowed capsule moving along the digestive organs. This SNR variation here is referred to as the shadow fading. In order to derive the average BER with this randomly varying SNR, we should know the PDF of the SNR. It has been derived that the shadow fading follows a log-normal distribution. According to the same derivation process described in Section 6.2.1, we can obtain the average BER performance for representative UWB modulation schemes.

Figures 6.18 and 6.19 show the comparisons between an ideal BER in an AWGN channel and a deteriorated average BER for IR-UWB with PPM and OOK, respectively. In both schemes noncoherent energy detection is employed at the receiver. As noted, the average BER performance of OOK is 3 dB lower in comparison with that

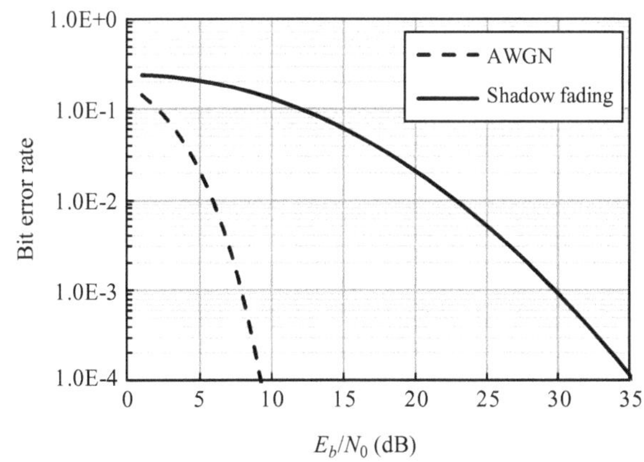

Figure 6.18 BER comparison in AWGN channel and in-body shadow fading channel for IR-UWB with PPM scheme and noncoherent detection

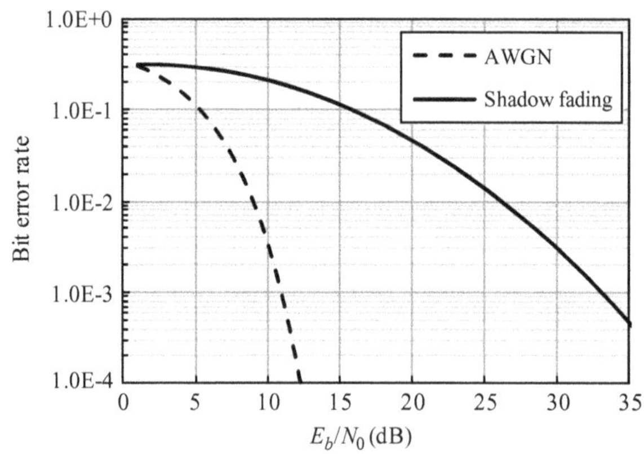

Figure 6.19 BER comparison in AWGN channel and in-body shadow fading channel for IR-UWB with OOK scheme and noncoherent detection

of PPM. This is because with the same data rate and noise power in the two considered schemes, the bandwidth per bit with PPM is twice that with OOK, which contributes to a half noise power spectral density for PPM. From the results, it can be noted that the average BER performance in the in-body to on-body UWB channel for capsule endoscopy has deteriorated significantly.

Diversity reception is an effective means of improving the BER performance. Diversity reception refers to a method of improving the reliability of a signal by combining two or more communication channels with different characteristics. This can be achieved by placing two or more antennas at the receiver, which is known as spatial diversity reception. The analysis model shown in Figure 4.6 agrees with this placement requirement. The combining method considered here is MRC in terms of spatial diversity. It represents a theoretically optimal combiner compared with the other diversity schemes in which the desired SNR can be maximized over fading channels in a communication system.

Based on the in-body to on-body channel characteristics derived in Sections 4.2.3 and 4.3.3, the combined signal with MRC diversity over two log-normal distributed channels can be denoted as a log-normal sum distribution. The derivation of this log-normal sum PDF is of great importance for evaluating the average BER in this in-body to on-body channel. Given the importance of the log-normal sum distribution in analyzing this in-body to on-body wireless system performance, effort has to be devoted to derive the PDF of the log-normal sum distribution. However, as stated in Section 5.4, closed-form theoretical expressions for log-normal sum PDF are unknown. In contrast, an approximation method using the MGF is an analytical and flexible approach in which the log-normal sum is still approximated as a log-normal distribution. This method matches a short Gauss–Hermite approximation of the MGF of the log-normal sum with that of a log-normal distribution. The mean μ_Y and standard deviation σ_Y in decibels in the log-normal sum PDF are obtained from

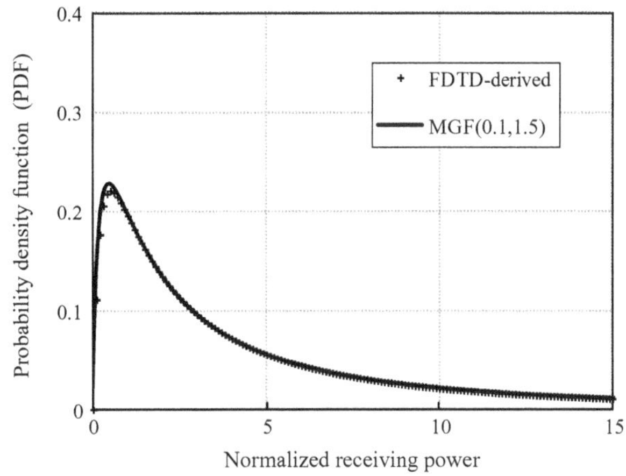

Figure 6.20 PDF of the received power based on FDTD-derived data and that approximated by the log-normal sum approximation with MGF

Equation 5.47. The Hermite integration order N and parameter S_m for adjusting the weighted integrals of the short Gauss–Hermit integration play an important role in an accurate log-normal sum approximation. In the in-body to on-body channel, based on the fitting accuracy to the FDTD-derived PDF of the received power, the parameter S_m is determined as $(S_1, S_2) = (0.1, 1.5)$, and N is determined as six. Then the nonlinear equation shown as Equation 5.47 with the unknowns μ_Y and σ_Y can be numerically solved, for example, using *fsolve* function in MATLAB®.

Figure 6.20 shows the PDF of the received power based on FDTD-derived data and that approximated by the log-normal sum approximation with MGF (Shi, Anzai, and Wang, 2011). The log-normal sum approximation corresponds to the MRC diversity reception of Rx1 and Rx2 in Figure 4.6. The approximated two parameters μ and σ in the log-normal sum for MRC diversity are shown in Table 6.5. As can be seen from Figure 6.20, this method gives a very good approximation to the FDTD-derived PDF. In addition, compared with the results in Table 6.5 for a single branch, we can observe that the parameter σ for MRC diversity becomes smaller.

By using the log-normal sum PDF with approximated parameters in Table 6.5 and the bit error probability for PPM or OOK in AWGN channel with noncoherent detection, we can calculate the average BER for MRC diversity of Rx1 and Rx2 with

Table 6.5 Parameters of the log-normal distribution

	μ	σ
Rx1	0	1.88
Rx2	0	1.76
Rx1 and Rx2	1.19	1.38

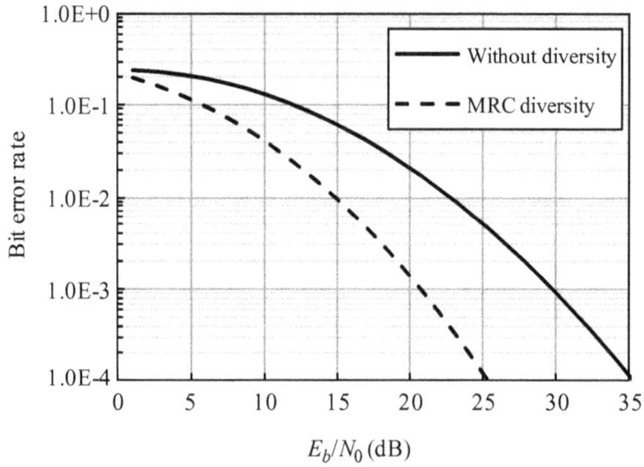

Figure 6.21 Average BER performances of MRC diversity for IR-UWB with PPM scheme and noncoherent detection

Equation 6.1. The average BER performances of MRC diversity are shown in Figures 6.21 and 6.22, respectively, for the IR-UWB with PPM and OOK schemes. Significant BER improvements are obtained around 10 dB for both PPM and OOK at BER of 10^{-3}, and a diversity gain of 2 is achieved.

It should be noted, however, that the theoretical derivation of the log-normal sum distribution assumes statistical independence between the two received signals. This is not really true because a correlation coefficient between two single received signals may exist. This reduces the diversity effect. Such two channels which have a smaller correlation coefficient should therefore be chosen to obtain an effective diversity improvement.

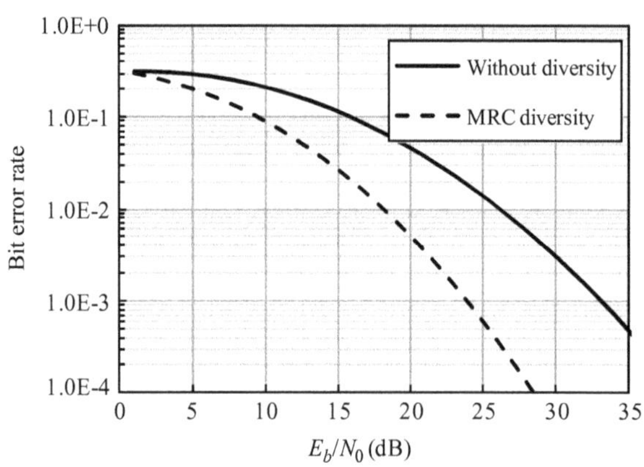

Figure 6.22 Average BER performances of MRC diversity for IR-UWB with OOK scheme and noncoherent detection

6.3.1.2 In-Body to Off-Body Transmission for Cardiac Pacemaker Application

The details of the in-body to off-body channel model for the cardiac pacemaker application are given in Sections 4.2.3 and 4.3.3. The SNR variation is mainly due to the shadowing of various tissues at different transmitter or receiver locations. A two-path impulse response model is sufficient to represent the channel characteristics. The corresponding parameters for the statistical distributions are shown in Table 4.13.

With the same methodology and approach as in the in-body to on-body case, we can derive the BER performance using Equation 6.1. However, instead of the theoretical analysis approach, we here employ computer simulation to derive the BER performance (Wang, Masami, and Wang, 2011). This approach gives the same BER result in principle as the theoretical analysis approach.

The transmitter still adopts the IR-UWB with PPM scheme, and the receiver employs correlation detection. Table 6.6 gives the simulation specifications for the communication performance evaluation.

The BER performance in the in-body to off-body chest channel is shown in Figure 6.23. It is an average BER for more than 20 discrete impulse responses induced with the parameters in Table 4.13. This is because the BER depends on the particular realization of the channel impulse response. Compared with the capsule endoscope channel, the average BER performance is much better because of the smaller path loss in the chest channel. However, an E_b/N_0 of about 18 dB is still required to obtain a BER of 0.001, which should be attributed to the multipath influence.

In an IR-UWB system, the RAKE receiver takes advantage of multiple correlators to cope with the multipath problems. To make a RAKE receiver work, different multipath components of the same transmitted pulse have to be analyzed separately and eventually combined. However, in the UWB low band system, the pulse width is around 2.1 ns, while the inter-path delay may be smaller. This means that it is almost impossible to separate each multipath from the received signals because the pulse width may be larger than the inter-path delay. However, the RAKE receiver can work effectively if it has prior knowledge of the multipaths.

Based on the characterization of the in-body chest channel, the first two paths dominate and the inter-path delay between them is around 1.43 ns with a very small

Table 6.6 Simulation specifications

Frequency band (GHz)	3.4–4.8
Pulse shape	Second derivative Gaussian pulse
Pulse width (ns)	2.1
Modulation scheme	IR-UWB with PPM
Data rate (Mbps)	10
Chip period (ns)	12.5
Chips per bit	8
Receiver structure	Correlation detection or RAKE reception

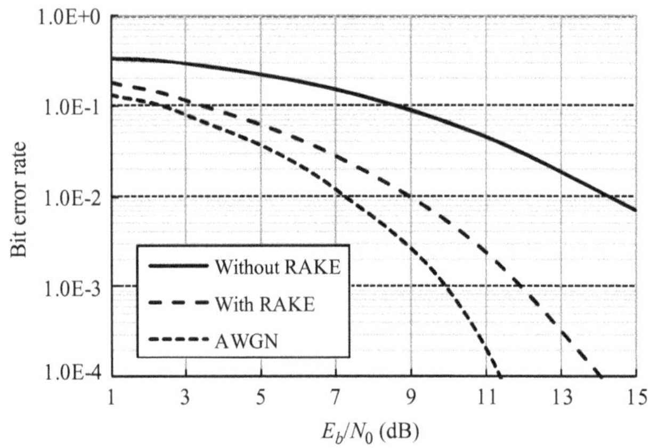

Figure 6.23 Average BER performances of 2-finger RAKE receiver for IR-UWB with PPM scheme and correlation detection

deviation. It is therefore reasonable to use a 2-finger RAKE receiver and fix the time delay of the second finger to 1.43 ns. Such a RAKE structure can avoid the channel estimation for the arrival time of the multipath components and significantly reduces the receiver complexity. The simulated average BER is shown in Figure 6.23 for the 2-finger RAKE receiver. The BER performance is effectively improved nearly 6 dB at a BER of 0.001 compared with that without RAKE reception. The degradation in performance of the 2-finger RAKE receiver is <2 dB with respect to the correlation receiver in an ideal AWGN channel.

6.3.2 *Link Budget*

6.3.2.1 In-Body to On-Body Transmission for Capsule Endoscope Application

In order to analyze the link budget for the in-body to on-body transmission in UWB low band for capsule endoscope application, in addition to the derived path loss, the possible signal power and noise characteristics also have to be calculated.

The UWB emission mask issued by the FCC imposes the limit on the maximum allowed transmitting power. For the UWB low band, for example, from 3.4 to 4.8 GHz, the maximum allowed transmitting power is 0.1 mW or −10 dBm. Given the allowed maximum power, we can evaluate the system margin when a predetermined probability of error must be guaranteed at the receiver.

As described in Section 6.2, we use Equation 6.30 to obtain the received power under the maximum transmitting power, and Equation 6.31 to obtain the link SNR or E_b/N_0. With BER = 0.001 as the predetermined BER threshold, we can get the required E_b/N_0 of IR-UWB with PPM and OOK for noncoherent energy detection. The corresponding E_b/N_0 values are 30 and 33 dB, respectively, without diversity, and 21 and 24 dB, respectively, with MRC diversity.

Table 6.7 Parameters for in-body to on-body UWB link budget analysis

Transmitter and receiver	
Frequency (GHz)	3.4–4.8
Transmitter output power P_t (dBm)	−10
Transmitter antenna gain G_{TX} (dBi)	0
Receiver antenna gain G_{RX} (dBi)	0
Standard temperature T (K)	300
Receiver noise figure N_F (dB)	6
Boltzmann constant k (J/K)	1.38E-23
Signal quality	
Bite error rate	10^{-3}
$[E_b/N_0]_{spec}$ (dB)	30 (PPM, noncoherent detection)
	33 (OOK, noncoherent detection)
	21 (PPM, 2-branch MRC diversity)
	24 (OOK, 2-branch MRC diversity)
Coding gain (dB)	0

Using the assumed system parameters in Table 6.7, Figures 6.24 and 6.25 show the dependence of system margin on distance with data rates of 0.1, 1 and 10 Mbps for IR-UWB with PPM and OOK, respectively. It can be found that, at a data rate of 1 Mbps, the system has a margin larger than 0 dB for PPM at a distance of 9 cm and for OOK at a distance of 8 cm both from the body surface. If the data rate is increased to 10 Mbps, the possible communication distance will be limited to only 6.5 cm for PPM and 6 cm for OOK. These results suggest that a further BER improvement is necessary to make the data transmission from the digestive organs possible. Figures 6.26 and 6.27 show the dependence of system margin on distance with the MRC diversity reception. Such an improved receiver structure provides a better system margin. As can be seen, in almost all of the transmitter locations inside

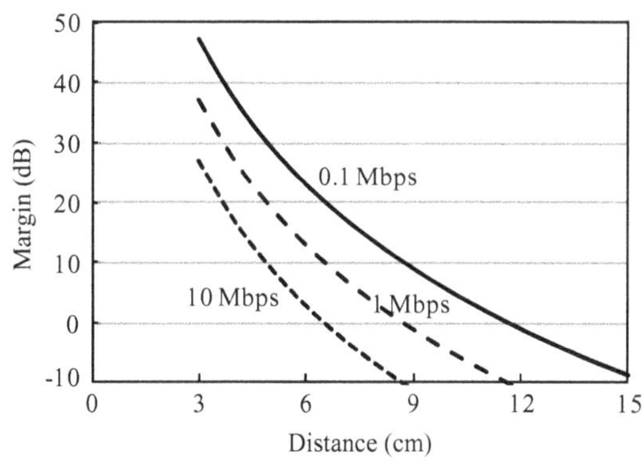

Figure 6.24 Dependence of system margin on distance for PPM-IR-UWB

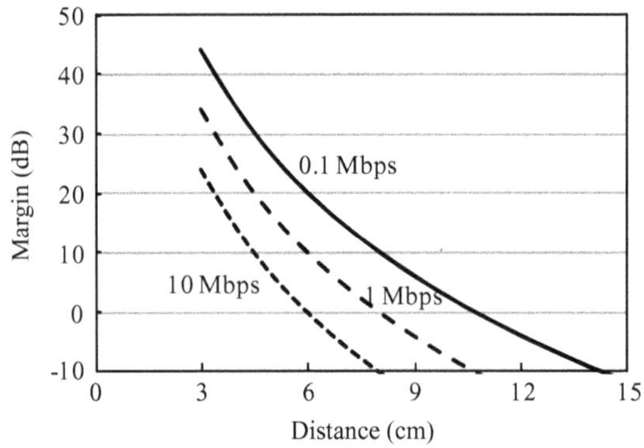

Figure 6.25 Dependence of system margin on distance for OOK-IR-UWB

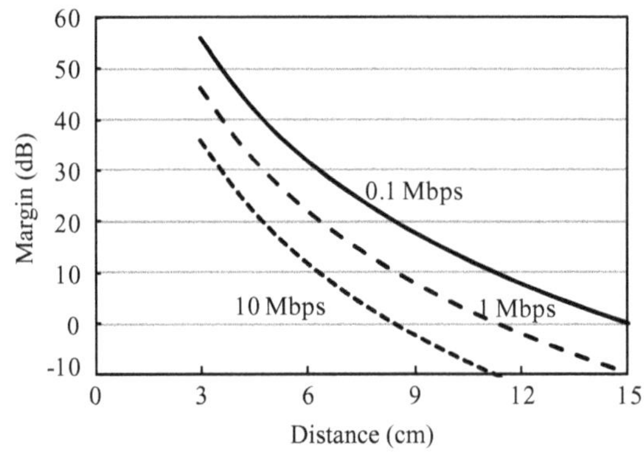

Figure 6.26 Dependence of system margin on distance for PPM-IR-UWB with MRC diversity

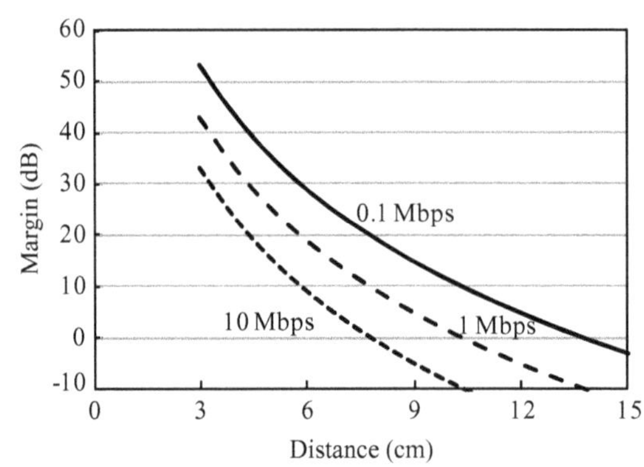

Figure 6.27 Dependence of system margin on distance for OOK-IR-UWB with MRC diversity

the digestive organs which are usually within 15 cm from the body surface, a data rate of 0.1 Mbps can always have a system margin larger than 0 dB for PPM with 2-branch MRC diversity. When the data rate is increased to 1 and 10 Mbps, however, the corresponding communication distances will be reduced to 11.5 and 8.5 cm, respectively. The possible communication distance for OOK is somewhat shorter compared with PPM. Although the 2-branch spatial diversity has improved the link budget effectively, it seems insufficient to secure a reliable wireless link for the capsule endoscope application. A higher order diversity or larger transmitting power is required for a high data rate transmission.

The above analysis is under the condition that the allowable maximum transmitting power is −10 dBm based on the FCC UWB regulation. However, this regulation is defined in an indoor environment. For a UWB transmitter inside the human body, the signals arriving at the body surface will be largely attenuated due to the lossy tissues. This means that a larger transmitting power may be allowed as long as the signals at the body surface are sufficiently weak. It is therefore meaningful to clarify the relationship between the data rate and the required transmitting power. Under the 2-branch diversity reception, we use Equation 6.33 to derive the data rate f_b. Here E_b/N_0 is set to be $[E_b/N_0]_{spec}$ to obtain a BER of 0.001, that is, 21 dB for IR-UWB with PPM and 24 dB for IR-UWB with OOK. Figure 6.28 shows the required transmitting power as a function of data rate at a communication distance of 15 cm, which covers almost all of the possible distance for a capsule endoscope application. The result indicates a required transmitting power of at least 10 dBm (10 mW) for achieving a data rate as high as 10 Mbps for the in-body to on-body transmission. Such a data rate can be expected to provide a high-quality image transmission for a capsule endoscope application. Moreover, if the transmitting power is allowed up to 20 dBm, the data rate may achieve 50 Mbps so that a real-time transmission for uncompressed capsule images becomes possible.

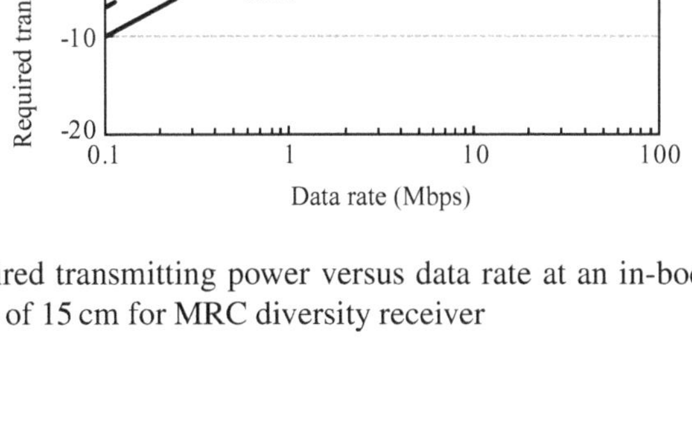

Figure 6.28 Required transmitting power versus data rate at an in-body to on-body communication distance of 15 cm for MRC diversity receiver

Table 6.8 Parameters for in-body to off-body UWB link budget analysis

Transmitter and receiver	
Frequency (GHz)	3.4–4.8
Transmitter output power P_t ([dBm)	−10
Transmitter antenna gain G_{TX} (dBi)	0
Receiver antenna gain G_{RX} (dBi)	0
Standard temperature T (K)	300
Receiver noise figure N_F (dB)]	6
Boltzmann constant k (J/K)	1.38E-23
Signal quality	
Bite error rate	10^{-3}
$[E_b/N_0]_{spec}$ (dB)	18 (PPM, correlation detection)
	12 (PPM, 2-finger RAKE)
Coding gain (dB)	0

6.3.2.2 In-Body to Off-Body Transmission for Cardiac Pacemaker Application

As described in the previous subsection, the maximum allowed transmitting power is 0.1 mW or −10 dBm for the UWB low band. From Equation 6.30 we can obtain the received power when the maximum permissible power is transmitted, and from Equation 6.31 we can obtain the link SNR or E_b/N_0. Also with BER = 0.001 as the predetermined threshold, we can get the required E_b/N_0 for the in-body to off-body transmission for a cardiac pacemaker application. Based on the BER performance in Figure 6.23, the corresponding E_b/N_0 values are 18 dB without RAKE reception and 12 dB with 2-finger RAKE reception, respectively, for the IR-UWB with PPM scheme and correlation detection.

Using the assumed system parameters in Table 6.8, Figure 6.29 shows the dependence of system margin on distance between the implant transmitter and the

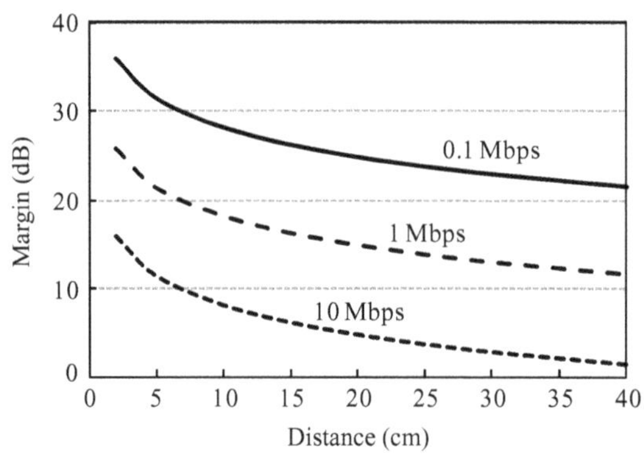

Figure 6.29 Dependence of system margin on distance for PPM-IR-UWB without RAKE reception

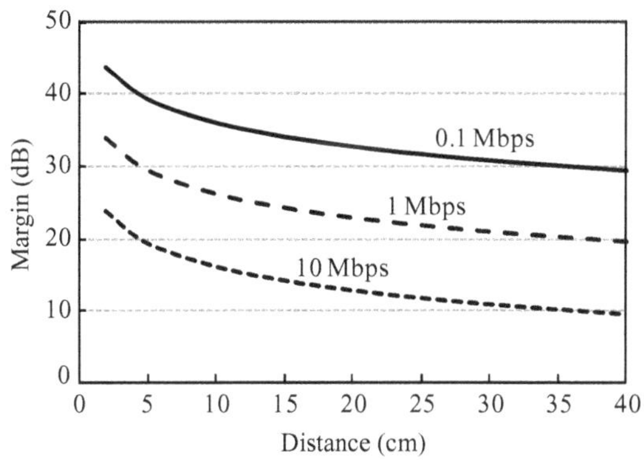

Figure 6.30 Dependence of system margin on distance for PPM-IR-UWB with RAKE reception

off-body receiver with data rates of 0.1, 1 and 10 Mbps for PPM without RAKE reception. It can be seen that the system always has a margin larger than 0 dB within an off-body distance of 40 cm for the transmission from the chest, even at a data rate as high as 10 Mbps. With the employment of 2-finger RAKE reception, as shown in Figure 6.30, the system margin is increased to 10 dB at 10 Mbps and the farthest communication distance of 40 cm. These link budget results demonstrate that a reliable wireless link in the UWB low band is feasible for a cardiac pacemaker application.

On the other hand, we give the relationship between the data rate and required transmitting power for possible transmission at higher data rate. This relationship can be derived from Equation 6.33 with the path-loss term replaced by Equation 4.14. Here E_b/N_0 is set to be $[E_b/N_0]_{spec}$ to obtain a BER of 0.001, that is, 18 dB for

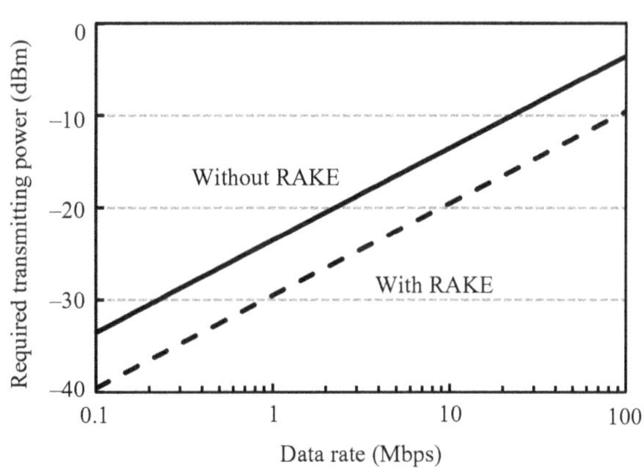

Figure 6.31 Required transmitting power versus data rate at an in-body to off-body chest channel with a communication distance of 40 cm for PPM-IR-UWB

without RAKE reception and 12 dB for with 2-finger RAKE reception. Figure 6.31 shows the required transmitting power as a function of data rate at a communication distance of 40 cm between the implant transmitter and the off-body receiver. The result indicates that, under the maximum allowed transmitting power of −10 dBm, a data rate up to 100 Mbps is possible with RAKE reception for the in-body to off-body UWB transmission for a cardiac pacemaker application.

6.4 In-Body MICS-Band Communication

In-body MICS-band communication covers the scenario of a wireless capsule endoscope application. In this section, with the same methodology as in the previous sections, we will derive the average BER performance for the in-body MICS-band channel. We will also conduct a link budget analysis based on the derived BER performances, and discuss the data rate versus required transmitting power.

6.4.1 Bit Error Rate

On account of the deeper penetration of signals compared with UWB, the MICS band should provide better performance for a real time picture or video transmission from the digestive organs, although the available bandwidth becomes narrow. The SNR variation with the capsule endoscope movement along the digestive organs is referred to as shadow fading here. Since it follows a log-normal distribution as shown in Section 4.2.4, we can use the same method as described in Section 6.2.1 to obtain the PDF of the SNR and then the average BER for representative narrow band modulation schemes.

Figure 6.32 shows a comparison between an ideal BER in an AWGN channel and an average BER for BPSK with optimum demodulation in the shadow fading

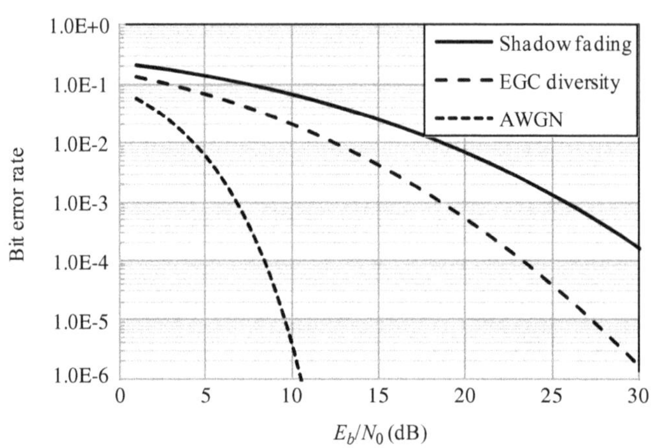

Figure 6.32 BER comparison in AWGN channel and in-body MICS band shadow fading channel for BPSK with optimum demodulation

channel. The BER performance is obviously deteriorated due to the shadow fading. Although it exhibits somewhat better E_b/N_0 versus BER performance than that in the UWB low band, an E_b/N_0 of nearly 26 dB is still required to obtain a BER of 0.001.

In order to improve the BER performance, a spatial diversity technique is also employed in the receiver. The diversity chooses two from the seven receivers placed around the human body as shown in Figure 4.11. The diversity combination adopts the EGC for simplicity because it does not need a phase detector and a shifter. The combined signal with EGC diversity over two log-normal distributed channels is approximated as a log-normal distribution. The MGF method, which matches a short Gauss–Hermite approximation of the MGF of the log-normal sum with that of a log-normal distribution, is then used to derive the PDF of the combined signal. Table 6.9 gives the two parameters μ and σ for two single received signal powers at Rx1 and Rx2 and the EGC combined signal power. The parameter σ is reduced significantly by the EGC diversity.

The PDF for E_b/N_0 is the same as the received signal power except for a shift of the mean value. With the derived PDF, we can calculate the theoretical BER performance for diversity reception. The result is shown in Figure 6.32. Compared with a single branch, the EGC diversity with two branches effectively improves the BER performance by an order of 8 dB. For example, the EGC diversity receiver only needs an E_b/N_0 of 18 dB to obtain an average BER of 0.001. Moreover, the diversity gain can also be confirmed as 2 from the BER curves in Figure 6.32.

6.4.2 Link Budget

In order to analyze the link budget for the MICS-band transmission for capsule endoscope application, we should first prescribe the maximum transmitting power. Since the MICS band allows a maximum transmitting power of 25 μW or −16 dBm, we here use this power level in the link budget analysis. However, from the point of view of biological safety, 20 mW or 13 dBm may be also acceptable since it will never induce an SAR in the human tissue >2 W/kg in any 10 g.

As described in Section 6.2, we use Equation 6.30 to obtain the received power under the maximum transmitting power, and Equation 6.31 to obtain the link SNR or E_b/N_0. With BER = 0.001 as the predetermined BER threshold for a BPSK optimum receiver, we can obtain the required E_b/N_0 to be 26 dB without diversity and 18 dB with two-branch EGC diversity.

Table 6.9 Parameters of the log-normal distribution

	μ	σ
Rx1	0	1.87
Rx2	0	1.70
Rx1 and Rx2	−0.45	1.38

Table 6.10 Parameters for in-body MICS band link budget analysis

Transmitter and receiver	
Frequency (MHz)	400
Transmitter output power P_t (dBm)	−16
Transmitter antenna gain G_{TX} (dBi)	0
Receiver antenna gain G_{RX} (dBi)	0
Standard temperature T (K)	300
Receiver noise figure N_F (dB)	6
Boltzmann constant k ([J/K)	1.38E-23
Signal quality	
Bite error rate	10^{-3}
$[E_b/N_0]_{spec}$ (dB)	26 (BPSK, correlation detection)
	18 (BPSK, 2-finger EGC diversity)
Coding gain (dB)	0

Using assumed system parameters in Table 6.10, Figures 6.33 and 6.34 show the dependence of system margin on distance with data rates of 0.1, 1 and 10 Mbps for a conventional optimum receiver without diversity and with EGC diversity, respectively. It can be seen that, without diversity, the system only has a margin larger than 0 dB within a distance of 12 cm at 0.1 Mbps and 5 cm at 10 Mbps for the transmission from the digestive organs. This is obviously insufficient for a real time capsule endoscope application. With the use of 2-finger EGC diversity, the communication distance may be increased to be more than 17 cm at 0.1 Mbps but is still smaller than 8 cm at 10 Mbps at a system margin larger than 0 dB. At a data rate of 1 Mbps, a communication distance of 12 cm, which basically covers the entire range from the digestive organs to the body surface, is achieved to have a system margin larger than

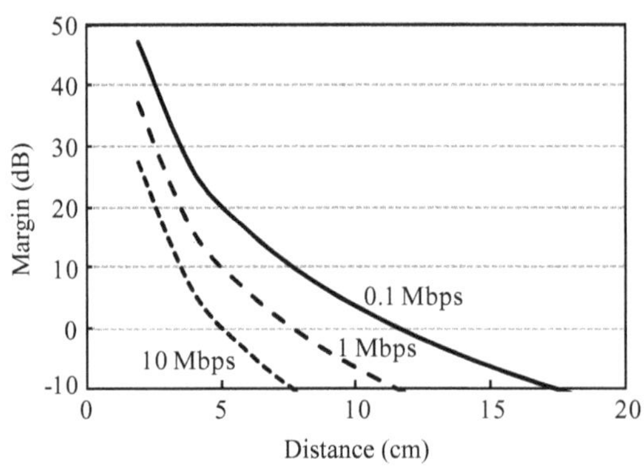

Figure 6.33 Dependence of system margin on distance for conventional BPSK correlation receiver

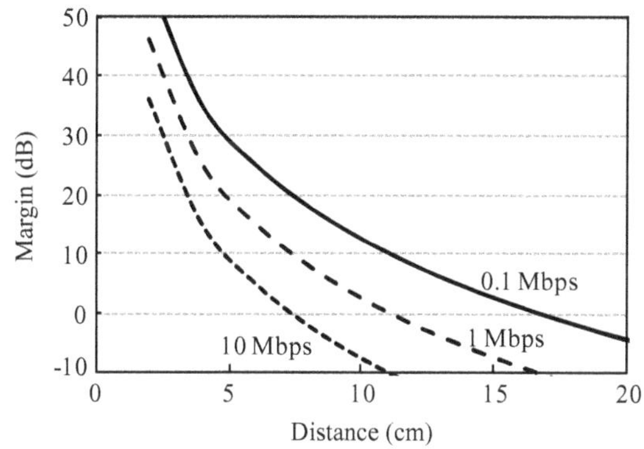

Figure 6.34 Dependence of system margin on distance for BPSK correlation receiver with EGC diversity

0 dB by EGC diversity. This link budget result shows the feasibility for capsule endoscope transmission at a low data rate but the difficulty at a high data rate in the MICS band.

On the other hand, we can give the relationship between the data rate and required transmitting power for possible transmission at higher data rates. This relationship can be derived from Equation 6.33. Here E_b/N_0 is set to be $[E_b/N_0]_{spec}$ to obtain a BER of 0.001, that is, 26 dB without EGC diversity and 18 dB with 2-branch EGC diversity. Figure 6.35 shows the required transmitting power as a function of data rate at a communication distance of 15 cm. The result indicates the feasibility of a high data rate of 10 Mbps under a transmitting power below 0 dBm. Such a transmitting power level does not cause any biological thermal effect in the human body.

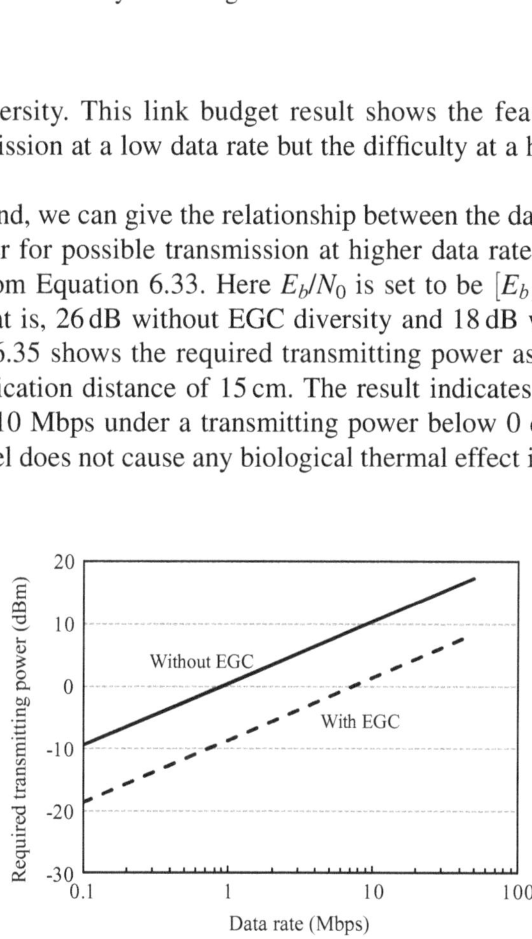

Figure 6.35 Required transmitting power versus data rate for in-body MICS band communication with a communication distance of 15 cm

6.5 Human Body Communication

HBC has an on-body path loss much smaller than that of UWB. Since the signal propagates mainly along the human body itself, the effect of human body posture is not so severe on the communication performance. In this section, with the same methodology as in the previous sections, we will derive the average BER performance using the HBC path loss model in Equation 4.34. We will also analyze the link budget based on the derived BER performances, and discuss the data rate versus required transmitting power.

6.5.1 Bit Error Rate

Figure 6.36 shows a typical example of transmitted and received pulse signals on the human body in the HBC band. Taking the correlation of the two signals we can get a correlation coefficient on the order of 0.9. This means that signal distortion is quite small due to the propagation characteristic of the human body except for an attenuation factor. In addition, the low data rate for HBC means no distinguishable multipath components exist. Thus the BER performance in a HBC channel is similar to that in an AWGN channel. For BPSK with correlation detection, the error probability is given as

$$P_b = Q(\sqrt{2E_b/N_0} \cdot \rho) = \frac{1}{2}\mathrm{erfc}(\sqrt{E_b/N_0} \cdot \rho) \tag{6.34}$$

where ρ is the correlation coefficient between the transmitted and received signals. Figure 6.37 shows the calculated BER in a HBC channel with $\rho = 0.9$. The E_b/N_0

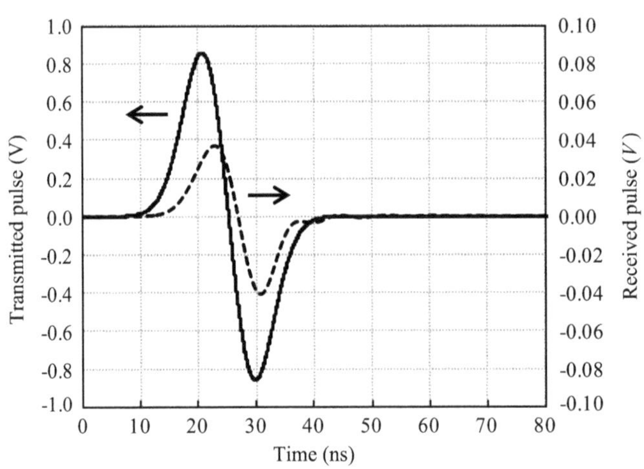

Figure 6.36 Example of transmitted and received pulse signals on the human body in the HBC band

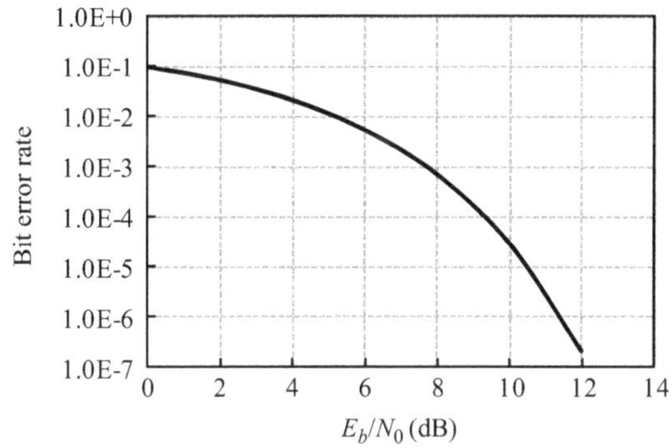

Figure 6.37 BER performance in a HBC channel for BPSK correlation detection

versus BER performance is much better than that of on-body UWB. Only an E_b/N_0 of 8 dB can provide a BER of 0.001.

6.5.2 *Link Budget*

As described in Section 6.4.2, since a transmitting power of 20 mW or 13 dBm will never induce an SAR in the human tissue >2 W/kg in any 10 g, we here use 13 dBm as the allowed maximum transmitting power in the link budget analysis. With BER = 0.001 as the predetermined BER threshold for a BPSK optimum receiver, we can obtain the required E_b/N_0 to be 8 dB from Figure 6.37.

Table 6.11 gives the assumed system parameters for link budget analysis. It should be noted that HBC usually employs electrodes in the transmitter and receiver instead of antennas. The G_{Tx} and G_{Rx} in Table 6.11 are actually the gains for the transmitting and receiving electrodes. Since the electrode structure in the HBC band is

Table 6.11 Parameters for HBC band link budget analysis

Transmitter and receiver	
Frequency (MHz)	10–50
Transmitter output power P_t (dBm)	13
Transmitter electrode gain G_{TX} (dBi)	−10
Receiver electrode gain G_{RX} (dBi)	−10
Standard temperature T (K)	300
Receiver noise figure N_F (dB)	6
Boltzmann constant k (J/K)	1.38E-23
Signal quality	
Bite error rate	10^{-3}
$[E_b/N_0]_{spec}$ [dB]	8 (BPSK, correlation detection)
Coding gain (dB)	0

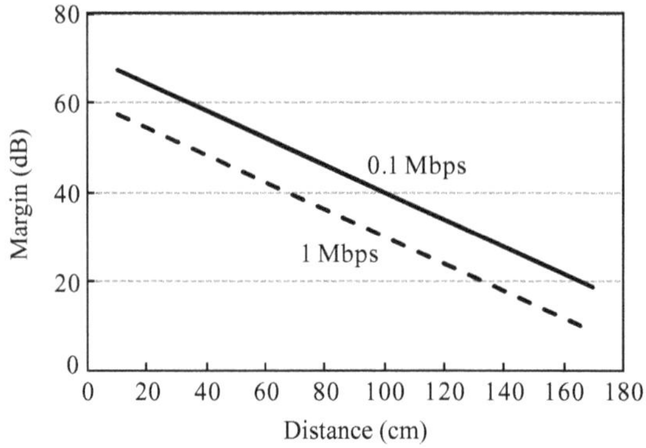

Figure 6.38 Dependence of system margin on distance for on-body HBC communication with BPSK correlation detection

difficult to have a 0 dB gain, we here assume a gain of −10 dB at both the transmitter and receiver. Figure 6.38 shows the dependence of system margin on distance with data rates of 0.1 and 1 Mbps, respectively. It can be seen that the system always has a margin larger than 0 dB for BER = 0.001 within a distance of 1.7 m for a data rate up to 1 Mbps. This link budget result demonstrates the feasibility of the on-body HBC link for covering the entire human body area.

On the other hand, we give the relationship between the data rate and required transmitting power for possible transmission at various data rates. This relationship can be derived from Equation 6.33 with the path-loss term replaced by Equation 4.34. Here E_b/N_0 is set to be $[E_b/N_0]_{spec}$ to obtain a BER of 0.001, that is, 8 dB. Figure 6.39 shows the required transmitting power as a function of data rate at an

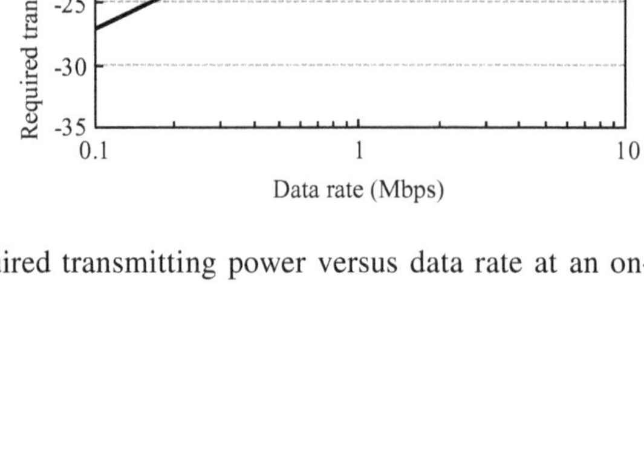

Figure 6.39 Required transmitting power versus data rate at an on-body HBC distance of 1 m

on-body communication distance of 1 m. The result indicates that a low transmitting power is usually sufficient for an on-body transmission with the HBC technique.

6.6 Dual Mode Body Area Communication

A BAN for healthcare and medical applications may need both on-body and in-body communication functions. The HBC band is a good candidate for on-body communication because the human body acts as a transmission medium in this frequency band. It is superior to the GHz band from the point of view of path loss and information security. A high security feature is also important in healthcare and medical BAN. On the other hand, in order to provide high-speed communication in implant BAN, the MICS band and UWB low band are promising candidates for in-body to on-body transmission. A dual model receiver structure can provide an effective network for both on-body and in-body BANs (Cho *et al.*, 2009).

Figure 6.40 shows the concept of dual mode transmission. The on-body transceiver has both the HBC function and the in-body to on-body communication function. In order to have low power consumption, the design of the system requires a structure that is as simple as possible. Transmitting a very short pulse with no RF modulation, that is, the IR-UWB, uses a typically simple structure in which the information data symbols can modulate the pulses with the PPM scheme.

Figure 6.41 shows a block diagram for the realization of a dual model IR-UWB system. The pulses to be transmitted may be sinusoidally modulated Gaussian pulses as shown in Figure 6.42, which are generated by an oscillator and a band pass filter (BPF). The pulse widths are different for different modes: 50 ns for the on-body mode and 2.1 ns for the in-body mode. In the on-body mode, the oscillator generates a sinusoidal signal at 30 MHz, which is then passed through a Gaussian-shaped BPF

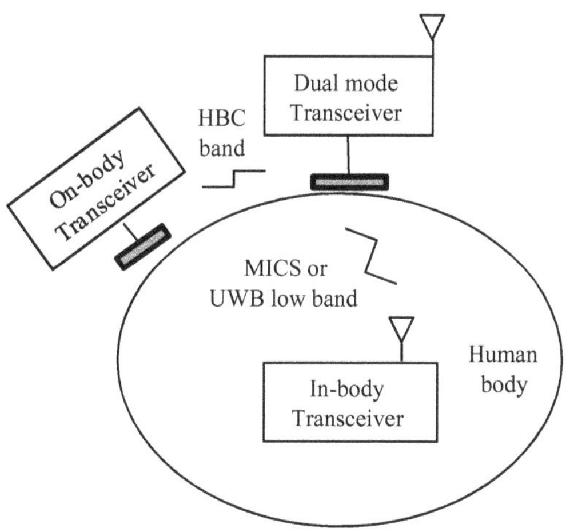

Figure 6.40 Concept of dual mode communication

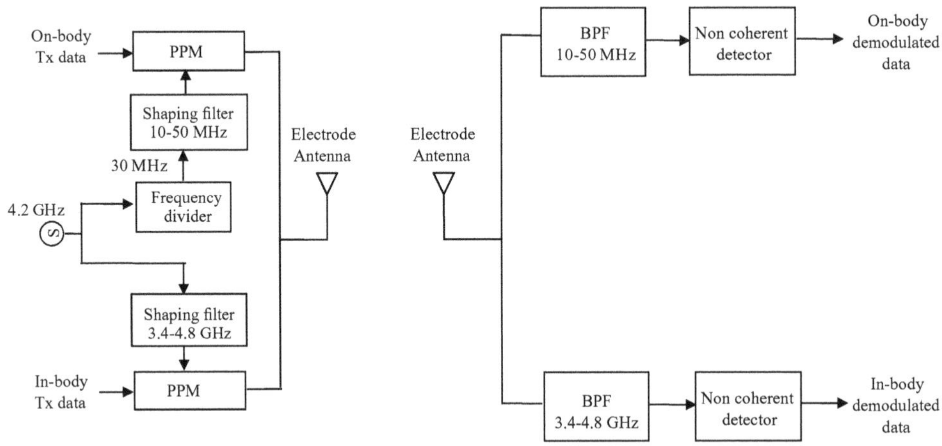

Figure 6.41 Block diagram for the realization of a dual model IR-UWB system

to generate a pulse with a −10 dB bandwidth between 10 MHz and 50 MHz. On the other hand, in the in-body mode, the signal at 4.2 GHz is passed through a Gaussian-shaped BPF to produce a pulse with a −10 dB bandwidth between 3.4 GHz and 4.8 GHz. Figure 6.43 shows two examples of power spectra produced by the sinusoidally modulated Gaussian pulses in the on-body and in-body modes. They basically meet the above requirements. Since both of the two pulses meet the condition that the ratio of the bandwidth to the center frequency is larger than 0.2, they are the UWB signals. Using the two types of pulses, one for on-body and the other for in-body communication, the information data are modulated by PPM and then transmitted as pulse trains.

For reception of the two mode signals, a common antenna/electrode structure is expected for system simplicity. After the antenna/electrode structure, the signals are sent to separate BPFs, and are then demodulated according to the mode. The dual mode design effectively meets the basic requirement for both on-body and in-body

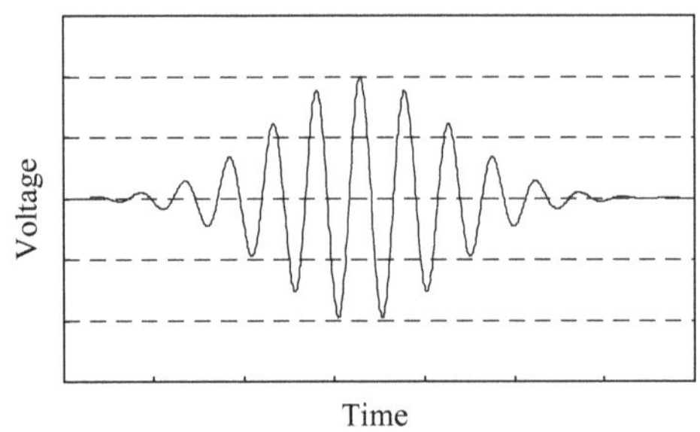

Figure 6.42 Sinusoidally modulated Gaussian pulses. The pulse widths are 50 ns for on-body mode and 2.1 ns for in-body mode

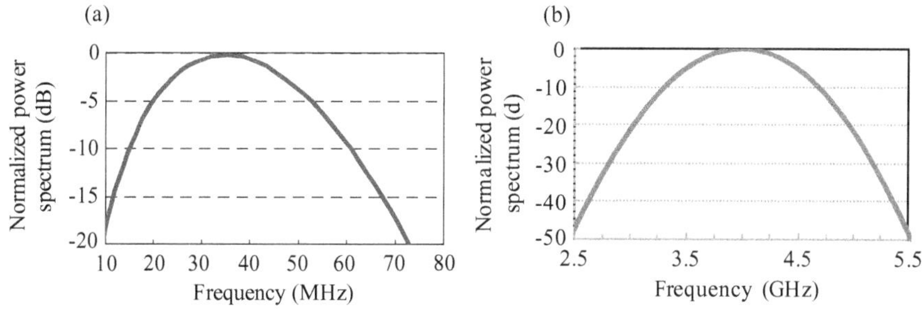

Figure 6.43 Normalized power spectra of the sinusoidally modulated Gaussian pulses (a) on-body mode; (b) in-body mode

data rates in healthcare and medical applications. Its communication performance in each mode can be determined from the previous sections.

References

Benedetto, D.M.-G. and Giancola, G. (2004) *Understanding Ultra Wide Band Radio Fundamentals*, Prentice Hall, New Jersey, pp. 294–300.

Bic, J.C., Duponteil, D., and Imbeaux, J.C. (1991) *Elements of Digital Communication*, John Wiley & Sons Ltd., Chichester, pp. 462–467.

Cho, N., Roh, T., Bae, J., and Yoo, H.-J. (2009) A planar MICS band antenna combined with a body channel communication electrode for body sensor network. *IEEE Transactions on Microwave Theory and Technique*, **57** (10), 2515–2522.

FCC (2002) Federal Communications Commission: Revision of Part 15 of the Commission's rules Regarding Ultra-Wideband Transmission System: First report and order, Technical Report FCC 02–48 (Adopted February, 14 2002; Released April 22, 2002).

Shi, J., Anzai, D., and Wang, J. (2011) Performance analysis of diversity effect for in-body to on-body wireless UWB link. Proceedings of the 4th International Symposium on Applied Science in Biomedical and Communication Techniques, Barcelona, Spain.

Wang, J., Masami, K., and Wang, Q. (2011) Transmission performance of an in-body to off-body UWB communication link. *IEICE Transactions on Communications*, **E93-B** (4), 150–157.

Wang, Q. and Wang, J. (2010) Performance of ultra wideband on-body communication based on statistical channel model. *IEICE Transactions on Communications*, **E93-B** (4), 833–841.

7

Electromagnetic Compatibility Considerations

7.1 Introduction

Body area communications makes an electromagnetic field source just on or in the human body. Electromagnetic radiation from a body area communication device may give rise to energy absorption in the human body as well as possible interference with medical devices. This implies a potential for possible biological effects as well as malfunction of medical devices that are worn by or implanted in a human body. Both of these electromagnetic compatibility (EMC) issues have to be considered in the design of a body area communication system.

Biological effects caused by exposure to RF fields are related to the electric and magnetic fields inside the human body. The distribution of the internal fields is related to a number of parameters, including the communication frequency, the dielectric properties of tissues, the geometrical shape of the human body, and the antenna structure of the transmitter of body area communications. The interaction of the induced electromagnetic fields with the human body results in the flow of electric current, the polarization of bound charge (formation of electric dipoles), and the reorientation of electric dipoles already present in human tissue. The electrical conductivity governs the flow of electric current, and the permittivity governs the magnitude of polarization effects.

Current knowledge tells us that there are mainly two types of biological effects of electromagnetic fields. One is the stimulation effect to nervous system due to the induced current in tissue. The other is the thermal effect due to energy absorption in tissue. Low frequency electromagnetic fields normally result in negligible energy absorption and no measurable temperature rise in the human body. However, when the frequency exceeds 100 kHz, the stimulation effect of current grows weak

Body Area Communications: Channel Modeling, Communication Systems, and EMC, First Edition.
Jianqing Wang and Qiong Wang.

because the cell membrane is approximately short-circuited, whereas a significant energy absorption and corresponding temperature rise may occur. In view of this feature, we only need to quantify the energy absorption in body area communications to consider the possible biological effects.

The quantification of energy absorption in the human body is known as dosimetry. The most important quantity of dosimetry at body area communication frequencies is SAR. SAR is the mass averaged rate of energy absorption in the human body

$$SAR = \frac{d}{dt}\left(\frac{dU}{dm}\right) \tag{7.1}$$

where U is the absorbed energy and m is the mass. In other words SAR is the absorbed power per unit mass in units of watts per kilogram. As described in Chapter 2, it is related to the internal electric field E (root mean square value) by

$$SAR = \frac{\sigma}{\rho}E^2 \tag{7.2}$$

where σ is the conductivity of tissue in S/m and ρ is the mass density in kg/m^3.

Since body area communications operate on or in the human body, the safety of the human body has a higher priority than in other wireless communications. This is also one of the major issues to be addressed by IEEE 802.15.6 standard. It has been known that RF energy absorption can result in heating of the human body and an increase in body temperature. If the absorbed energy is greater than the heat released by the human body into the environment, the temperature of the interior of the human body will rise. Tissue damage may emerge primarily because of the body's inability to cope with or dissipate the excessive heat. When the body temperature rises from its normal value, an adverse biological effect may occur. It is therefore essential to limit the transmitting power to as low as possible in order to ensure the safety of humans in body area communications.

On the other hand, electromagnetic interaction from body area communication signals may interfere with body-worn or implanted medical devices to cause a malfunction. A typical implanted medical device is the cardiac pacemaker. The cardiac pacemaker is normally connected to the heart by an electrode to read the ECG signal and to stimulate the heart beat by voltage pulses if necessary. External electromagnetic fields from body area transmitters can couple with the cardiac pacemaker to cause an interference voltage in the internal electronic circuit. If the interference voltage exceeds a threshold in the cardiac pacemaker, the pulse voltage to stimulate the heart beat may be triggered to yield a malfunction. Both electromagnetic interference (EMI) evaluation and the EMC design for medical devices are important in body area communications.

This chapter consists of two main parts which will present the energy absorption in the human body and EMI analysis for a cardiac pacemaker for body area communication signals. For the first issue, we will describe methods to calculate the SAR for both sinusoidal signals and UWB pulse signals. Moreover, we will also discuss the SAR addition effect because multiple transmitters may be used simultaneously in body area communications. For the second issue, we will describe the basic EMI mechanism in the cardiac pacemaker and present a two-step approach to analyze the produced EMI voltage at the cardiac pacemaker. The first step is to calculate the input voltage at the pacemaker circuit using a full-wave electromagnetic field analysis tool by modeling the pacemaker as a receiving antenna, whereas the second step is to analyze the output voltage of the pacemaker circuit to quantitatively evaluate the EMI level. Based on the understanding of EMI mechanism, we will give some guidelines for the EMC design of cardiac pacemakers.

7.2 SAR Analysis

7.2.1 Safety Guidelines

Safety guidelines for RF electromagnetic field exposure are based on the results of critical evaluations and interpretations of the relevant scientific research. From the evaluations, a threshold SAR is established for the most sensitively confirmed biological effect which could be considered harmful to humans. To account for uncertainties in the data and to increase confidence that the guideline is below the levels at which adverse effects could occur, the resulting threshold is usually lowered by a safety factor of 10–50 times below the observed threshold.

The International Commission on Non-Ionizing Radiation Protection (ICNIRP) publishes safety guidelines for RF exposure (ICNIRP, 1998). The physical quantities used to specify the basic restriction are SAR and current density. Protection against adverse biological effects requires that these basic restrictions are not exceeded. According to ICNIIRP, the basic restriction is provided on current density to prevent effects on nervous system functions between 1 Hz and 10 MHz, and on SAR to prevent whole-body thermal stress and excessive localized tissue heating between 10 MHz and 10 GHz. The latter frequency range is of interest for body area communications.

Scientific literature has shown that most confirmed biological effects are associated with significant temperature rise in tissue. Laboratory research with rodents has demonstrated the range of adverse biological effects, from physiological changes to disruption of learned behavioral tasks, resulting from either whole-body or partial-body temperature rises in excess of 1–2 °C. A body temperature rise of 1 °C may be induced by RF exposure for about 30 min at a whole-body average SAR of between 1 W/kg and 4 W/kg. At lower exposure levels, however, there is no convincing evidence for the adverse effects, although sensitive studies can detect adaptive responses such as increased sweating or decreased metabolic rate. In the absence of any convincing evidence for so-called "nonthermal" effects at long-term low-level exposure, modern safety guidelines are based on short-term thermal effects. The threshold for the thermal effects is found to be greater than 4 W/kg even in the most sensitive tissues. A basic restriction for the whole-body average SAR is therefore established at 0.4 W/kg with a safety factor of 10 times for occupational exposure. An additional safety factor of 5 is further introduced for exposure to the general public, giving an average whole-body SAR of 0.08 W/kg.

In contrast to the whole-body average SAR, however, the localized spatial peak SAR is more meaningful in body area communications because the corresponding RF exposure by a transmitter is a highly localized one. This means that the high SAR concentrates in the vicinity of the transmitter, whereas the whole-body SAR is so low as almost negligible. The local heating will be carried to other parts of the human body by blood flow. Hence, the body temperature rise does not only depend on the localized SAR but also on the blood flow. Since the blood flow varies in a complicated manner with body temperature, accurate estimation of temperature rise for a localized exposure is difficult. The ICNIRP also gives a basic restriction for localized SAR as averaged over 10 g of tissue. However, the basis for the threshold of localized SAR is not so clear. Under local exposure conditions, significant thermal damage can occur in sensitive tissues such as the eye and the testis. An SAR from 100 to 140 W/kg at rabbits' eyes was observed to yield a temperature as high as 41–43 °C which caused cataracts. This suggests that a localized SAR of 10 W/kg in eyes will not cause a temperature rise greater than 1 °C. It is thus reasonable to choose 10 W/kg as a basic restriction for localized SAR for occupational exposure. For the general public, moreover, an additional safety factor of 5 yields a localized SAR of 2 W/kg as the basic restriction.

Table 7.1 summarizes the basic restrictions for whole-body SAR and localized SAR between 10 MHz and 10 GHz. All SAR values are averaged over any 6-min period in order to reach a steady state of temperature. For localized SAR, the averaging mass is any 10 g of tissue, and the maximum SAR so obtained is the value used for the estimation with the basic restriction. In addition, for pulse signals of duration T, the equivalent frequency to apply in the basic restrictions is suggested to be calculated as $f = 1/(2T)$. Moreover, for pulsed exposure in the frequency range of

Table 7.1 Basic restrictions for whole-body SAR and localized SAR between 10 MHz and 10 GHz

Exposure characteristics	Whole-body average SAR (W/kg)	Localized SAR (head and trunk) (W/kg)	Localized SAR (limbs) (W/kg)
Occupational exposure	0.4	10	20
General public exposure	0.08	2	4

0.3–10 GHz and for localized exposure to the head, an additional basic restriction is recommended in order to limit or avoid the microwave hearing effect (Lin, 1978). The microwave hearing effect is attributed to a thermo-elastic interaction in the auditory cortex of the brain. The auditory sensation is described as a buzzing, clicking or popping sound, depending on the modulation characteristics of the pulsed field. The basic restriction for pulsed electromagnetic field employs the specific energy absorption (SA) in units of joules per kilogram. The corresponding SA should not exceed 10 mJ/kg for occupational exposure and 2 mJ/kg for the general public, also averaged over 10 g of tissue. In addition, beyond our interested frequency range, between 100 kHz and 10 MHz, which may be also used in HBC, an additional basic restriction for the current density in the head and trunk is required. The corresponding root mean square value of the current density is $f/100\ \text{mA/m}^2$ where f is the frequency in hertz.

Furthermore, body area communications may simultaneously employ multiple frequencies to form a network. In this case, the total SAR is the sum of the SAR at each frequency because it is related to the power.

7.2.2 Analysis and Assessment Methods

It is usually difficult to measure the SAR directly in a living human body, and therefore dosimetry efforts are forced to rely on computer simulation with numerical human body models or experimental simulation with tissue-equivalent phantoms in order to evaluate human safety for body area transceivers.

7.2.2.1 Numerical Techniques

An anatomically based human body model is essential for numerical dosimetry. Such numerical models have been described in Section 3.4 and are especially useful in the FDTD method. For narrow band body area communication signals, the SAR can be evaluated at the carrier frequency. It is straightforward to calculate the SAR at a specific location in the human body for sinusoidal signals because the FDTD method gives the three components of the electric field in each discretized cell. However, the electric fields are given at the edges of each cell in the FDTD method, whereas the tissue types are defined in terms of the cell itself. In order to obtain the SAR in each cell, we need to place all the electric field components in the center of the cell. Figure 7.1 shows the concept for summation of the 12 electric field components, which correctly place all field components in the cell's center to obtain E_{xc}, E_{yc} and E_{zc}, which are the amplitudes of the three electric field components. Then

$$\text{SAR} = \frac{\sigma}{2\rho}\left(E_{xc}^2 + E_{yc}^2 + E_{zc}^2\right). \tag{7.3}$$

This approach should be the most reasonable for calculating the SAR in each cell.

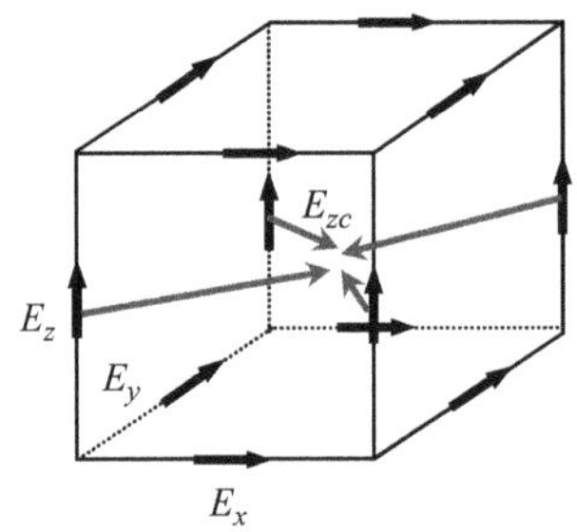

Figure 7.1 Summation of the field components with the 12-component approach

In order to assess the safety for the localized exposure by the body area transceivers, we need to calculate the 10-g averaged spatial peak SAR as defined in the basic restrictions of safety guidelines. The whole-body average SAR usually does not need to be considered in body area communications in view of its highly localized exposure and consequent low average of energy absorption in the whole body.

Local SAR assessment sometimes leads to a question of interpretation of the appropriate volume over which the SARs should be averaged in the FDTD method. The specified volume is 10 g of tissue in the shape of a cube or 10 g of contiguous tissue. Since the human body is irregular in shape and is made up of different tissues, finding a precise 10-g cube of tissue around the peak SAR values is often impossible. In addition, due to the difference in mass density of tissues, even cubic volumes of the same size may still yield different mass. Moreover, the FDTD cell size is not always divisible into exactly 10-g cubes.

A possible method to cope with this problem is to use interpolation or extrapolation of the SAR data. For tissue in the body, the spatial peak SAR should be calculated in a cubic volume which contains a mass within ±5% of 10 g. The cubic volume is first centered at each location in the body, and is then expanded in all directions until the desired value for the required mass is reached with no surface boundary of the averaging volume extending beyond the most exterior surface of the body, as shown in Figure 7.2 (IEEE, 2002). For the case of body surface locations, the averaging volume should be constructed with the location centered at one surface of the cube and the other five surfaces of the cube should be expanded evenly in all directions until the required 10 g of tissue in the body is enclosed within this volume.

With the local SAR in each cell, a linear interpolation algorithm can be used to derive the 10-g averaged SAR for safety assessment. According to the above-described expanding rule for cubic volumes, a sequence of cubic volumes of increasing size is obtained. The mass-average SAR is then calculated based on the power deposited in two consecutive cubic volumes, from the sequences which have lower and higher mass than the required 10 g, as shown in Figure 7.3. A weighted average

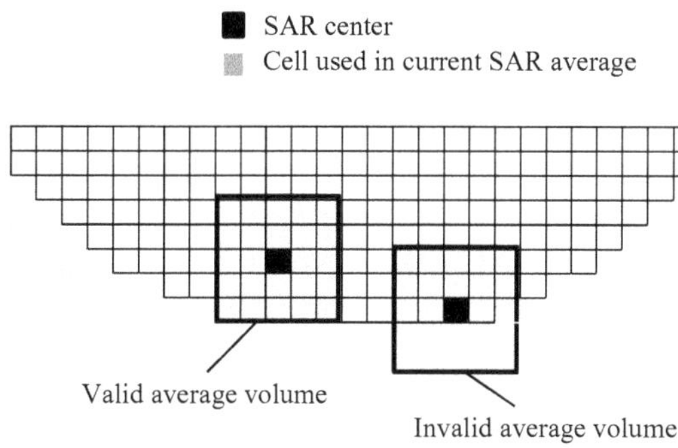

Figure 7.2 The left cube shows an averaging volume centered on the highlighted cell. The average SAR value is obtained from the enclosed cells and assigned to the highlighted location. The right cube shows an invalid averaging volume (IEEE, 2002). Reproduced with permission from IEEE Std. C95.3-2002 (2002): IEEE recommended practice for measurements and computations of radio frequency electromagnetic fields with respect to human exposure to such fields, 100 kHz–300 GHz

between the two cubic volumes is calculated by implementing the following formula (Caputa, Okoniewski, and Stuchly, 1999)

$$\mathrm{SAR}_{10\mathrm{g}} = \frac{P_c + P_e \frac{m_g - m_c}{m_e}}{m_g} \tag{7.4}$$

where SAR_{10g} is the 10-g averaged SAR, m_g is the required mass of 10 g, P_c is the power deposited in the volume of the lower mass, m_c is the mass of that volume, P_e is the power deposited in the extra part of cubes in the next larger volume of the sequence, and m_e is the mass of that part.

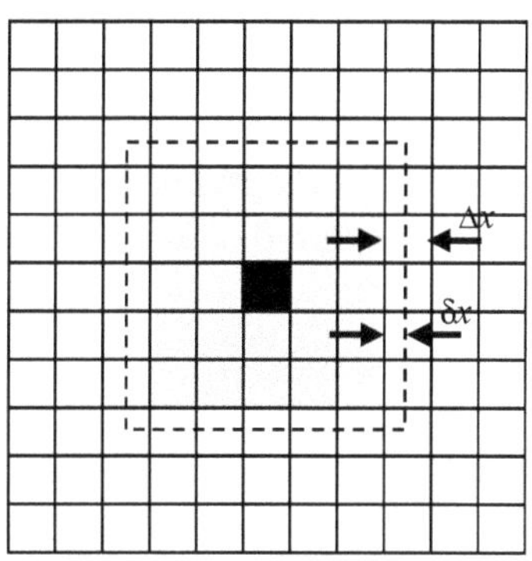

Figure 7.3 Interpolation for obtaining an accurate 10-g averaged SAR. The gray area is the core volume with a mass m_c smaller than 10 g. The extra volume with a mass m_e is between the dotted lines and the gray area (Caputa, Okoniewski, and Stuchly, 1999). Reproduced with permission from Caputa K., Okoniewski M. and Stuchly M.A., "An algorithm for computations of the power deposition in human tissue," *IEEE Antennas and Propagation Magazine*, **41**, 4, 102–107, 1999. © 1999 IEEE

The maximum value of the 10-g averaged SAR obtained in such a way can be used as the required quantity in the safety assessment with the basic restrictions.

On the other hand, for wide band communication signals such as UWB signals, the frequency-dependent FDTD method is usually used in numerical analysis. Since the UWB signals produce pulsed fields, both the SA and the SAR are required for the safety assessment. Here we show two approaches for the SA and SAR calculation.

The first approach is a time-domain approach. Referring to Section 3.1.5, it calculates the current density in the time domain in the frequency-dependent FDTD method by

$$J(t) = \sigma_0(t) + \varepsilon_0 \frac{d}{dt}[\chi(t) * E(t)]. \tag{7.5}$$

Then the SA can be obtained from

$$\mathrm{SA} = \int_0^T \frac{J(t)E(t)}{\rho} dt \tag{7.6}$$

where T is the pulse duration. This calculation is straightforward in the frequency-dependent FDTD method, and it does not need additional calculation burden.

The second approach is essentially a frequency-domain expression of the first approach. According to Parserval theorem,

$$\mathrm{SA} = \int_{-\infty}^{\infty} \sigma(\omega)|E(\omega)|^2/\rho \cdot df \tag{7.7}$$

where

$$\sigma(\omega) = -\omega\varepsilon_0 \mathrm{Im}[\varepsilon_r(\omega)] \tag{7.8}$$

is the lossy component of the dielectric properties and $E(\omega)$ is the Fourier transfer of electric field $E(t)$. We can get $E(\omega)$ in the frequency-dependent FDTD method as a running summation at each time step based on the discrete Fourier transfer such as

$$E(m\Delta\omega) = \sum_{n=0}^{N-1} E(n\Delta t)e^{-j2\pi mn/N} \tag{7.9}$$

where Δt is the time step, N is the number of total time steps, and $\Delta\omega = 2\pi/N\Delta t$. This approach requires additional calculation burden.

Although the corresponding numerical algorithms are different, the two approaches are mathematically equivalent. Let us consider a simple example for comparison of the two approaches. A disk dipole antenna is placed in the front of a homogenous cube, and is excited by a UWB pulse signal. The cube has a length of 200 mm and the dielectric properties of muscle. Figure 7.4 shows the SA profile,

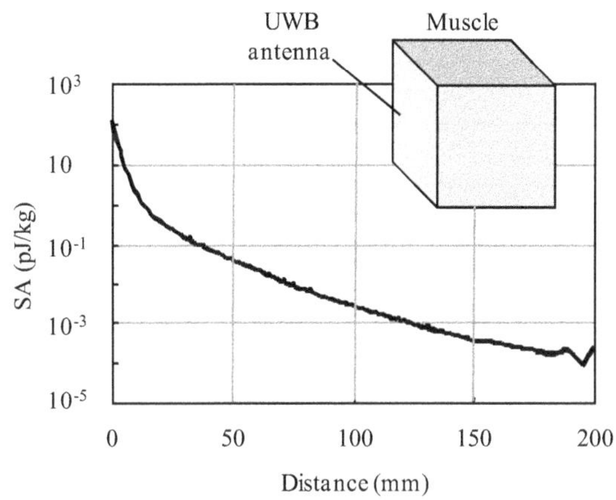

Figure 7.4 Comparison of the calculated SA between the time-domain approach and the frequency-domain approach. The two approaches give the same result so that the two curves completely overlap

taken from the front to the back of the muscle cube, calculated by using the two different approaches. As can be seen, the two approaches give the same SA values. In view of the calculation efficiency, the first approach in the time domain is more appropriate for the SA calculation in the frequency-dependent FDTD method. As for the SAR, it can be simply obtained from the ratio of the SA to the pulse duration T, that is,

$$\mathrm{SAR} = \frac{\mathrm{SA}}{T}. \tag{7.10}$$

7.2.2.2 Measurement Techniques

Experimental dosimetry is focused mainly on homogeneous, tissue-equivalent phantoms, because of the difficulty in developing a heterogeneous human body model. Phantoms generally simulate a single tissue such as muscle, or a mixed-tissue such as the body, by having the same permittivity and conductivity as the actual tissue. Phantoms can be mainly classified into three types depending on the phantom form: liquid, solid, or gel.

A liquid phantom permits scanning of the electric fields inside it using an electric field probe, which can provide high-precision SAR measurement. The main materials in most liquid phantoms are deionized water, sugar, and sodium chloride. Also other materials such as hydroxyethyl cellulose are often used to adjust the permittivity and conductivity. The permittivity is adjusted mainly by varying the percentage of sugar, and the conductivity is adjusted mainly by altering the percentage of

Table 7.2 Composition of liquid phantom (wt%) for muscle at 400 MHz

Water	Sugar	Sodium chloride	Cellulose	Preservative
52.4	45	1.5	1	0.1

sodium chloride. Because it is based on water, a liquid phantom is generally good at simulating a high water content tissue. A mixture of a physiological salt solution and ethylene glycol enables the liquid phantom to also simulate a low water content tissue. The drawback of liquid phantoms is the evaporation of water, which changes the permittivity and conductivity. This, however, can be managed by the appropriate replenishment of water. A possible composition at 400 MHz for muscle tissue is shown in Table 7.2 (Hartsgrove, Kraszewski, and Surowiec, 1987).

In liquid phantoms, the SAR assessment is usually conducted by measuring the electric field. The SAR is calculated from the measured electric fields with Equation 7.2. The main requirements for an electric field probe in SAR measurement are: (1) high sensitivity and linear response over a broad frequency range; (2) high spatial resolution; (3) isotropy in different media; (4) low interaction with the measured field; and (5) small size. Most existing electric field probes are based on Schottky diode detectors. The measured signal at the probe output is a voltage proportional to the electric field E or the squared electric field E^2.

Because the SAR assessment requires all three components of the electric field, most electric field probes consist of three small dipoles, with detector diodes at their center gaps. A triangular-beam structure, as shown in Figure 7.5 (Schmid, Egger,

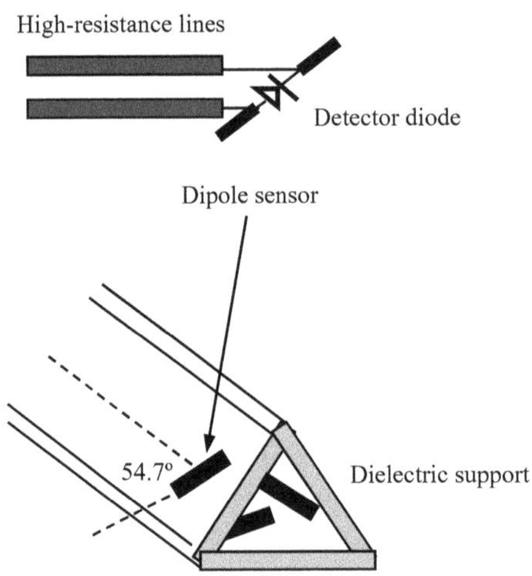

Figure 7.5 A typical electric field probe with a triangular beam structure. Schmid, Egger, and Kuster, 1996

and Kuster, 1996), is a typical design because of its small size and the possibility of placing the detector in the center of the probe. The probe consists of three sensors. Each sensor consists of (1) a short dipole antenna, (2) a diode detector at the dipole feed-gap, (3) a dielectric mechanical support, and (4) a highly resistive feed-line to extract the signal detected by the diode to the measuring unit. The probe's tip length and tip diameter are on the order of 2 cm and 1 cm, respectively. This permits a high spatial resolution for scanning electric fields within a liquid phantom. Such probes may typically have a frequency range from 10 MHz to 10 GHz, a dynamic range of 5 μW/g to 100 mW/g, and an orthogonal directivity pattern for the three sensors.

Probe calibration for SAR measurement in liquid phantoms will give either an SAR or an electric field conversion factor. Because SAR is proportional to liquid conductivity, a direct calibration in terms of SAR is valid only for liquids with exactly the same conductivity. The electric field sensitivity depends more on both the liquid permittivity and the liquid conductivity and therefore is less sensitive to the conductivity alone. Calibration in terms of electric field, rather than SAR, should have a broader range of validity. Possible calibration methods include both waveguide calibration and thermal calibration. In a waveguide calibration, the measured fields are compared with analytical solutions. In the thermal calibration, the SAR values are compared with the temperature rise. The former is preferable, because it gives higher precision.

In addition to liquid phantoms, solid and gel phantoms are also useful in SAR assessment. One representative solid phantom is the TX-151 phantom (Ito *et al.*, 1998). A TX-151 phantom is applicable for simulation of a high water content tissue in the microwave band. The permittivity is adjusted mainly by the amount of polyethylene powder mixed in agar, and the conductivity is adjusted mainly by the amount of sodium chloride. TX-151 acts as an adhesive in the agar liquid; thus, it forms a solid phantom. The materials required are easily obtained, and it is easy to make TX-151 phantoms of differing shapes. If the phantom is wrapped with vinyl film, it is possible to maintain constant permittivity and conductivity for 1 month at room temperature. Glycerin-based phantoms can provide a longer life span than TX-151. Their life span can be extended to 6 months by wrapping the phantom with vinyl film because glycerin acts to maintain humidity. The permittivity in a glycerin-based phantom is adjusted mainly with deionized water, and the conductivity is adjusted with sodium chloride. Polyethylene powder is used in fine adjustment of the permittivity and conductivity. Compared with the TX-151 phantom, the drawback of a glyceric phantom is the narrower frequency range for a single composition. At different frequencies the percentages of the various materials have to be changed.

Gel phantoms are typically manufactured from a physiological salt solution and polyethylene powder mixed with TX-150 or agar. They are used to simulate a high water content tissue. With a gel phantom, it is difficult to have a fixed shape, unless a container is used.

Scanning the electric fields in a solid or gel phantom for SAR assessment is impractical. They are more appropriate for measuring the temperature rise. Under the assumption of linear energy deposition over a fixed period of time, the SAR can be determined from

$$\mathrm{SAR} = C_p \frac{dT}{dt} \approx C_p \frac{\Delta T}{\Delta t} \tag{7.11}$$

where C_p is the specific heat of the phantom, ΔT is the temperature rise caused by the exposure, and Δt is a very short exposure time.

There are two methods for the measurement of temperature. One method is to use a flour-optic temperature probe. These probes utilize temperature-dependent fluorescent decay or interferometric micro shifts of cavity resonators. One can insert a flour-optic temperature probe into a phantom and measure the temperature over a period of time; this allows a history of temperature rise over time to be recorded. The second method is to use an infrared image camera. The solid or gel phantom, before exposure, is first set at the ambient temperature, and the corresponding infrared image is recorded. Immediately after the electromagnetic exposure, another infrared image is taken. Then the temperature rise at the phantom surface is obtained from the temperature difference, and the SAR is determined from Equation 7.11. However, the exposure time for thermal measurement has to be short enough to prevent heat transfer from phantom to air. In order to keep a linear temperature rise over a short period of time requires use of a high-output power amplifier. More than 10 W is usually required to produce an efficient temperature rise of 1–2 °C within 1 min. The thermal method is therefore a weak approach when measuring low SAR values, which is actually a typical situation in body area communications. The precision of thermal measurement is also poorer than the electric field probe approach.

In view of these observations, a numerical analysis approach with realistic-shaped human body models or a measurement approach with electric field probe scans in a liquid phantom is the appropriate choice for SAR assessment in body area communications.

7.2.3 *Transmitting Power versus SAR*

7.2.3.1 On-Body UWB

UWB communication is a candidate for on-body applications. A typical UWB antenna may be an elliptic disk dipole with a major axis radius of 12 mm and a minor axis radius of 10 mm. As shown in Figure 7.6, such a structure yields a VSWR of nearly 2.0 between 3.1 GHz and 10.6 GHz. In order to make a quantitative analysis for the SA and SAR in a human body for an actual UWB transmitter, we employ this antenna in an on-body communication scenario.

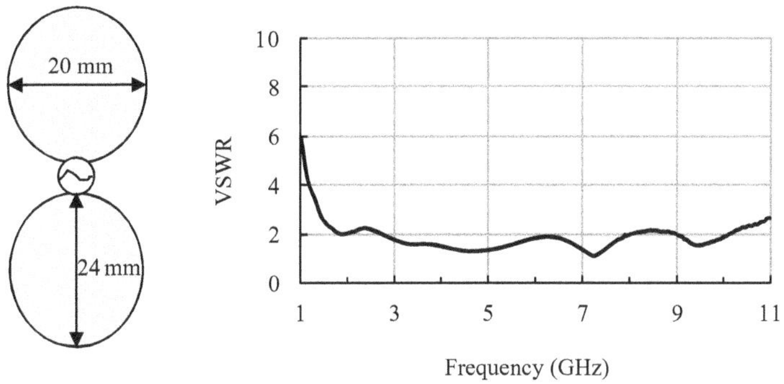

Figure 7.6 An elliptic disk dipole antenna and its VSWR

The antenna is assumed to have a spacing of 2 mm from the human body surface. The UWB pulse to be transmitted is considered as a Gaussian derivative because it can be generated in the easiest way by a real pulse generator and be radiated in an efficient way. Figure 7.7 shows a fifth-derivative Gaussian pulse waveform with a width of nearly 500 ps, and its EIRP. The EIRP is the amount of power that a theoretical isotropic antenna (which evenly distributes power in all directions) emits to induce the peak power density observed in the direction of maximum antenna gain. It is derived as follows

$$\text{EIRP} = P_a \cdot G_a \tag{7.12}$$

where P_a is the power supplied to the UWB antenna and G_a is the maximum absolute gain of the antenna. The determination of P_a depends on the UWB modulation scheme. For IR-UWB, we can first calculate the antenna power density

$$P_{SD}(f) = Re\left[\frac{V(f)V^*(f)}{Z_{in}(f)}\right]/T_p \tag{7.13}$$

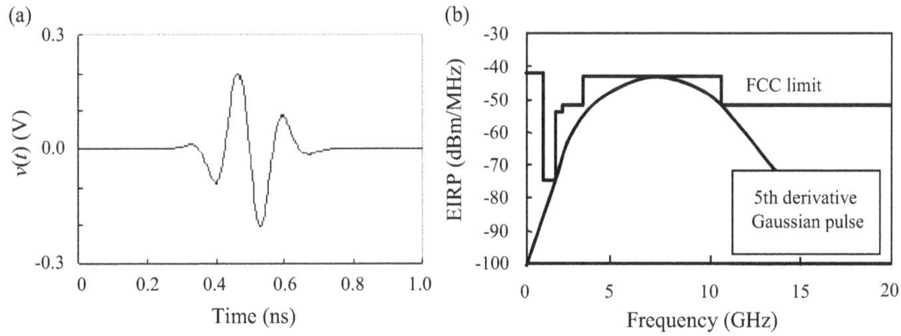

Figure 7.7 (a) Fifth-derivative Gaussian pulse waveform and (b) its EIRP versus frequency

where $V(f)$ is the Fourier transform of the UWB pulse voltage in units of V/Hz, $Z_{in}(f)$ is the antenna input impedance and T_p is the pulse duration. Then we have the EIRP as

$$\mathrm{EIRP} = 10\log_{10}[2P_{SD}(f)] + 10\log_{10} G_a + 90 \qquad (\mathrm{dBm/MHz}) \tag{7.14}$$

where the number 90 is introduced by the units of mW and MHz. It can be found from Figure 7.7 that the maximum power density of the employed fifth-derivative Gaussian pulse is limited to −43.1 dBm/MHz between 3.1 GHz and 10.6 GHz, that is, it meets with the FCC indoor emission spectrum mask. Careful adjustment of pulse shape factors for other orders of Gaussian derivatives is also possible to produce a power density which approximately meets with the FCC UWB emission limit. However, the fifth-derivative Gaussian pulse is a superior choice because it closely approximates the FCC emission mask so that more permissible power can be radiated. This is desirable in view of the excitation source requirement in the SA/SAR analysis.

In view of typical on-body communication scenarios, we consider that the UWB antenna is placed at the chest, ear, eye and waist, respectively, parallel to the body surface. Moreover, since in actual on-body applications many transmitters may communicate at the same time, it is necessary to also consider a multiple exposure scenario in order to investigate the possible addition of energy absorption. As a multiple exposure scenario, two UWB antennas are simultaneously placed at the chest and waist, or four UWB antennas are simultaneously placed at the above four locations. Figure 7.8 shows the SA distributions on the body surface for the UWB pulse exposure under the FCC emission limits (Wang and Wang, 2009). The 0 dB corresponds to 10 pJ/kg. As can be seen, the SA concentrates on a very small area on the body surface near the UWB antenna location, and the attenuation is more than 30 dB at a location 10 cm away from the UWB antenna.

For an IR-UWB pulse signal, the SAR can be simply obtained from the ratio of SA to the pulse duration T (Equation 7.10). It is reasonable because the transmitted signal is in a form of successive pulses in the IR-UWB scheme. Table 7.3

Table 7.3 10-g averaged peak SA and SAR under FCC UWB pulse emission limit (approximately 0.3 mw transmitting power)

Antenna location	10-g peak SA (pJ/kg)	10-g peak SAR (mW/kg)
Chest	0.473	0.946
Ear	0.037	0.074
Eye	0.268	0.536
Waist	0.232	0.464
Chest + waist	0.474	0.948
Chest + ear + eye + waist	0.476	0.952

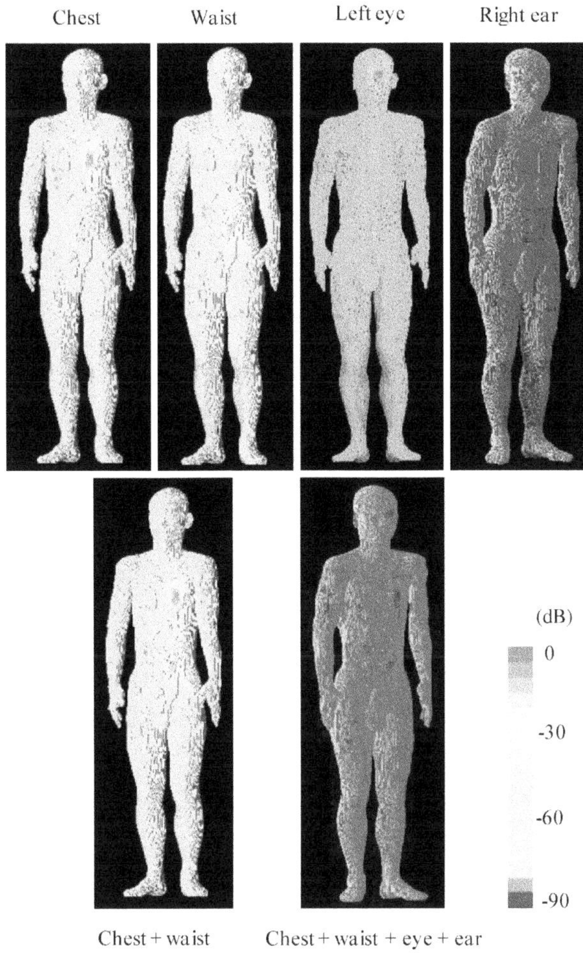

Figure 7.8 SA distributions under the FCC UWB limit (approximately 0.5 mW transmitting power) for various antenna locations. The 0 dB corresponds to 10 pJ/kg (Wang and Wang, 2009). Reproduced with permission from Wang Q. and Wang J., "SA and SAR analysis for wearable UWB body area applications," *IEICE Transactions on Communications*, **E92-B**, 2, 425–430, 2009

summarizes the 10-g averaged spatial peak SA and SAR in all of the considered transmitter locations under the FCC emission limits. It should be noted that the full UWB band permits a maximum transmitting power of 0.556 mW as described in Chapter 1. Since the actual UWB pulse as shown in Figure 7.7 does not completely fit the shape of the FCC spectrum emission limit, the transmitting power is somewhat smaller than 0.556 mW, typically 3 dB smaller than the maximum power. Here we approximate it as 0.3 mW. As can be seen, under this transmitting power, the 10-g averaged SA is on the order of pJ/kg which is much smaller than the ICNIRP

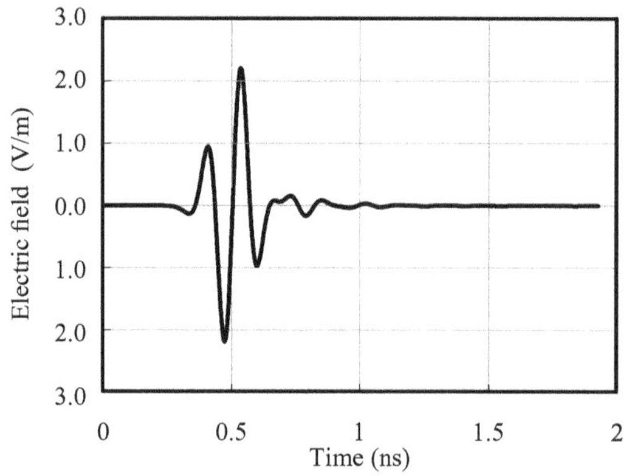

Figure 7.9 Electric field waveform at the chest under multiple exposures

safety guideline of 2 mJ/kg. Moreover, the 10-g averaged SAR is on the order of mW/kg which is smaller than 1/2000th of the ICNIRP safety guideline of 2 W/kg. Besides, the energy absorbed by the whole body is found to be 0.01 pJ which is about a quarter of the energy radiated from the antenna which is around 0.04 pJ.

Moreover, compared with the SAR for single exposure, almost no obvious increase is observed in the SA and SAR for the multiple exposures from Table 7.3. This can also be observed from the SA distributions in Figure 7.8. In order to further verify this conclusion, let us consider the time waveform of electric field at the chest under multiple exposures. Since the separation distance between two on-body transmitters is about or larger than 20 cm, the transmission time is at least 0.7 ns. This means that if the effect of signals from other transmitters is not negligible at the chest, there should be some significant components after 0.7 ns in the observed electric field waveform. This does not depend on whether the transmitters are working simultaneously or not. From Figure 7.9, however, no significant component is observed over there. This means that as long as the separation between two UWB devices is larger than 20 cm, which is usually satisfied in most cases of on-body communications, the additional effect of signals from other transmitters can be ignored because of the rapid surface attenuation of UWB signals.

In addition, the SA and SAR vary with the distance between the antenna and the human body. Table 7.4 shows the 10-g averaged peak SA and SAR corresponding to

Table 7.4 10-g averaged peak SA and SAR with different antenna distance from the body

Antenna spacing from body	10-g peak SA (pJ/kg)	10-g peak SAR (mW/kg)
2 mm	0.473	0.946
1 cm	0.034	0.068
2 cm	0.013	0.026

distances of 2 mm, 1 cm and 2 cm when the UWB antenna is placed on the chest. It can be seen that the 10-g averaged peak SA and SAR values will decrease by 11 and 16 dB, respectively, when the distances are 1 and 2 cm compared with the initial distance of 2 mm. In practice, the antenna is placed as close as possible to the body surface. The SA and SAR at 2-mm spacing should represent an almost worst case. The further the distance becomes, the larger the safety margin will be.

7.2.3.2 In-Body MICS Band

The MICS band attracts much attention especially in in-body communication such as the capsule endoscope and cardiac pacemaker. Since the implant transmitters are mainly used for medical treatment, the permissible SAR level may be higher than that in daily situations. However, there are no established restrictions for SAR in medical treatment. The safety limit of 2 W/kg for the general public or 10 W/kg for occupational exposure, as averaged over any 10 g of tissue, may be used in in-body communication.

The approach for in-body SAR evaluation may be summarized as follows:

1. Determine the required transmitting power for ensuring a BER performance via a link budget analysis.
2. Calculate the 10-g averaged spatial peak SAR for various typical transmitting antenna locations and directivities.
3. Derive the maximum localized SAR and the SAR statistics value for safety evaluation.

In order to have a quantitative understanding of the SAR at the in-body MICS band, let us consider a capsule endoscope scenario. We assume the implant antenna to be a 2-cm-long dipole. Since the induced SAR will be different for different antenna types, the following result is just an example to demonstrate this analytical approach. In addition, it should be noted that the SAR is defined within 6 min of reaching a thermal equilibrium. A capsule endoscope may stay in one location in the digestive organs more or less than 6 min. In either case, the calculated SAR can be considered as a worst case estimation.

As described in Section 4.2.4, the dipole is moved along the digestive organs, from the esophagus and stomach to the small intestines and large intestines, with a spacing of several centimeters and three directivities of x, y and z, respectively. With the transmitting power as the input to the antenna, the 10-g averaged spatial peak SAR is calculated using the FDTD method in conjunction with the anatomical human body model. Figure 7.10 shows an example of SAR distribution when the antenna is located in the stomach. It can be seen that the higher SAR area is limited in a small region. This is because of the small size of antenna which yields a high field concentration.

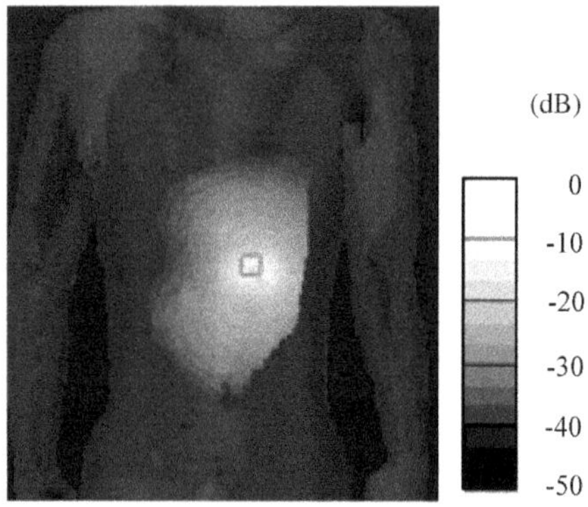

Figure 7.10 Normalized SAR distribution when the antenna is located in the stomach in the MICS band

Based on a link budget analysis as described in Section 6.4.2, at each receiver location in Figure 4.11, the required transmitting power is shown in Table 7.5 for all transmitting dipole locations and directivities in the digestive organs to ensure a BER of 10^{-3} at a data rate of 1 Mbps with BPSK modulation. The required mean transmitting power denotes the power which yields an average BER of 10^{-3}, while the required maximum transmitting power denotes the power which yields the BER always smaller than or equal to 10^{-3}. From Table 7.5, the required mean transmitting power at BER $= 10^{-3}$ is almost smaller than 20 mW. Although such a transmitting power is larger than that in Figure 6.35 because all the capsule locations are considered, it can never induce a 10-g averaged SAR exceeding 2 W/kg. In this sense the safety guideline is always satisfied. However, if we require a higher communication quality to ensure the BER is always smaller than 10^{-3} in any capsule location, the SAR will be determined by the maximum transmitting power in Table 7.5.

Moreover, when the receiving antenna is located in the front of the human body, the path loss is relatively small so that the required transmitting power is also small. In order to have a statistical observation for the SAR characteristics, we show the probabilities of the 10-g averaged spatial peak SAR in Figure 7.11 for all

Table 7.5 Required maximum and mean transmitting power at BER $= 10^{-3}$ for all capsule locations

	Maximum transmitting power (mW)	Mean transmitting power (mW)
Rx_1	120.71	20.85
Rx_2	20.83	2.96
Rx_3	52.04	5.04
Rx_4	115.10	18.25
Rx_5	105.12	7.70

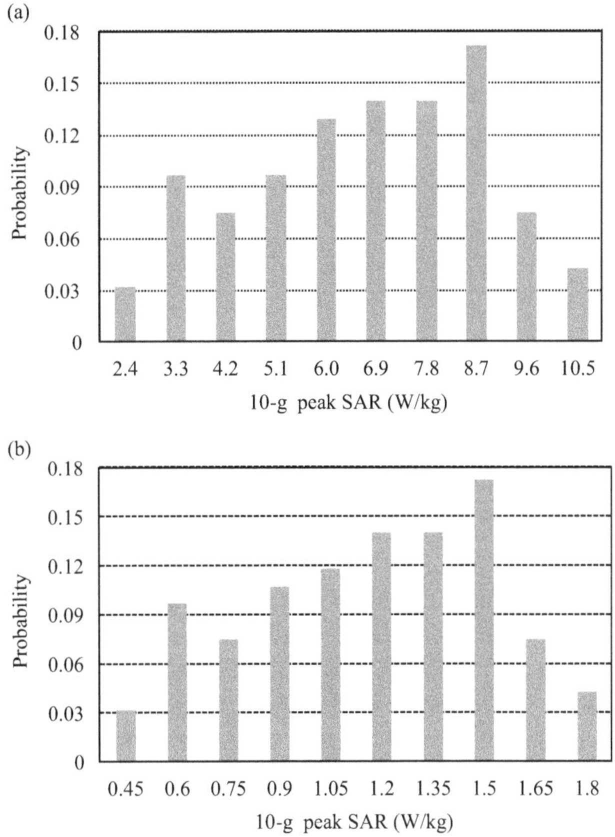

Figure 7.11 Probabilities of 10-g averaged peak SAR in digestive organs for receiver locations (a) Rx1 and (b) Rx2 to have a BER always below 10^{-3}

transmitting antenna locations and directivities (90 data in total) in the digestive organs. The vertical axis in the figure denotes the percentage of SAR values on the horizontal axis. The SAR values are calculated at the maximum transmitting power for Rx1 and Rx2, respectively. As can be seen, if the receiving antenna is set at the Rx2 location, a transmitting power of 20.83 mW is required to ensure a BER not exceeding 10^{-3}. Such a transmitting power yields a local peak SAR ranging from 0.45 to 1.8 W/kg, which does not exceed the safety guideline of 2 W/kg. On the other hand, if the receiving antenna is set at the Rx1 location, a transmitting power of 120.71 mW is required to ensure a BER not exceeding 10^{-3}. In this case the local peak SAR may always exceed 2 W/kg at any transmitter locations.

However, in view of the fact that the capsule endoscope is a medical treatment, it should be acceptable to use 10 W/kg as the safety guideline. So the induced local peak SAR satisfies the safety guideline for occupational exposure. Moreover, the results in Figure 7.11 also suggest that the SAR is dependent on the receiver

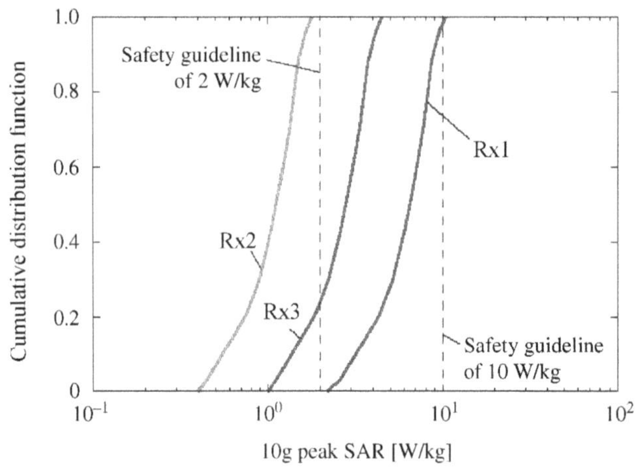

Figure 7.12 Cumulative distribution functions of 10-g averaged peak SAR in digestive organs for securing a BER always below 10^{-3}

locations. If we can optimally choose a receiver location, for example Rx2, both a BER smaller than 10^{-3} and a local SAR smaller than 2 W/kg are obtainable.

Another useful expression for SAR statistics is the cumulative distribution function or CDF. Figure 7.12 shows the CDFs of the 10-g averaged peak SAR when the transmitting antenna moves along the digestive organs to have 30 locations and three different directivities. The SAR values are calculated at the maximum transmitting power for receiver locations Rx1, Rx2 and Rx3, respectively. The plots obviously show that the 10-g peak SAR will satisfy the safety guideline of 2 W/kg at Rx2, and 10 W/kg at Rx3. However, if the receiving antenna is set at Rx1, a transmitting power of 120.71 mW is required so that even the safety guideline of 10 W/kg cannot be satisfied in some transmitter locations.

An effective technique to improve the BER performance is known as spatial diversity reception. Such a technique can also be used to reduce the SAR. Here we consider a two-branch EGC diversity. As described in Chapter 6 in detail, we first obtain the E_b/N_0 versus BER performance for the diversity reception via a theoretical analysis or computer simulation, then we derive the required transmitting power to secure a BER of 10^{-3} based on the established path loss channel model. The required mean transmitting power and maximum transmitting power are shown in Table 7.6. Compared with the results for the single receiver case, we can see that applying spatial diversity reception to the receivers can significantly reduce both the required mean transmitting power and the required maximum transmitting powers. Furthermore, similar to the single receiver case, the required mean transmitting power is also less than 10 mW at all selections of the receiving antennas, that is, the 10-g averaged SAR will never exceed 2 W/kg. If we can choose a good combination of receiving antenna positions, for example Rx2 and Rx3, the achievable required

Table 7.6 Required mean and maximum transmitting power at BER = 10^{-3} for diversity reception

	Mean transmitting power (mW)				
	Rx1	Rx2	Rx3	Rx4	Rx5
Rx1	—				
Rx2	0.70	—			
Rx3	0.85	0.35	—		
Rx4	1.50	0.60	0.70	—	
Rx5	1.05	0.42	0.55	0.90	—
	Maximum transmitting power (mW)				
	Rx1	Rx2	Rx3	Rx4	Rx5
Rx1	—				
Rx2	1.58	—			
Rx3	2.99	1.54	—		
Rx4	2.98	1.49	2.71	—	
Rx5	5.19	2.54	5.08	4.81	—

maximum transmitting power is only 1.54 mW, which is only about 7% of the required maximum transmitting power in the case of a single receiver.

Figure 7.13 shows the CDFs of the 10-g averaged peak SAR in the case with the spatial diversity reception at the required maximum transmitting power for the combination of Rx2 and Rx3 and the combination of Rx1 and Rx5. From this figure, in addition to the required transmitting power, the SAR is also remarkably

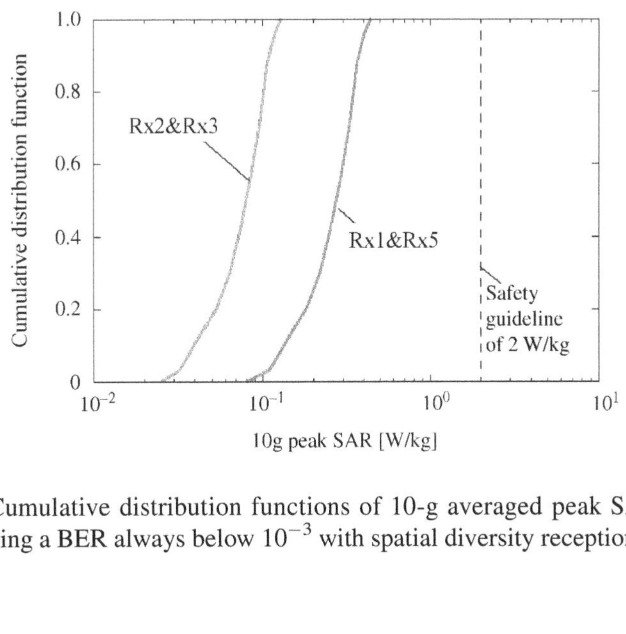

Figure 7.13 Cumulative distribution functions of 10-g averaged peak SAR in digestive organs for securing a BER always below 10^{-3} with spatial diversity reception

Table 7.7 Threshold transmitting power at a local SAR of 2 or 10 W/kg

Local peak SAR (W/kg)	Transmitting power (mW)
2	24.36
10	121.80

reduced compared with that for the single branch case. Even if the two receiving antennas are set to Rx1 and Rx5 (the worst selection), a transmitting power of only 5.19 mW is required to ensure a BER not exceeding 10^{-3}. Such a transmitting power yields a 10-g averaged peak SAR below 0.5 W/kg. This result means that the maximum transmitting power can never yield a 10-g averaged SAR larger than 2 W/kg with a spatial diversity reception. Moreover, if we can optimally select the diversity branches, for example the selection of Rx2 and Rx3, the 10-g averaged peak SAR will have a margin more than 10 dB with respect to the safety guideline of 2 W/kg.

Table 7.7 gives the permissible transmitting power for securing the local SAR not exceeding 2 W/kg or 10 W/kg, which can be used as a threshold transmitting power in safety evaluation for the capsule endoscope transceiver with dipole antenna. Then, if an implant transceiver needs to be evaluated, we can insert it into a liquid phantom of the human body and measure its transmitting power. By comparing the measured transmitting power with the threshold transmitting power, it is possible to evaluate the safety of the transceiver for capsule endoscope application.

Of course for different transmitter antennas the in-body SAR values may vary. However, the basic approach of SAR evaluation is the same as above. It should be noted that the improvement of receiver performance, for example, to adopt a high-gain receiving antenna or to adopt a spatial diversity reception technique, can effectively reduce the required transmitting power inside the body. This will contribute to the reduction of the peak SAR. As long as the transmitting power is below 20 mW, the 10-g averaged peak SAR will never exceed 2 W/kg.

7.2.3.3 HBC Band

HBC is considered between 10 MHz and 50 MHz in IEEE 802.15.6 standard. A typical HBC transmitter employs an electrode structure on the surface of the human body to transmit signals. In view of a 60 dB average path loss in the HBC band, A 1 V transmitting voltage is required to produce at least 1 mV receiving voltage at the receiver for demodulation. We therefore evaluate the SAR when the electrode is excited with a 1 or 10 V voltage source at 30 MHz.

Consider the human body model in Figure 4.19. The electrode is assumed to be on the left chest or back pocket location. The SAR is then calculated using the FDTD method in conjunction with a homogeneous human body model. The homogeneous human body model is assumed to have a conductivity and permittivity two-thirds

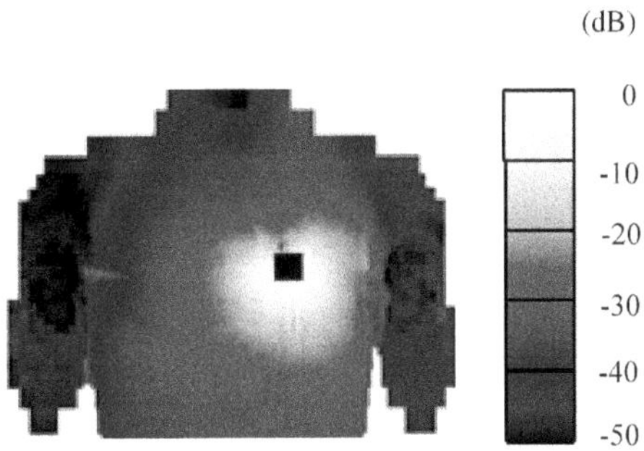

Figure 7.14 SAR distribution on the human body surface in the HBC band

Table 7.8 10-g averaged peak SAR (μW/kg) for 1 or 10 V transmitting voltage in the HBC band

	1 V	10 V
Chest	0.99	99
Back pocket	0.10	10

that of muscle. This is because a human body has an approximated average composition of two-thirds muscle-like tissue. Figure 7.14 shows the SAR distribution on the human body surface with the transmitting electrode on the chest. The energy absorption is also quite concentrated due to the small size of the electrode. Table 7.8 summarizes a calculation example of the 10-g averaged spatial peak SAR for 1 V as well as 10 V transmitting voltage. Compared with that in the back pocket case, the higher 10-g SAR level is found in the chest due to its flat shape. In total, the 10-g averaged spatial peak SAR is on the order of μW/kg at the usual 1 V transmitting voltage, with a safety factor of more than 60 dB compared with the safety limit of 2 W/kg. The result suggests that it is not difficult to satisfy the SAR guidelines in human body communications.

7.3 Electromagnetic Interference Analysis for the Cardiac Pacemaker

7.3.1 Cardiac Pacemaker Model and Interference Mechanism

In addition to the electromagnetic energy absorption issue due to body area communication signals, the electromagnetic interaction of body area communication signals with the human body implies a potential for EMI with implanted medical devices such as a cardiac pacemaker. The cardiac pacemaker consists of a shielded housing with electronic circuits inside and an electrode. It is connected to the heart by the

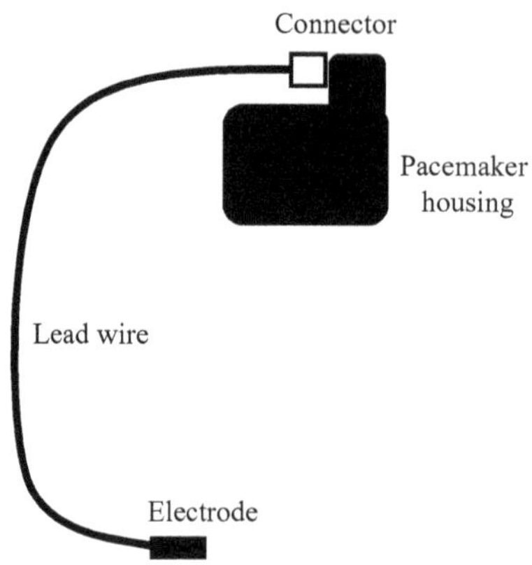

Figure 7.15 Basic configuration of a cardiac pacemaker

electrode to read the ECG signal and to simulate the heart beat by voltage pulses if necessary. External electromagnetic fields can couple into the pacemaker to cause an interference voltage at the input of the internal sensing circuit. The interference voltage produced at the input of the sensing circuit of pacemaker will be amplified and low-pass filtered. If the output voltage of the amplifier and low-pass filter exceeds a threshold, the pulse voltage to simulate the heart beat may be triggered and a malfunction of the cardiac pacemaker may occur. Although the body area communication signal is generally at higher frequencies (greater than MHz) compared with the working frequency (in the order of kHz) of the pacemaker, the nonlinearity of the internal sensing circuit may produce a demodulation effect so that the RF signals fall into the working frequency band of the pacemaker.

In this section, we will introduce a detailed two-step approach to analyze the EMI voltage induced at the cardiac pacemaker. In the first step, the input voltage of the pacemaker circuit is calculated using the FDTD method by considering the pacemaker as a receiving antenna. In the second step, a Volterra series is employed to analyze the output voltage of the nonlinear amplifier and low-pass filter circuit in the pacemaker for evaluating the EMI effect.

The pacemaker may act as a receiving antenna with respect to the external electromagnetic fields (Wang, Fujiwara, and Nojima, 2000). It is reasonable to apply this behavior to any body area communications frequency (Wang *et al.*, 2009). Figure 7.15 shows a basic configuration for an implanted cardiac pacemaker. The cardiac pacemaker consists of a shielded housing with electronic circuits inside and one (unipolar) or two (bipolar) electrodes as well as the lead wires. Without losing generality, we here consider a unipolar electrode case. By considering the internal impedance seen from the connector to the internal circuit as a load, and the metal portions consisting of the pacemaker housing and the lead wire of the electrode as two elements of a receiving antenna, the resultant equivalent circuit for the pacemaker can be

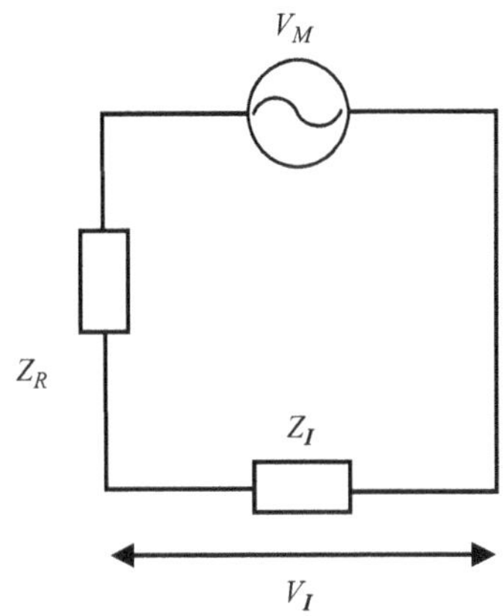

Figure 7.16 Equivalent circuit for EMI predication

shown as in Figure 7.16. Here, Z_R is the radiation impedance of the pacemaker, V_M is the open voltage induced between the pacemaker housing and the lead wire due to the electromagnetic fields from the external communication devices, Z_I is the internal impedance of the pacemaker seen from the connector, and V_I is the voltage produced through the connector onto the internal sensing circuit, which is referred to here as the input interference voltage to the internal sensing circuit of the pacemaker. Then the input interference voltage V_I of the sensing circuit can be obtained as

$$V_I = \frac{Z_I}{Z_R + Z_I} V_M. \tag{7.15}$$

Figure 7.17 shows the block diagram of the analog sensing circuit of the pacemaker (Tarusawa *et al.*, 2005; Schenke, Fichte, and Dickmann, 2007). The input voltage V_I is amplified and low-pass filtered. The resulting output voltage V_O is compared with a sensing threshold V_t. When the voltage exceeds the sensing threshold, it will switch on the pulse output, and then create a malfunction in the pacemaker.

Whether a body area communication signal can pass through the low-pass filter depends on its characteristic parameters such as the signal magnitude, frequency component, modulation scheme as well as the parameters of the sensing circuit.

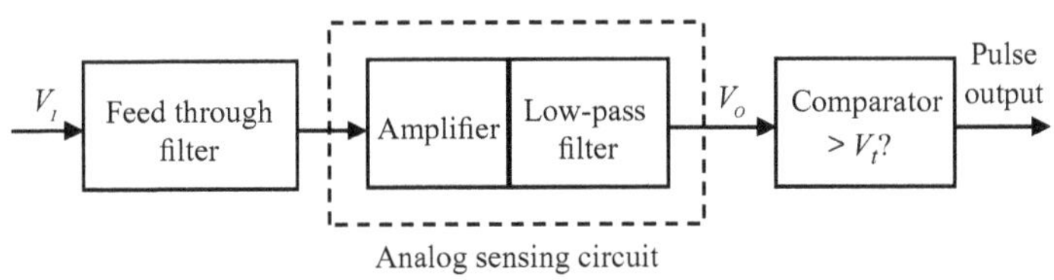

Figure 7.17 Block diagram of the internal pacemaker circuit

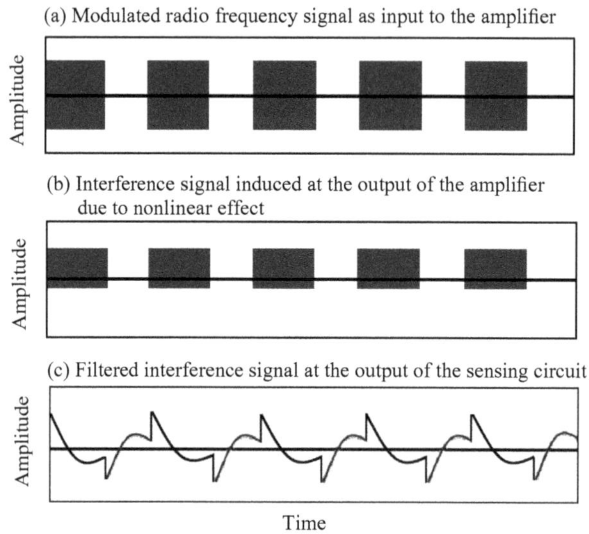

Figure 7.18 Basic concept of EMI caused by a body area communication signal

Figure 7.18 illustrates the basic concept of EMI caused by a RF communication signal. An ECG signal usually ranges from 1 Hz to 1 kHz. The electric circuit of a pacemaker is thus principally designed to work in this low frequency band so that it is difficult for a RF interference signal to appear in the input to the comparator. However, if the RF interference signal is sufficiently strong, the pacemaker circuit may behave nonlinearly. The nonlinearity comes from either the feed through filter or the amplifier (Barbaro *et al.*, 2003). After passing through the feed through filter, the RF signal is impressed to the amplifier for ECG sensing. Since the interference signal frequency is usually beyond the working range of the amplifier, the interference signal will be weakened but the amplifier's nonlinearity may produce a new DC offset and low frequency components as shown in Figure 7.18(b). The DC and low frequency components can pass through the low-pass filter and become the input of the comparator as shown in Figure 7.18(c). If the comparator cannot efficiently reject this interference signal V_O, the pacemaker may malfunction when V_O is larger than the threshold voltage V_t.

The nonlinearity of an analog sensing circuit can be analyzed using the Volterra series method which is a classical and powerful tool to describe a weakly nonlinear circuit. Based on the nonlinear circuit model, the output interference voltage V_O can be derived to predict the potential interference to the pacemaker.

This two-step approach for analyzing pacemaker EMI can be outlined as follows:

1. **Electromagnetic field approach**
   Determine the open-voltage V_M at the connector using a full-wave electromagnetic field simulation tool such as the FDTD method and then get the input

interference voltage V_I of the sensing circuit from Equation 7.15. The open-voltage V_M is obtained by modeling the pacemaker as a receiving antenna, in which both the pacemaker housing and the lead wire/electrode are considered as a conductor, and the connector between them is considered as an open load.

2. **Electric circuit approach**
   Determine the output interference voltage V_O based on the nonlinear Volterra series model of the analog sensing circuit.

Now let us introduce the electromagnetic field approach and electric circuit approach in Sections 7.3.2 and 7.3.3, respectively.

7.3.2 Electromagnetic Field Approach

The open voltage V_M at the connector can be obtained by modeling the pacemaker as a receiving antenna with an open load at the connector using the FDTD method. The receiving antenna has a dipole structure in which the two radiating elements are the metal housing and the metal lead line of electrode. Using the FDTD method, the open voltage V_M between the metal housing and the lead wire, that is, at the connector, can be obtained when the transmitter electrode is excited, and then the interference voltage at the input of the internal pacemaker sensing circuit can be estimated.

The detailed steps for the electromagnetic field approach are as follows:

1. **Determine the radiation impedance of the pacemaker**
   This impedance is obtained by simulating the pacemaker as a transmitting antenna, that is, applying a source voltage at the connector to drive the electrode and lead wire against the metal pacemaker housing. The current flowing through the connector is then obtained from the FDTD-calculated circumferential magnetic fields according to Ampere's law, and the radiation impedance is determined from the ratio of the source voltage to the current at the connector.
2. **Determine the open voltage at the connector**
   This voltage is based on the consideration that the pacemaker acts as a receiving antenna with an open load at the connector. By exciting the transmitting antenna of a body area communication device, the voltage induced between the metal housing and the lead wire is calculated as the open voltage also using the FDTD method.
   For calculating the open voltage using the FDTD method, a lumped resistor R is modeled in one FDTD cell. Consider the resistor to be x-directed and let V_R be the voltage at the resistor. The current flowing along it at the time step $(n-1/2)$ is then

$$I_R^{n-(1/2)} = \frac{V_R^{n-(1/2)}}{R} = \frac{\Delta x}{R}\frac{E_x^{n-1}+E_x^n}{2} \tag{7.16}$$

where E_x is the electric field component at the resistor location and Δx is the FDTD cell size in the x direction. From Maxwell's equations, the corresponding time-stepping relation for electric field E and magnetic field H at the resistor is then given by

$$E_x^n = \frac{1 - \dfrac{\Delta t \Delta x}{2R\varepsilon \Delta y \Delta z}}{1 + \dfrac{\Delta t \Delta x}{2R\varepsilon \Delta y \Delta z}} E_x^{n-1} + \frac{\dfrac{\Delta t}{\varepsilon}}{1 + \dfrac{\Delta t \Delta x}{2R\varepsilon \Delta y \Delta z}} (\nabla \times H^{n-(1/2)})_x. \tag{7.17}$$

When $R \to \infty$, the voltage V_R in Equation 7.16 can be considered as the open voltage V_M.

3. **Calculate the input interference voltage V_I**
This voltage goes through the connector and acts as the input of the internal pacemaker circuit. It can be calculated from Equation 7.15. In order to apply Equation 7.15, however, we need to know the internal impedance Z_I of the pacemaker seen from the connector. The magnitude of Z_I is mainly determined by the pacemaker sensing circuit, which will be discussed next.

7.3.3 Electric Circuit Approach

A pacemaker sensing circuit is composed of an amplifier and a low-pass filter. A typical circuit can be considered as an operational amplifier (opamp) combined with external elements. An opamp is generally a high-gain electronic voltage amplifier with a differential input and a single-ended output. Figure 7.19 shows a representative negative feedback opamp configuration with differential inputs. As described above, although the body area communication signal is generally at higher frequencies compared with the heart-beat-stimulating pulse signal, the opamp's nonlinearity may convert the high frequency signal to lower frequencies and enable it to pass through the low-pass filter. Measured results in Schenke, Fichte, and Dickmann

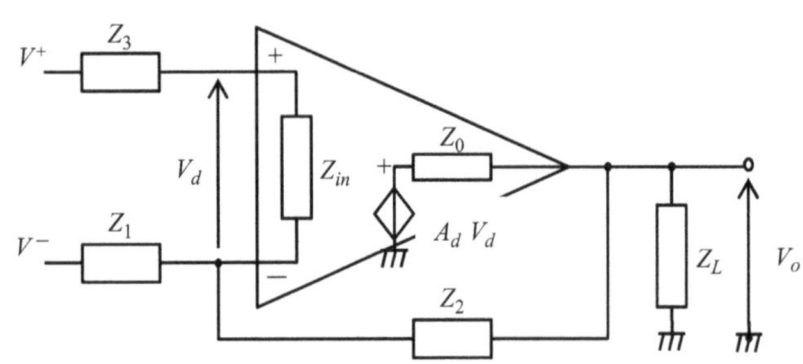

Figure 7.19 General negative feedback opamp configuration with differential inputs

(2007) also show that the ideal linear opamp model is no longer applicable at higher frequency due to the demodulation property of the opamp. A nonlinear model is therefore necessary for an opamp with low-pass filter character in order to accurately predict the interference output. Although the feed through filter before the opamp may also yield a nonlinear result, we here focus the nonlinear analysis on the opamp.

The Volterra series is known as a powerful and accurate tool for weakly nonlinear time-invariant analog circuit analysis (Schetzen, 1980). It is nowadays extensively used to calculate small, but nevertheless troublesome, distortion terms in transistor amplifiers and systems. Compared with the Taylor series, it is characterized by the ability to capture "memory" effect which implies the output of the nonlinear system depends on the input to the system at all other times. This effect is beneficial to model the nonlinear behavior in "memory" effect devices such as capacitors and inductors. In mathematics, the Volterra series represents a functional expansion of a dynamic, nonlinear, time-invariant function and is a series of infinite sum of multi-dimensional convolution integrals.

It is known that a linear time-invariant system with memory can be described by the convolution representation

$$y(t) = \int_{-\infty}^{\infty} h(\tau)x(t-\tau)d\tau \tag{7.18}$$

where $x(t)$ is the input signal, $y(t)$ is the output signal and $h(t)$ is the impulse response of the system.

According to the Volterra series method, the output signal $y(t)$ of a nonlinear time-invariant system with memory can be expressed in the form (Schetzen, 1980)

$$y(t) = \sum_{i=1}^{\infty} y^{(i)}(t) \tag{7.19}$$

where

$$\begin{aligned} y^{(i)}(t) = & \int_{-\infty}^{+\infty} \cdots \int_{-\infty}^{+\infty} H_{k_1,\cdots,k_i}(f_1,\cdots,f_i) \\ & \times X_{k_1}(f_1)\cdots X_{k_i}(f_i)e^{j2\pi(f_1+\cdots+f_i)t}df_1\cdots df_i. \end{aligned} \tag{7.20}$$

$X_{k_i}(f_i)$ is the Fourier transform of the system input and $H_{k_1,\cdots,k_i}(f_1,\cdots,f_i)$ with $1 \leq k_i \leq d$ (d is the potential highest order of the system) are the ith order frequency-domain Volterra kernels. If the system is considered as a second-order system ($d = 2$), the $H_{k_1,\cdots,k_i}(f_1,\cdots,f_i)$ item will contain $H_1(f)$, $H_2(f)$, $H_{11}(f_1,f_2)$, $H_{12}(f_1,f_2)$, $H_{21}(f_1,f_2)$ and $H_{22}(f_1,f_2)$. The Volterra series method depicts that any nonlinear system, in principle, can be modeled through multi-dimensional convolution. However, in practice, the kernels are usually only considered up to order 3 or

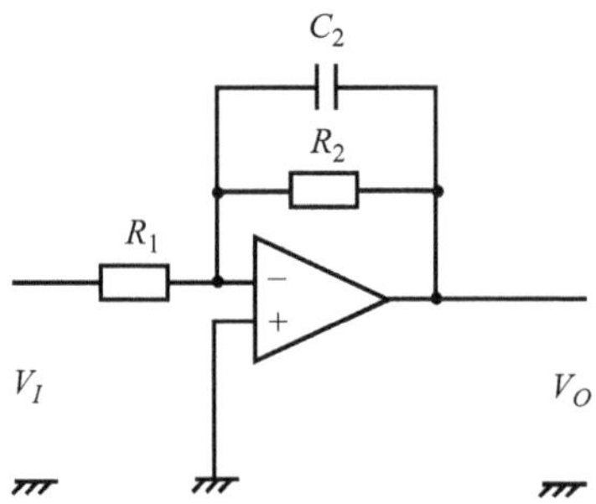

Figure 7.20 An opamp model for an internal pacemaker sensing circuit

rarely 5 due to complex kernels. As a result of this limitation, this method is only used for weakly nonlinear systems.

For a weakly nonlinear negative feedback opamp driven by RF interference, the output voltage expression for two arbitrary input signals has been derived using the second-order Volterra series (Fiori and Crovetti, 2003). Based on the derived Volterra kernels in Fiori and Crovetti (2003) and with input voltage $V^+ = 0$ in the pacemaker case, we can get the opamp output voltage as follows

$$\begin{aligned} v_o(t) = &\int_{-\infty}^{\infty} H_2(f_1)V^-(f_1)e^{j2\pi f_1 t}df_1 \\ &+ \int_{-\infty}^{\infty}\int_{-\infty}^{\infty} H_{22}(f_1,f_2)V^-(f_1)V^-(f_2)e^{j2\pi(f_1+f_2)t}df_1 df_2 \end{aligned} \tag{7.21}$$

where $H_2(f_1)$ and $H_{22}(f_1,f_2)$ are the frequency domain Volterra kernels, which depend on the opamp parameters and the external components. With $V^+ = 0$ in the pacemaker case, the negative feedback opamp in Figure 7.19 can be simplified as in Figure 7.20. Therefore, Equation 7.21 is a general second-order nonlinear opamp output voltage expression for an internal pacemaker sensing circuit. In terms of the weak nonlinearity of the Volterra series, we assume the opamp of the pacemaker internal circuit to be a second-order nonlinear system.

Next we give the frequency domain Volterra kernels $H_2(f)$ and $H_{22}(f_1,f_2)$ in Equation 7.21.

1. **Assumption for various impedances of opamp**
   By comparing Figures 7.19 and 7.20, it can be noted

$$Z_3(f) = 0, Z_{in}(f) = \infty, Z_0(f) = 0, Z_L(f) = \infty.$$

   Therefore, the corresponding impedance relationship is as follows

$$Z_1(f) = R_1 \tag{7.22}$$

$$Z_2(f) = \frac{R_2}{1 + j\omega R_2 C_2} \tag{7.23}$$

and the voltage transmission function is

$$A_d(f) = A_0 \frac{1}{1 + jf/f_1} \tag{7.24}$$

where A_0 is the amplification factor and f_1 is the cut-off frequency of the low pass filter.

2. **Derivation of $H_2(f)$**

$$H_2(f) = \frac{[1 - B(f)]A_d(f)}{1 + B(f)A_d(f)} \tag{7.25}$$

$$\because \quad B(f) = -\frac{Z_0(f)}{Z_0(f) + Z'_L(f)} \cdot \frac{Z'_i(f)}{Z'_i(f) + Z_2(f)} \cdot \frac{Z_{in}(f)}{Z_{in}(f) + Z_3(f)} \tag{7.26}$$

but

$$Z'_i(f) = \frac{Z_1(f)[Z_3(f) + Z_{in}(f)]}{Z_1(f) + Z_3(f) + Z_{in}(f)} = Z_1(f) \tag{7.27}$$

$$Z'_L(f) = \frac{Z_L(f)\left[Z_2(f) + Z'_i(f)\right]}{Z_L(f) + Z_2(f) + Z'_i(f)} = Z_1(f) + Z_2(f) \tag{7.28}$$

$$\therefore\ B(f) = 0 \tag{7.29}$$

$$\therefore\ H_2(f) = A_d(f). \tag{7.30}$$

3. **Derivation of $H_{22}(f_1,f_2)$**

$$H_{22}(f_1,f_2) = G_{12}(f_1)G_{22}(f_2)H_0(f_2,f_1) + G_{22}(f_1)G_{12}(f_2)H_0(f_1,f_2) \tag{7.31}$$

where

$$H_0(f_1,f_2) = \frac{1}{2}\frac{1}{2I_0}\frac{A_d(f_1+f_2)}{1+B(f_1+f_2)A_d(f_1+f_2)} \times \frac{2g_m j2\pi f_1 C_T}{j2\pi f_1(2C_{gs}+C_T)+2g_m}$$
$$= \frac{1}{4I_0}A_d(f_1+f_2)\cdot\frac{2g_m j2\pi f_1 C_T}{j2\pi f_1(2C_{gs}+C_T)+2g_m} \tag{7.32}$$

$$H_0(f_2,f_1) = \frac{1}{4I_0}A_d(f_1+f_2)\cdot\frac{2g_m j2\pi f_2 C_T}{j2\pi f_2(2C_{gs}+C_T)+2g_m} \tag{7.33}$$

I_0 is the bias current at the transistors, g_m is the transfer conductance of the opamp, C_{gs} is the gate-to-source capacitance of each transistor, C_T is the sum of the parasitic capacitances related to the ground and the power supply, and

$$G_{12}(f) = \frac{-D(f)Y_1(f)}{Y_1(f)+Y_2'(f)\left[1+A_d'(f)D(f)\right]+Y_3'(f)} \tag{7.34}$$

$$G_{22}(f) = \frac{[2-D(f)]Y_1(f)}{2\{(f)+Y_2'(f)\left[1+A_d'(f)D(f)\right]+Y_3'(f)\}}. \tag{7.35}$$

$$\because D(f) = \frac{Z_{in}(f)}{Z_3(f)+Z_{in}(f)} = 1 \tag{7.36}$$

$$Y_1(f) = \frac{1}{Z_1(f)} \tag{7.37}$$

$$Y_2'(f) = \frac{1}{Z_2(f)+\dfrac{1}{Y_0(f)+Y_L(f)}} = \frac{1}{Z_2(f)} \tag{7.38}$$

$$Y_3'(f) = \frac{1}{Z_3(f)+Z_{in}(f)} = 0 \tag{7.39}$$

$$A_d'(f) = A_d(f)\frac{Z_L(f)}{Z_L(f)+Z_0(f)} = A_d(f) \tag{7.40}$$

$$\therefore \quad G_{12}(f) = -\frac{\dfrac{1}{Z_1(f)}}{\dfrac{1}{Z_1(f)} + \dfrac{1}{Z_2(f)}[1 + A_d(f)]} = -\frac{Z_2(f)}{Z_2(f) + Z_1(f)[1 + A_d(f)]} \tag{7.41}$$

$$G_{22}(f) = \frac{\dfrac{1}{Z_1(f)}}{2\left\{\dfrac{1}{Z_1(f)} + \dfrac{1}{Z_2(f)}[1 + A_d(f)]\right\}} = \frac{Z_2(f)}{2\{Z_2(f) + Z_1(f)[1 + A_d(f)]\}}. \tag{7.42}$$

The derivation of the Volterra kernels used for the second-order nonlinear opamp has been presented in detail. In short, the Volterra kernels depend on the electrical parameters of the opamp components. If the input interference to the internal pacemaker sensing circuit is also known, we can then get the output interference signal based on the Volterra nonlinear model in Equation 7.21.

Based on the above two approaches, the prediction of EMI voltage at the analog sensing circuit output of a pacemaker can be now summarized as follows:

1. Calculate the open voltage V_M using the FDTD method by modeling the pacemaker as a receiving antenna. Since the radiation impedance Z_R is much smaller than the input impedance Z_I of the opamp circuit, based on Equation 7.15, it is reasonable to use the open voltage V_M as the input voltage V_I, which actually considers a worst case.
2. Calculate the output voltage V_O using Equation 7.21 and compare it with the sensing threshold V_t. If $V_O > V_t$, the heart-beat-simulating pulse will be triggered and a malfunction may occur.

7.3.4 *Transmitting Signal Strength versus Interference Voltage*

With the above-described method, in this section, we will focus on two types of on-body communication signals as the external interference sources to analyze the potential interference voltage at the implanted pacemaker output. From the perspective of the interference signal type, the carrier signal and pulse signal are two representative transmission signals. For the carrier signal, we will give an example in the HBC band, while for the pulse signal, the UWB band is naturally used. They actually correspond to narrow band and wide band communication signals, respectively.

7.3.4.1 Narrow Band HBC Signal

It has been described in Chapters 2 and 4 that the propagation in the HBC band is mainly based on an electrostatic coupling or approximate surface wave. The dielectric properties of human tissue determine that such a transmission should be expected in a frequency below dozens of MHz. On the other hand, however, the electromagnetic interaction of on-body communication signals with the human body is significant in this frequency range because of the easier penetration of electromagnetic fields into the human body. This implies a potential EMI with an implanted cardiac pacemaker.

For a narrow band on-body HBC communication signal at carrier frequency f_c, the input interference voltage can be approximated in a delta-function form $V_I = V_{IA}\delta(f_c)$. Based on Equation 7.21, the output voltage at the sensing circuit can be written as

$$v_O(t) = V_{IA}H_2(f_c)\cos[2\pi f_c t + \angle H_2(f_c)] + 0.5V_{IA}^2\mathrm{Re}[H_{22}(f_c, -f_c)] \qquad (7.43)$$

where V_{IA} is the amplitude of the input narrow band signal, and $H_2(f_c)$ and $H_{22}(f_c, -f_c)$ are the frequency-domain Volterra kernels, which depend on the opamp parameters and the external components. Since the term at f_c (in MHz) is beyond the opamp circuit bandwidth (in kHz), only the second term in Equation 7.43 is effective after the low-pass filter. This indicates that the EMI effect is actually caused by an offset voltage in the output. The interference output voltage produced by the narrow band HBC signal can then be rewritten as

$$V_O = 0.5V_{IA}^2\, Re[H_{22}(f_c, -f_c)]. \qquad (7.44)$$

Based on the derivation of $H_{22}(f_c, -f_c)$ from Equations 7.31 to 7.42, the following is obtained

$$H_{22}(f_c, -f_c) = A_0 \frac{g_m}{2I_0} \frac{j2\pi f_c C_T}{j2\pi f_c(2C_{gs} + C_T) + 2g_m} \times \left[\frac{Z_2(f_c)}{R_1 + Z_2(f_c)}\right]^2. \qquad (7.45)$$

Then the interference output voltage V_O is

$$\begin{aligned} V_O &= 0.5V_{IA}^2\, Re[H_{22}(f_c, -f_c)] \\ &= V_I^2 A_0 \frac{g_m}{4I_0} Re\left\{\frac{j2\pi f_c C_T}{j2\pi f_c(2C_{gs} + C_T) + 2g_m} \times \left[\frac{Z_2(f_c)}{R_1 + Z_2(f_c)}\right]^2\right\}. \end{aligned} \qquad (7.46)$$

According to Schenke, Fichte, and Dickmann (2007), the low-pass filter of the pacemaker circuit has a cut-off frequency of 1 kHz which means $f_0 = 1$ kHz. Assuming a 10 dB gain of the negative feedback opamp circuit, the component parameters for the circuit in Figure 7.20 can be determined as $R_1 = 1\,\mathrm{k\Omega}$, $R_2 = 3\,\mathrm{k\Omega}$ and $C_2 = 53.05$ nF. Furthermore, with typical values of A_0, I_0, g_m, C_{gs} and C_T, we can

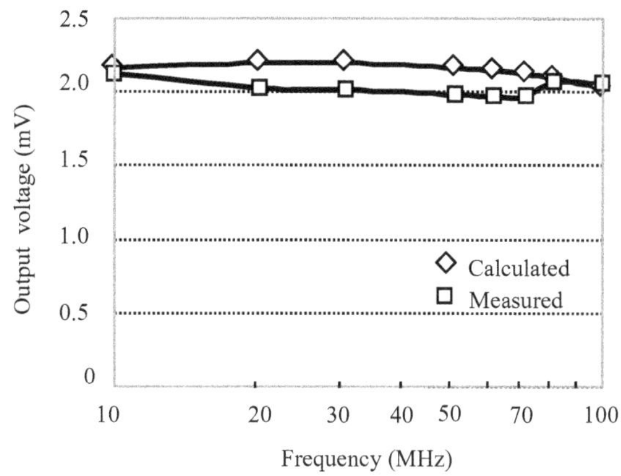

Figure 7.21 Calculated and measured output voltages

obtain the output voltage V_0 for a known interference input voltage at the pacemaker sensing circuit.

The measured values given in Schenke, Fichte, and Dickmann (2007) show a critical input voltage of almost 1 V from 10 to 100 MHz. This means that when the input interference voltage is 1 V, the output interference voltage will be the threshold voltage V_t. When the input interference voltage exceeds 1 V, the pulse output switches on and this is rated as the heart beat. According to Irnich *et al.* (1996) and (Barbaro *et al.* (2003), the mean value for the threshold voltage V_t of a pacemaker is around 1–4 mV. Here we choose a typical value of 2 mV as the threshold voltage. Figure 7.21 shows the threshold output voltage, derived from the measured data in Schenke, Fichte, and Dickmann (2007), as a function of frequency between 10 MHz and 100 MHz. It is found that the threshold output voltage is almost flat within this frequency band and can have an almost constant level of 2 mV for appropriately chosen circuit parameters.

On the other hand, using the formula in Equation 7.46, we can calculate the output voltage V_O for the same input voltage as given in Schenke, Fichte, and Dickmann (2007). The results are also plotted in Figure 7.21. The calculated interference output voltage is found to agree well with the measured one, which supports the validity of the nonlinear circuit model. The circuit parameters used in the model are listed in Table 7.9.

This approach is applied to a HBC scenario as shown in Figure 7.22. In this scenario the pacemaker user has an on-body HBC transmitter on the chest. The transmitter is designed to communicate with the receiver at the finger position. The human body model is a homogeneous one with dielectric properties two-thirds of muscle. The on-body transmitter is an electrode structure consisting of two metal plates on the human body surface. Since the transmitter is on the chest surface, the

Table 7.9 Opamp circuit parameters

Amplification factor, A_0	1 000 000
Bias current, I_0 (μA)	10
Transfer conductance, g_m (mS)	1.2
Gate-to-source capacitance, C_{gs} (fF)	100
Parasitic capacitance, C_T (pF)	1

EMI voltage produced should have a significant effect on the implanted pacemaker. Figure 7.23 shows the FDTD-calculated open voltage V_M as a function of frequency from 10 to 100 MHz for an exciting voltage of 10 V at the transmitter. This exciting voltage should be the maximum possible level in normal on-body HBC applications. As can be seen, the produced interference input voltage decreases with frequency, which may be attributed to the more difficult penetration into the body tissue at

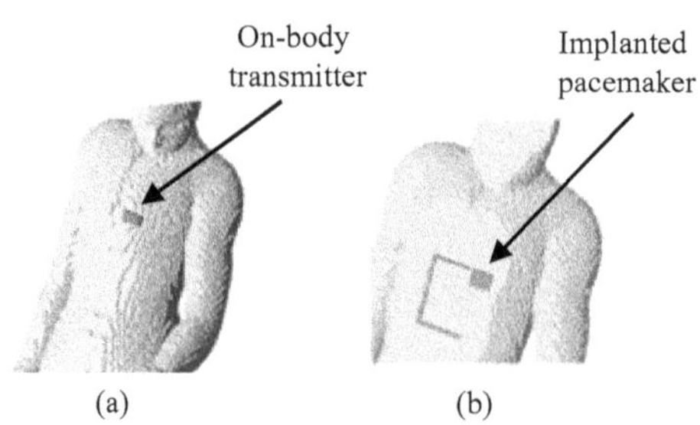

Figure 7.22 (a) A realistic-shaped human body model with a transmitter on the left-chest surface and (b) an implanted pacemaker inside the chest

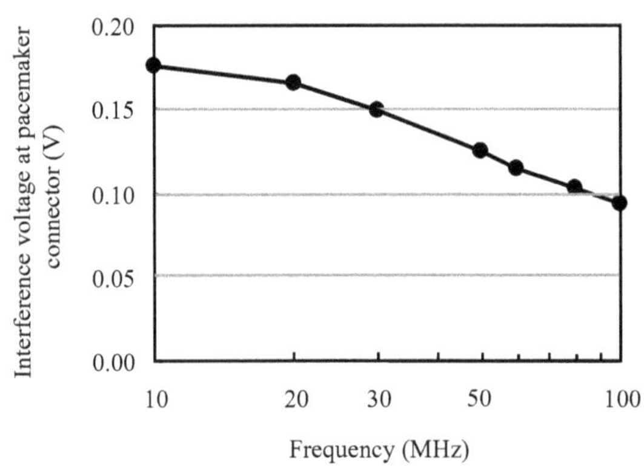

Figure 7.23 FDTD-calculated interference voltage V_M at the pacemaker connector. (This voltage is almost equal to the input voltage V_I)

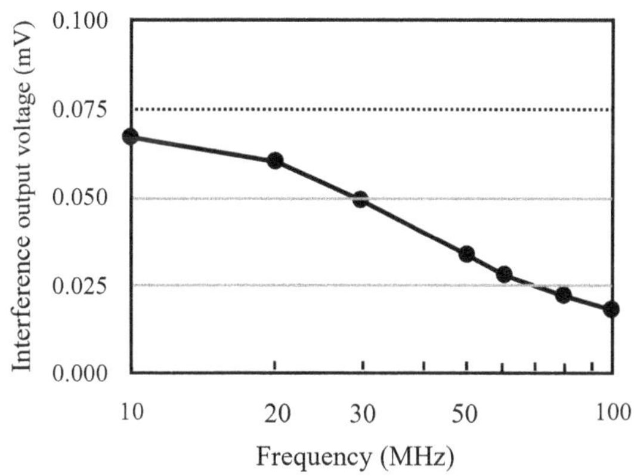

Figure 7.24 Calculated output interference voltage at the pacemaker circuit in an on-body HBC scenario

higher frequencies. Within the interested frequency band, the open voltage V_M ranges from 0.18 to 0.09 V. These voltages are added to the sensing circuit input as V_I approximately. Figure 7.24 shows the predicted output voltage of the analog sensing circuit using the nonlinear opamp model, that is, Equation 7.46. It is found that the output voltage ranges from 0.07 to 0.02 mV within this frequency band. Compared with the sensing threshold V_t of 2 mV, there is a safety margin of at least 30 dB between 10 MHz and 100 MHz. This result suggests that it is unlikely to malfunction for the pacemaker under usual transmitting voltage level in HBC applications.

7.3.4.2 UWB Pulse Signal

IR-UWB transmission makes use of short pulses to produce wide band frequency occupancy. In on-body UWB communication scenarios, communication devices on the left chest area may couple with the implanted pacemaker due to the opamp nonlinearity. In this application scenario, similar to Figure 7.22, the pacemaker user is assumed to have an on-body UWB transmitter on the chest for some biomedical purpose. Since the transmitter is on the chest surface, the EMI voltage produced may have a significant effect on the implanted pacemaker.

The UWB pulse is generally generated in the form of an *n*th-derivative Gaussian pulse. The expression of the *n*th-derivative Gaussian pulse has been given in Equation 4.5 and the corresponding Fourier transform is given in Equation 4.6. Then we have $V^-(f) = X_n(f)$ in Equation 7.21. For the EMI evaluation, we choose a second-derivative Gaussian pulse as the transmitted UWB pulse. The pulse peak is kept at 0.3 V and the pulse shape factor has been adjusted to $\alpha = 1.18 \times 10^{-10}$ so as to have a spectrum close to the FCC emission requirement as much as possible. The pulse width is about 280 ps as shown in Figure 4.3.

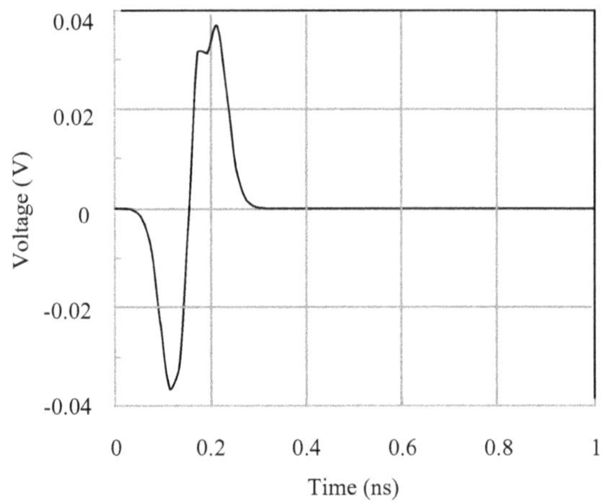

Figure 7.25 Calculated interference output voltage waveform of the pacemaker circuit for maximum permissible UWB pulse input

In this analysis, the implanted pacemaker is at a depth of about 1.5 cm from the left-chest surface. Since the pacemaker can be considered as a receiving antenna, we can use the frequency-dependent FDTD method to calculate the produced voltage at the pacemaker connector as the input to the sensing circuit. On the other hand, as another alternative, based on the path loss analysis result in Section 4.2.3, the UWB pulse will undergo a 54 dB attenuation between the transmitter on the left-chest surface and the implanted pacemaker location. In view of the fact that the transmitted UWB pulse peak is 0.3 V under the FCC emission limit, we can then determine the peak voltage produced at the input of the pacemaker to be 0.6 mV. This voltage will be the maximum UWB input voltage at the sensing circuit of the implanted pacemaker.

Based on the general expression in Equation 7.21, the interference output voltage is no longer a DC component for a wideband input signal. The output voltage is calculated using Equation 7.21 and the opamp circuit parameters in Table 7.9. The result is shown in Figure 7.25. It can be noted that the voltage waveform is smoothed which is attributed to the decreased high-frequency components. Compared with the sensing threshold voltage $V_t = 2$ mV, the peak value of the interference output voltage is only 0.037 mV. This corresponds to a safety margin of 35 dB for the pacemaker circuit for UWB interference signals. The DC component of the interference output voltage is only 0.06 μV, much smaller than the sensing threshold voltage.

Figure 7.26 shows the relationship between the input and output voltages for the peak voltage and DC voltage components, respectively. It can be concluded that when the peak voltage of the input pulse is less than 1 mV, both the peak and DC voltage components of the output pulse have linear relationships with the peak voltage of the input pulse. This is reasonable since when the input voltage is very small, the second term in Equation 7.21 which contains a quadratic of the input signal will

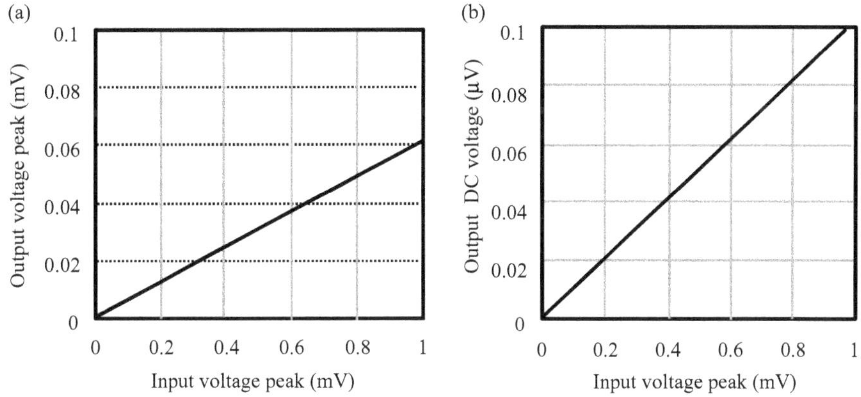

Figure 7.26 (a) Interference output voltage peak and (b) DC component versus input pulse voltage peak

become extremely small. Moreover, the frequency domain second-order Volterra kernel $H_{22}(f_1, f_2)$ is much smaller than the first-order Volterra kernel $H_2(f)$ at the UWB frequency band. Therefore the nonlinear effect on input–output voltage is negligible and a linear relationship can be maintained as shown in Figure 7.26.

However, the demodulating property of the opamp changes the output spectral components. The interference output spectrum is not proportional to the input spectrum within the interested frequency band because of the nonlinearity. Low frequency components will be amplified while high frequency components will be attenuated. The transmission characteristics at different frequencies are shown in Figure 7.27 using the spectral amplitude ratio of output to input voltages. The high frequency components in the input voltage are demodulated and then significantly contribute to the low frequency components of the output voltage. Figure 7.28 shows the spectral amplitude ratio of the output to input voltage at 1 kHz (the cut-off

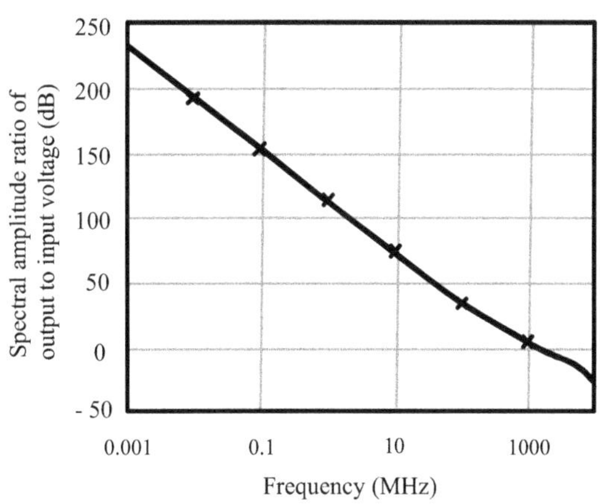

Figure 7.27 Spectral amplitude ratio of interference output to input voltages

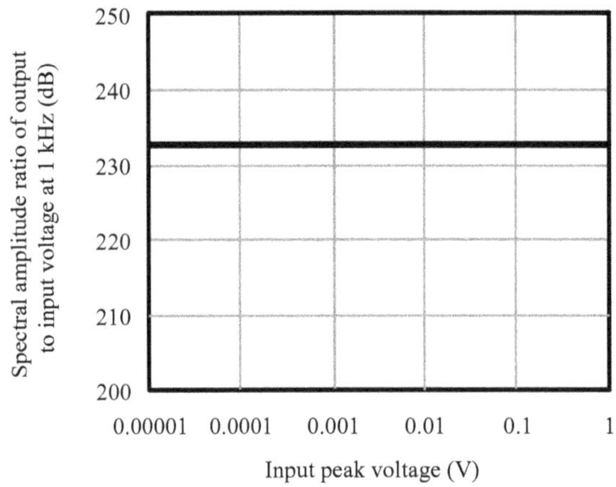

Figure 7.28 Spectral amplitude ratio of output to input voltages at 1 kHz versus input pulse peak voltage

frequency of the opamp) with the peak of input pulse voltage ranging from 10 μV to 1 V. The spectral component ratio at 1 kHz is as high as 230 dB which indicates a significant amplification of low frequency components. The high gain of the low frequency component is far beyond the opamp amplification ability. It should be attributed to the demodulating property of the opamp. The same result and conclusion can be drawn around the heartbeat frequency of 0.1 kHz.

Although the above analysis has not considered the nonideal behavior of the resistors and capacitors in Figure 7.20 and has also assumed the constant parameters of opamp up to GHz frequencies, the derived finding is helpful to understand the basic mechanism and get a preliminary EMI evaluation for cardiac pacemakers. In conclusion, the mechanism of cardiac pacemaker malfunction in body area communications is mainly due to the nonlinearity of the internal pacemaker circuit. The DC and low frequency components resulting from the demodulation of the body area communication signals should be paid much attention and may be used as an index for EMI evaluation. Except for the analytical approach demonstrated above, if we have a nonlinear electric circuit model, we can also use a circuit simulator such as SPICE to derive the DC or low frequency interference voltage.

7.3.5 Experimental Assessment System

In view of the importance of EMI to the cardiac pacemaker, experimental assessment is desirable for various pacemaker products. An *in vitro* test system has been developed for EMI assessment of the pacemaker by mobile phones and RFIDs (Irnich *et al.*, 1996; Tarusawa *et al.*, 2005; Futatsumori *et al.*, 2009). Its application to EMI assessment of the pacemaker in body area communications should be straightforward.

Figure 7.29 shows the configuration of the EMI test system for a cardiac pacemaker. The test system is composed of an electromagnetic wave source, an ECG

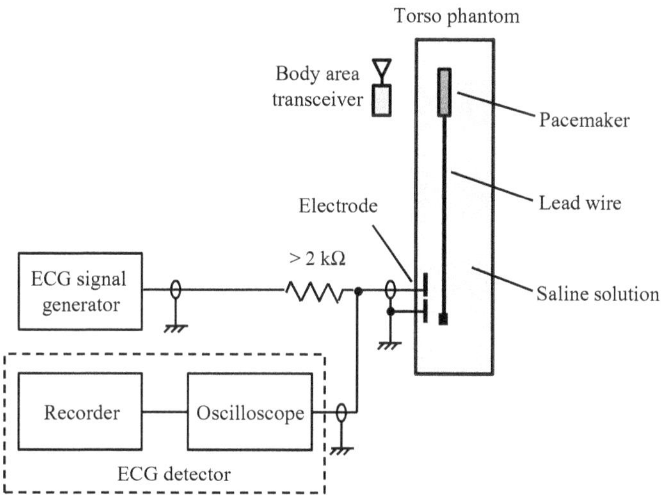

Figure 7.29 Configuration of test system for EMI assessment of a cardiac pacemaker. (Modified from Tarusawa *et al.*, 2005). Reproduced with permission from Tarusawa Y., Ohshita K., Suzuki Y., Nojima T. and Toyoshima T., "Experimental estimation of EMI from cellular base-station antennas on implantable cardiac pacemakers," *IEEE Transactions on Electromagnetic Compatibility*, **47**, 4, 938–950, 2005. © 2005 IEEE

signal generator, an ECG signal detector and a human torso phantom. The electromagnetic wave source may be a body area transceiver or a signal generator plus antenna to produce a body area communication signal. The ECG generator provides a simulated cardiac pulse to the pacemaker through an atrial or ventricle electrode. The ECG detector consists of an oscilloscope and a chart recorder. The oscilloscope is used to measure pace pulses generated from the pacemaker. The pace pulses are fed into the oscilloscope with 1 MΩ input impedance through the electrode for the atrium or ventricle, and is monitored by using a recorder equipped with sufficient digital memory to store the data.

The human torso phantom consists of a saline tank and electrodes. The saline tank is constructed from acrylic boards and is filled with a saline solution (NaCl 1.8 g/l). The saline solution provides similar dielectric properties as those of the human body. The positions of the pacemaker and the lead wires are fixed by using acrylic stays in the saline tank. The size of the saline tank is adjustable to some extent. Figure 7.30 shows a detailed human torso phantom saline tank with dimensions of 34 × 36 × 3.5 cm. As can be seen, two electrodes are installed in the saline tank as either a unipolar pacemaker or bipolar pacemaker equipped with atrial and ventricle lead wires. The electrodes are constructed from 3-cm diameter stainless steel rings and 0.6-cm diameter stainless steel patches. They are used to introduce a simulated ECG signal into the pacemaker and receive the pace pulses generated from the pacemaker. Moreover, they are adjusted to isolate each chamber's ECG signal level by more than 20 dB.

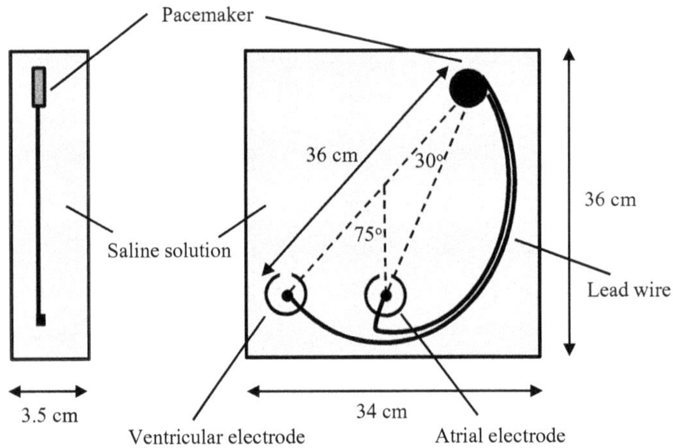

Figure 7.30 Human torso model used in the test system. Tarusawa *et al.*, 2005

The test procedure for the pacemaker EMI, shown in Figure 7.31, consists of five steps:

1. Set the operating modes and ECG sensitivity of the pacemaker under test. The maximum sensitivity or the minimum threshold voltage corresponds to a worst case.
2. Set the body area communication signal parameters such as the transmitting frequency, the transmitting power, the on-body or in-body position, and so on.
3. Conduct an inhibition test for assessing the pacemaker EMI. In this step the simulated ECG signal is off. The pacemaker under test does not exhibit any deviation in its pace-to-pace interval of the generated pacing pulses. If an interference signal from the body area transceiver inhibits a pacing pulse, this is considered as an EMI occurrence.
4. Conduct an asynchronous test for assessing the pacemaker EMI. In this step the simulated ECG signal is on. The pacemaker under test does not exhibit any pacing pulse. If an interference signal from the body area transceiver exhibits any pacing pulse, this is considered as an EMI occurrence.
5. In the inhibition test and the asynchronous test, the pacemaker EMI is assessed based on the parameters such as the transmitting power and the transceiver position on or in the torso phantom. If EMI occurs, recode the corresponding transceiver parameters, and then go back to Step 2 to set new parameters and repeat the same procedure.

Figure 7.32 shows two examples of EMI occurrence in the inhibition test and asynchronous test situations, respectively. The level of interference signal is set to be much higher than that of usual body area communication signals in order to

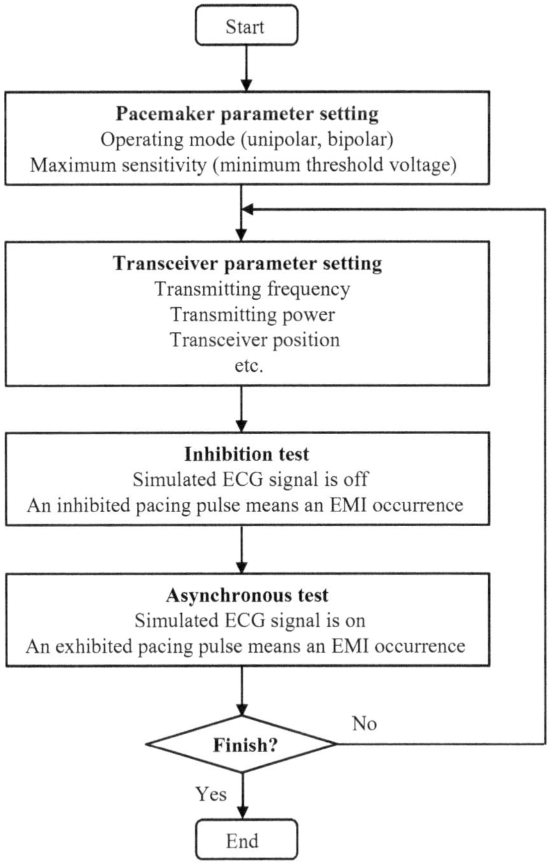

Figure 7.31 Procedure for testing pacemaker EMI

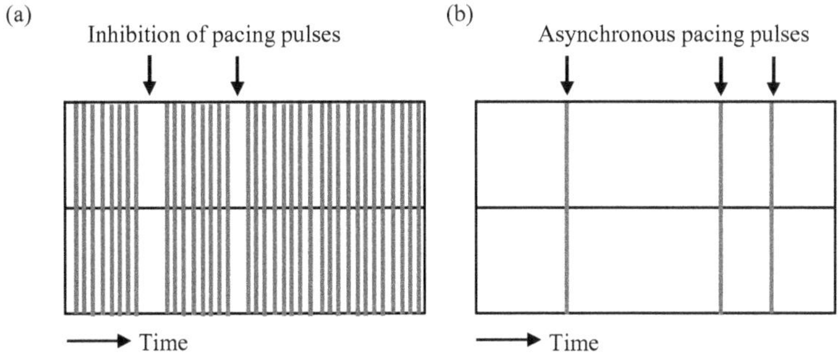

Figure 7.32 Example of EMI test results in (a) inhibition test and (b) asynchronous test

observe an obvious EMI with the pacemaker. As a result, two inhibited pacing pulse durations are observed in (a), and three exhibited pulses are observed in (b). Compared with the analytical and numerical assessments, such an experimental approach provides a more reliable and more intuitive EMI assessment for cardiac pacemakers.

References

Barbaro, V., Bartolini, P., Censi, F. *et al.* (2003) On the mechanisms of interference between mobile phones and pacemakers: parasitic demodulation of GSM signal by the sensing amplifier. *Physics in Medicine and Biology*, **48**, 1661–1671.

Caputa, K., Okoniewski, M., and Stuchly, M.A. (1999) An algorithm for computations of the power deposition in human tissue. *IEEE Antennas and Propagation Magazine*, **41** (4), 102–107.

Fiori, F. and Crovetti, P.S. (2003) Prediction of EMI effects in operational amplifier by a two-input Volterra series model. *IEE Proceeding of Circuits Devices System*, **150** (3), 185–193.

Futatsumori, S., Kawamura, Y., Hikage, T. *et al.* (2009) *In vitro* assessment of electromagnetic interference due to low-band RFID reader/writers on active implantable medical devices. *Journal of Arrhythmia*, **25** (3), 142–152.

Hartsgrove, G., Kraszewski, A., and Surowiec, A. (1987) Simulated biological materials for electromagnetic radiation absorption studies. *Bioelectromagnetics*, **8**, 29–36.

ICNIRP (1998) Guidelines for limiting exposure to time-varying electric, magnetic and electromagnetic fields (up to 300 GHz). *Health Physics*, **74**, 494–522.

IEEE (2002) IEEE recommended practice for measurements and computations of radio frequency electromagnetic fields with respect to human exposure to such fields, 100 kHz–300 GHz. IEEE Std C95.3-2002.

Irnich, W., Batz, L., Muller, R., and Tobisch, R. (1996) Electromagnetic interference of pacemakers by mobile phones. *PACE*, **19**, 1431–1446.

Ito, K., Furuya, K., Okano, Y., and Hamada, L. (1998) Development and the characteristics of a biological tissue-equivalent phantom for microwaves (in Japanese). *Transactions on IEICE*, **J81-B-I**, 1126–1135.

Lin, C. (1978) *Microwave Auditory Effects and Applications*, Charles C. Thomas, Springfield, IL.

Schenke, S., Fichte, L.O., and Dickmann, S. (2007) EMC modeling of cardiac pacemakers. Proceedings of the 2007 International Zurich Symposium on Electromagnetic Compatibility, Munich, Germany.

Schetzen, M. (1980) *The Volterra and Wiener Theories of Nonlinear Systems*, John Wiley & Sons, Ltd, New York.

Schmid, T., Egger, O., and Kuster, N. (1996) Automated E-field scanning system for dosimetric assessments. *IEEE Transactions on Microwave Theory and Techniques*, **44**, 105–113.

Tarusawa, Y., Ohshita, K., Suzuki, Y. *et al.* (2005) Experimental estimation of EMI from cellular base-station antennas on implantable cardiac pacemakers. *IEEE Transactions on Electromagnetic Compatibility*, **47** (4), 938–950.

Wang, J., Fujiwara, O., and Nojima, T. (2000) A model for predicting electromagnetic interference of implanted cardiac pacemaker by mobile telephones. *IEEE Transactions on Microwave Theory and Techniques*, **48** (11), 2121–2125.

Wang, Q. and Wang, J. (2009) SA and SAR analysis for wearable UWB body area applications. *IEICE Transactions on Communications*, **E92-B** (2), 425–430.

Wang, Q., Sanpei, T., Wang, J., and Plettemeier, D. (2009) EMI modeling for cardiac pacemaker in human body communication. Proceeding of the 2009 International Symposium on Electromagnetic Compatibility, Kyoto, Japan, pp. 629–632.

8

Summary and Future Challenges

Body area communications is a short range wireless communication technique in the vicinity of, or inside, a human body. The different operating environments mean that the body area communications is classified into on-body, in-body or in-body/on-body to off-body communication scenarios. It can provide a wide range of applications such as medical and healthcare services, assistance to people with disabilities, and consumer electronic connectivity and user identification.

As an introduction to body area communications, we have started with basic electromagnetic characterization and modeling methodology for the human body in various possible frequency bands. Based on the basic knowledge, we have focused on three major areas: channel modeling, modulation/demodulation performance, and EMC consideration in body area communications.

Body area communications is being considered to operate in UWB, MICS, ISM and HBC bands. Since electrical properties of human tissue are frequency-dependent, a Debye-type approximation expression is a useful means for modeling them. The basis for modeling the frequency dependence uses the dispersion phenomena in the dielectric spectrum of tissue. The dielectric spectrum is characterized by several dispersion regions. Each of them can be characterized by a single relaxation time constant in the Debye expression. The dielectric properties of the human body are then expressed as a summation of terms corresponding to the main dispersion mechanisms. For a frequency range from several Hz to 10 GHz, four Debye-type dispersion region expressions provide a good model for most human body tissues.

Body Area Communications: Channel Modeling, Communication Systems, and EMC, First Edition.
Jianqing Wang and Qiong Wang.

The frequency-dependent electrical properties mean that propagation along the human body is via different mechanisms at different frequency bands. For on-body communication:

- At HBC band such as 10 MHz, almost 80% of the received electric field component is contributed by an electrostatic field term at a unit communication distance. The signal transmission is actually realized by electrostatic coupling.
- At MICS or WMTS band around 400 MHz, almost 80% of the received electric field component is contributed by the surface propagation term, which acts as a main on-body propagation mechanism.
- In UWB band, more than 95% of the received electric field component is contributed by the surface propagation term. Actually it completely dominates the on-body propagation.

On the other hand, on-body communication mainly suffers from a path loss fluctuation or shadowing due to the body shape and structure and a multipath fading due to the body movement, while in-body communication mainly undergoes severe signal decay during the transmission through the lossy human tissue. These features mean that care has to be taken in choosing a frequency band for the on-body or in-body communication. In general, UWB and HBC bands provide some advantages for on-body communication. The former's advantages are based on its very low PSD and robustness to multipath fading, and the latter's are based on its small on-body path loss compared with other frequency bands. On the other hand, the MICS band and UWB low band may be more adequate for in-body communication because of the relatively small penetration depth in human tissues or the possibility of high data rates for real time transmission.

Based on the basic propagation mechanisms, we have generalized the channel model into a path loss model and an impulse response multipath model. The path loss model mainly describes the channel loss, including propagation loss, absorption loss as well as diffraction loss. It has been shown that the empirical power decay law can give a reasonable expression for all of the considered frequency bands. The path loss exponent may range from 2 to 5 for on-body UWB full band propagation, and is about 10 for in-body UWB low band and 6 for in-body MICS band propagation. The HBC propagation exhibits a smaller path loss around 30 dB per meter after 10 cm away from the transmitter. Moreover, in any frequency bands, the path loss fluctuation from the average path loss, that is, the shadowing, is found to always follow a log-normal distribution.

The multiple paths between transmitter and receiver, produced by body movement or transmitter movement, introduce complexity into the channel model. To take into account the time-varying channel properties, we have established a discrete time impulse response channel model based on the classical Saleh–Valenzuela model. First, the time delays of the multipaths are generated as follows: the first path is generated at a fixed arrival time, after which a temporal delay between two successive paths is generated according to the inverse Gaussian distribution and added to the arrival time of the previous path. Next, the gain coefficient for each path is determined acording to the log-normal distribution, whose mean follows an exponential power decay formula. In this discrete time impulse response model, four paths are usually necessary for representing an on-body UWB multipath channel, while two paths are sufficient for representing an in-body to on-body multipath channel in a capsule endoscope application.

The communication performance of body area communication systems is directly influenced by the channel it operates in. Based on the generalized body area channels in the forms of a static shadow fading channel and a dynamic multipath fading channel, we have evaluated the BER performance in terms of frequency bands and communication channels, that is, according to the following four types: on-body UWB communication, in-body UWB communication, in-body MICS band communication and human body communication, respectively. Link budget analysis and RAKE or diversity reception improvement are also carried out. The results can be summarized as follows:

- For on-body UWB communication, it is sufficient to realize a data rate not exceeding 10 Mbps within a 1 m communication distance in the multipath fading environment. If the RAKE reception is adopted, a system margin larger than 0 dB can be obtained almost on the whole body.
- For in-body UWB communication in the capsule endoscope scenario, diversity reception provides effective improvement on BER performance. In almost all of the transmitter locations inside the digestive organs, a data rate of 0.1 Mbps can always yield a system margin larger than 0 dB. When the data rate is increased to 1 Mbps or 10 Mbps, however, the corresponding communication distance will be reduced to about 10 cm. In order to make the communication possible in all the digestive organs, more than two-branch diversity reception is necessary. On the other hand, for in-body UWB communication in the cardiac pacemaker scenario, even without RAKE reception, a system margin can still be obtained larger than 0 dB at a data rate as high as 10 Mbps.
- For in-body MICS band communication in the capsule endoscope scenario, the permissible transmitting power is only −16 dBm. The conventional correlation receiver only supports a communication distance of 8 cm at a data rate of 1 Mbps. With the aid of diversity reception, the communication distance can be extended to basically cover all the digestive organs. However, transmission at a data rate of 10 Mbps seems still difficult.
- For HBC, the path loss is relative small and a low transmitting power can therefore be expected. It is basically available to communicate over the entire body area at a data rate up to 1 Mbps.

Finally, we have addressed the EMC issues in body area communications. One is SAR analysis for human safety evaluation, and the other is the EMI with a cardiac pacemaker. The SAR is basically low enough due to the low transmitting power in most of the body area transmitters. On the other hand, the EMI with a cardiac pacemaker is mainly due to the nonlinearity of the internal analog sensing circuit. Based on this mechanism of malfunction, we have presented a two-step approach for EMI evaluation with a pacemaker. In the first step, the input voltage of the sensing circuit of the pacemaker is calculated using a numerical electromagnetic analysis method by considering the pacemaker as a receiving antenna. In the second step, a Volterra series for a nonlinear system is employed to analyze the output voltage of the nonlinear sensing circuit for EMI evaluation. This two-step approach has provided an effective means of EMC consideration in the design of body area communication systems.

Although body area communications has shown obvious potential and rapid progress in medical, healthcare and consumer electronics areas, there are still many topics to be studied as well as many problems to be solved. Some topics for consideration are summarized as follows:

- Body area communications processes vital signals for patients and elderly people in situations that may be a matter of life and death. This feature requires that the communication link must be established at a highly reliable level, which does not allow a failure of communication and a loss of information. How to realize a body area communication at an error-free level should be a big challenge.
- Because of the short range and near-field coupling feature, body area channel models usually include the effects of the transmitting and receiving antennas used in measurement. This makes the derived channel models lack generality. How to develop an antenna-independent channel model is important for a new application of body area communications.
- On-body and in-body antenna design plays an important role in realizing high-quality body area communications. These antennas are obviously different from conventional ones, and a new design methodology is required. In view of the near-field feature, the antenna radiation pattern in far-field is meaningless in body area communications. In on-body communication, special attention should be paid to make the propagation along the body surface and the energy radiation not towards the human body. For in-body communication, how to make the radios penetrate into/through the lossy tissue is a challenge to antenna designers. Moreover, the transmitting antennas, the human body and the receiving antenna should be considered to be linked by an impedance matrix. Its optimization for energy transmission over the human body is the basic methodology for body area antenna design.
- Unifying the sensor and transceiver into a size as small as possible is a continuous challenge. Low consumption power and long battery time is also a key factor for the spread of applications. A wireless power transmission for in-body transceivers is especially expected.

- EMC issues will always rise in body area communications. The issues not only include the interference from other communication systems but also the interference to medical equipments. Due to the miniaturization and low-voltage operation of body area devices, how to solve the mutual coupling and interference problems inside the devices will also be a challenge to realize a highly reliable body area communication system.

Index

Body Area Communications: Channel Modeling, Communication Systems, and EMC, First Edition.
Jianqing Wang and Qiong Wang.

www.ingramcontent.com/pod-product-compliance
Ingram Content Group UK Ltd.
Pitfield, Milton Keynes, MK11 3LW, UK
UKHW050944290926
15470UKWH00005B/1624